W0037759

Security Informatics and Law Enforcement

Series Editor
Babak Akhgar, CENTRIC (Centre of Excellence
in Terrorism, Resilience, Intelligence and Organised
Crime Research)
Sheffield Hallam University
Sheffield, UK

The primary objective of this book series is to explore contemporary issues related to law enforcement agencies, security services and industries dealing with security related challenges (e.g., government organizations, financial sector insurance companies and internet service providers) from an engineering and computer science perspective. Each book in the series provides a handbook style practical guide to one of the following security challenges:

Cyber Crime – Focuses on new and evolving forms of crimes. Books describe the current status of cybercrime and cyber terrorism developments, security requirements and practices.

Big Data Analytics, Situational Awareness and OSINT – Provides unique insight for computer scientists as well as practitioners in security and policing domains on big data possibilities and challenges for the security domain, current and best practices as well as recommendations.

Serious Games – Provides an introduction into the use of serious games for training in the security domain, including advise for designers/programmers, trainers and strategic decision makers.

Social Media in Crisis Management – explores how social media enables citizens to empower themselves during a crisis, from terrorism, public disorder, and natural disasters

Law enforcement, Counterterrorism, and Anti-Trafficking – Presents tools from those designing the computing and engineering techniques, architecture or policies related to applications confronting radicalisation, terrorism, and trafficking.

The books pertain to engineers working in law enforcement and researchers who are researching on capabilities of LEAs, though the series is truly multidisciplinary – each book will have hard core computer science, application of ICT in security and security / policing domain chapters. The books strike a balance between theory and practice.

Ilias Gkotsis
Dimitrios Kavallieros • Nikolai Stoianov
Stefanos Vrochidis • Dimitrios Diagourtas
Babak Akhgar
Editors

Paradigms on Technology Development for Security Practitioners

 Springer

Editors
Ilias Gkotsis
Satways Ltd
Irakleio, Greece

Nikolai Stoianov
Bulgarian Defence Institute
Sofia, Sofiya, Bulgaria

Dimitrios Diagourtas
Satways Ltd
Irakleio, Greece

Dimitrios Kavallieros
M4D
CERTH-ITI
Ilioupoli, Greece

Stefanos Vrochidis
M4D
CERTH-ITI
Thessaloniki, Greece

Babak Akhgar
CENTRIC
Sheffield Hallam University
SHEFFIELD, UK

(cc) (i) BY

ISSN 2523-8507   ISSN 2523-8515   (electronic)
Security Informatics and Law Enforcement
ISBN 978-3-031-62082-9   ISBN 978-3-031-62083-6   (eBook)
https://doi.org/10.1007/978-3-031-62083-6

© The Editor(s) (if applicable) and The Author(s) 2025
This book is an open access publication.

Open Access This book is licensed under the terms of the Creative Commons Attribution 4.0 International License (http://creativecommons.org/licenses/by/4.0/), which permits use, sharing, adaptation, distribution and reproduction in any medium or format, as long as you give appropriate credit to the original author(s) and the source, provide a link to the Creative Commons license and indicate if changes were made.
The images or other third party material in this book are included in the book's Creative Commons license, unless indicated otherwise in a credit line to the material. If material is not included in the book's Creative Commons license and your intended use is not permitted by statutory regulation or exceeds the permitted use, you will need to obtain permission directly from the copyright holder.
The use of general descriptive names, registered names, trademarks, service marks, etc. in this publication does not imply, even in the absence of a specific statement, that such names are exempt from the relevant protective laws and regulations and therefore free for general use.
The publisher, the authors and the editors are safe to assume that the advice and information in this book are believed to be true and accurate at the date of publication. Neither the publisher nor the authors or the editors give a warranty, expressed or implied, with respect to the material contained herein or for any errors or omissions that may have been made. The publisher remains neutral with regard to jurisdictional claims in published maps and institutional affiliations.

This Springer imprint is published by the registered company Springer Nature Switzerland AG
The registered company address is: Gewerbestrasse 11, 6330 Cham, Switzerland

If disposing of this product, please recycle the paper.

FOREWORD

We are pleased to introduce this collection of papers, linked to the results of the Security and Defense R&D projects, examining several aspects of security and defense challenges our world faces nowadays and offering insights into the related ground-breaking technologies and strategies being developed to address these challenges.

In today's rapidly evolving geopolitical landscape, the imperatives of security and defense are constantly evolving, necessitating a nuanced understanding of emerging threats and innovative approaches to mitigate them. This collection offers a panorama of the E.U. research and innovation potential, being a valuable resource for anyone seeking to understand the latest developments and best practices in crisis management and defense technology.

We would like to thank all the contributors to this collection for their hard work and dedication to advancing E.U. security research and innovation, and we look forward to seeing the impact of their efforts on the world at large.

Athens, Greece

Ilias Gkotsis
Dimitris Kavallieros

PREFACE

In today's swiftly evolving global landscape, security threats are becoming increasingly intricate, frequently traversing international boundaries. From cyber threats to terrorism, and from illicit trafficking to hybrid warfare, these challenges necessitate a united and coordinated approach. Within the European Union (EU), EU security and defence research serves as a fundamental pillar in combating these multifaceted threats effectively.

EU security and defence research acts as a catalyst for innovation, knowledge cultivation, and capability enhancement, all crucial elements in addressing current security challenges and anticipating future ones. By fostering collaboration across borders and disciplines, EU security research not only strengthens Europe's resilience and security but also fosters a competitive European security industry.

In this direction, Research and Innovation Symposium for European SECURITY and Defense 2023 (RISE-SD2023) brought together experts from across the crisis management, physical and cyber security, critical infrastructure protection, border management, and defence technology spectrum to present, discuss, and showcase research results and some of the most innovative solutions developed in the context of relevant European R&D projects.

As we embark on this intellectual journey, we encourage readers to approach these papers with curiosity, critical inquiry, and a spirit of collaboration. It is our sincere hope that this volume will serve as a catalyst for continued dialogue, innovation, and cooperation in the pursuit of a more secure and peaceful world.

Irakleio, Greece	Ilias Gkotsis
Ilioupoli, Greece	Dimitrios Kavallieros
Sofia, Sofiya, Bulgaria	Nikolai Stoianov
Thessaloniki, Greece	Stefanos Vrochidis
Irakleio, Greece	Dimitrios Diagourtas
Sheffield, UK	Babak Akhgar

CONTENTS

Part V   Effective Management of EU External Borders

PART I

Disaster-Resilient Society for Europe

CHAPTER 1

The Data Governance Process Within the Digital Chain of Custody

Gabriel Pestana, Luís M. Carvalho, and Sebastian Chmel

INTRODUCTION

CBRNE incidents involve collecting and transporting samples for accurate identification, necessitating interactions among multiple stakeholders [1]. The integrity of the chain of custody must be maintained, and all actions must be documented to enable information flow auditing, especially at Custody Transfer Points (CTPs) where stakeholders exchange sample

G. Pestana (✉)
INOV – Institute of Systems and Computer Engineering Innovation, Lisbon, Portugal
e-mail: gabriel.pestana@inov.pt

L. M. Carvalho
Unidade Militar Laboratorial de Defesa Biológica e Química, Exército Português, Lisbon, Portugal

CINAMIL, AM, Instituto Universitário Militar, Lisbon, Portugal

S. Chmel
Fraunhofer INT (Fraunhofer Institute for Technological Trend Analysis), Euskirchen, Germany

© The Author(s) 2025

I. Gkotsis et al. (eds.), *Paradigms on Technology Development for Security Practitioners*, Security Informatics and Law Enforcement, https://doi.org/10.1007/978-3-031-62083-6_1

custodianship. Paper-based forms are used, highlighting the need for a digitalised approach to the chain of custody process [2]. Implementing a digital chain of custody (dCoC) strategy ensures traceability and security throughout the entire process, from collection to disposal. However, the management of dCoC for collected samples lacks standardisation, leading to inconsistencies among stakeholders within and across EU member states. To ensure evidence consistency and admissibility in court, a standardised dCoC process must be established across EU member states.

A literature survey was conducted to analyse existing standards' guidelines for the chain of custody framework, aiming to identify standardisation gaps and opportunities related to non-reputation of custodianship throughout the chain. Afterwards, an Experts by Experience Group (EEG) was set up to contribute to developing and standardising the dCoC process. The EEG was open to experts from entities that play significant roles in CBRNE incident response and management, as from the following sectors:

- First Responders:

    - Firefighters and Hazardous Materials (HazMat) teams responding to CBRNE incidents.
    - Emergency Medical Services (EMS) personnel providing medical assistance in CBRNE scenarios.
    - Law enforcement officers involved in securing the scene and assisting in evidence collection.

- Crisis Management Agencies:

    - National and regional crisis management centres responsible for coordinating responses to CBRNE incidents.
    - Government agencies overseeing and supporting disaster management and emergency response efforts, such as environmental, meteorological, health, etc.
    - Civil protection and emergency response teams at various levels of government.

- Other stakeholders:

  - Military units involved in CBRNE incident response and management, such as intelligence, reconnaissance, sampling and decontamination teams.
  - Non-governmental organisations focused on disaster relief and humanitarian assistance, especially in CBRNE scenarios.
  - Environmental organisations involved in monitoring and addressing CBRNE-related environmental risks.
  - Human rights organisations concerned with the impact of CBRNE incidents on vulnerable populations.

During the survey, intermediate actions included creating a web-based prototype in co-creation with the EEG to demonstrate a possible practical application of a dCoC in plausible scenarios where the collection of samples is required. These co-creation sessions enabled end users to contribute to identifying rules for authenticity, integrity and relevant metadata considerations.

Table 1.1 resumes the survey, classifying the achievements into three categories. Existing standards lack addressing challenges related to metadata standardisation to digitally monitor the custodianship of CBRNE

Table 1.1   Overview of existing standards

Standard scope	Normative overview
CBRNE sample	Guidelines focused on characterising what is a CBRNE sample and the packaging procedure of the collected sample for secure transportation AEP-66 [3] EN17173 [4] ISO/IEC 27042 [5]
Sample collection process	Guidelines on data recommendation related to control zones as well as data related to the identification of the collected sample AEP-66 ISO 22095 [6] ISO/IEC 27042
Digital evidence of the sample	Information regarding forensic procedures, including electronic discovery (i.e. the physical device holding the digital evidence) & digital evidence (i.e. digital data regarding the sample) ISO/IEC 27050 [7, 8] ISO/IEC 27042

evidence items, raising the pertinence of a consistent methodology for designing, implementing and managing the dCoC process.

Starting with and based on this literature survey and the EEG input, the following Technical Specifications (TS) were developed:

- Digital Chain of Custody for CBRNE Evidence—Part 1: Overview and Concepts.
- Digital Chain of Custody for CBRNE Evidence—Part 2: Data Governance and Audit

These TS offer guidelines to establish rules for implementing Custody Transfer Point (CTP) actions, ensuring the chain of evidence's integrity in administrative, disciplinary and judicial proceedings. Part 1 provides an overview of the data governance workflow for automating digital custody transfers in the chain of custody for CBRNE digital evidence items. Part 2 offers formal guidelines, using Business Process Model and Notation (BPMN) to implement the specified processes, ensuring a proper understanding of the requirements for auditing the digital metadata custody (DMC) and its admissibility in court [9].

The paper discusses results obtained in the STRATEGY project (N° 883520), specifically the results targeting potential standardisation aspects related to the CBRNE context. These aspects involve business scenarios, presented as high-level descriptions from the user's viewpoint, outlining the features that should be considered while characterising the information workflow and associated data governance for intelligent automation of the dCoC. The paper also highlights the collaboration with CEN/TC 391 in developing the TS.

THE DATA GOVERNANCE PROCESS

The primary aim of establishing a standard dCoC process is to ensure the traceability and security of CBRNE evidence items [10, 11], promoting transparency and accountability by documenting all actions in the chain of custodianship [12]. Achieving a unified approach to non-repudiation custodianship requires global agreement on digital custody metadata to maintain the integrity of the dCoC and enable digital evidence auditing [2].

The data governance process should provide clear guidelines for managing and auditing DCM, allowing stakeholders to identify and audit custody ownership of CBRNE digital evidence items throughout the chain of custody [13], particularly in its CTP lifecycles, and detect inconsistencies.

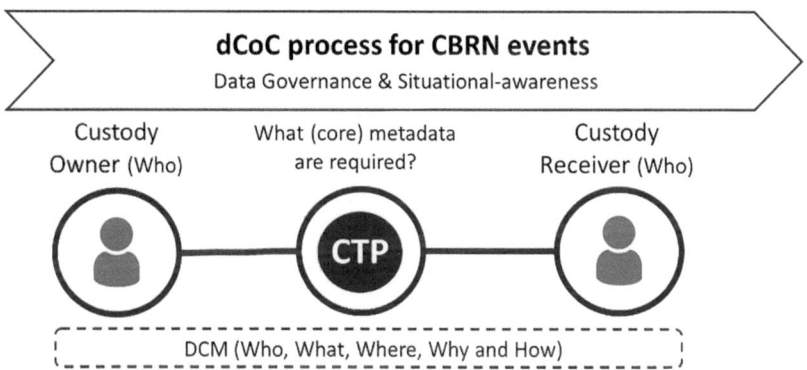

Fig. 1.1 Custody transfer schema within the dCoC process

Adopting a standardised approach with a well-defined DCM structure makes it possible to uniquely characterise each digital evidence item and effectively track custodians at each CTP [9]. To facilitate this, the CTP data model should include all the necessary metadata to describe the data package's transfer from stakeholder A to stakeholder B. Figure 1.1 outlines the essential components of the data governance workflow, wherein the Mission Command Team specifies the stakeholders involved in each CTP, along with their corresponding DCM.

The CTP lifecycle serves as a basis for gathering, interpreting, presenting and analysing stakeholder concerns from their respective viewpoints. These viewpoints are instrumental in establishing conventions that facilitate applying informal or tacit experiential knowledge. The goal is to assess whether the information reported by the DCM effectively addresses the concerns of stakeholders through role-play scenarios. The stakeholders encompass the Mission Commander, Reconnaissance, Sampling, Carrier and Laboratory Teams (Fig. 1.2).

The innovative concepts and viewpoints presented in the TS are summarised as follows:

- Custody Transfer Point (CTP)

  - List of metadata to identify a specific CTP within the dendrogram.
  - List of metadata to characterise the information within a CTP.
  - Audit the digital chain of custody, in particular, to analyse situations with inconsistencies.

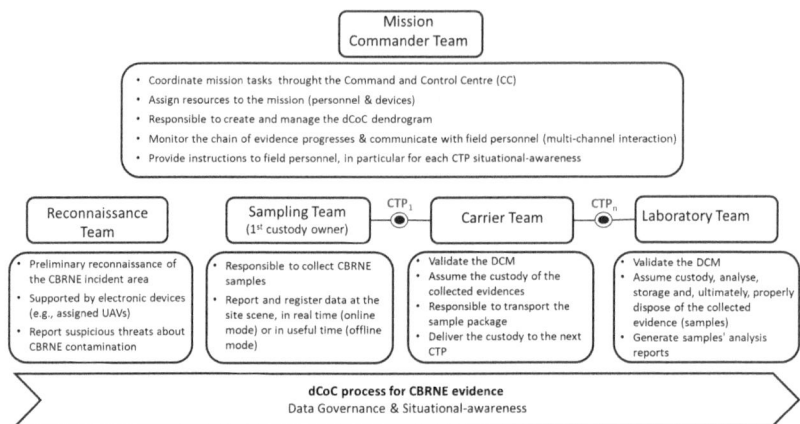

Fig. 1.2 Overview of the stakeholder's roles within the dCoC process

- Digital Custody Metadata (DCM)

    – Metadata guidelines to describe the CTP lifecycle within the dendrogram.
    – Digital log of information related to custodianship interactions (e.g. keep a historical record of the custody transfer actions performed, for each CTP, within the dendrogram).
    – Establish a metadata-centric approach for a non-repudiation digital log, creating a standard data structure for data management and auditing.

The CTP dendrogram, as exemplified in Fig. 1.3, outlines the dCoC process's information workflow and illustrates the relation between various CTPs. Once the Sampling Team collects samples at the scene, appropriate packaging is necessary for transporting them to intermediate or final destinations. Different samples may follow various paths, requiring specific transportation conditions, which the Carrier Teams must comply with. Since a package can contain multiple samples, standardised metadata is essential for consistent data sample descriptions. The DCM ensures this consistency by establishing a standardised data format structure, allowing for result comparability across the CTP dendrogram.

The mission Command Team manages the configuration of the dendrogram structure (i.e. CTP nodes and the corresponding paths within

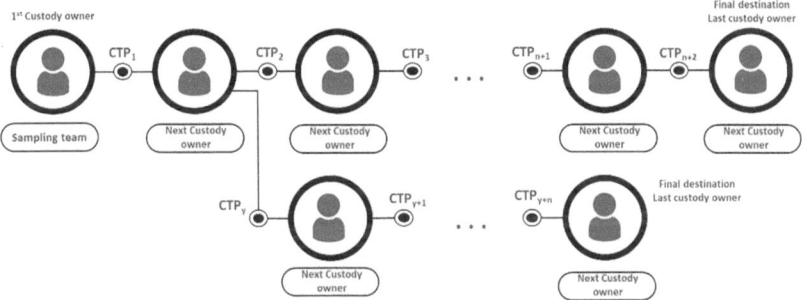

Fig. 1.3 A CTP dendrogram for a specific mission

the dCoC process). They should be able to make any changes to the dendrogram structure as operational aspects occur; for instance, whenever a stakeholder reports a data inconsistency for a specific CTP, the CTP status changes, requiring instructions on how to proceed. This makes it easier for stakeholders to understand the process and identify potential areas for improvement.

The model developed was the outcome of several co-sessions with the EEG, in which diverse scenarios and contexts were established to thoroughly test and evaluate the TS, incorporating multiple storylines that trigger actions from role-players.

A Mock-Up Demonstrating the Metadata Structure of a CTP

Figure 1.4 depicts a web-based prototype developed to help the EEG get familiarised with the essential TS concepts and the recommended metadata structure for each DCM in the dCoC process.

The layout is organised into three main sections:

- The top section provides data characterising the CTP and identifying the assigned mission, meaning data describing the mission to which the CTP is posted.
- The middle section provides a table with the essential data regarding package identifier. This area includes a line for each package assigned to the CTP. A graphical diagram on the right side visually represents the CTP status with a timestamp of the stakeholder's interactions with the system.

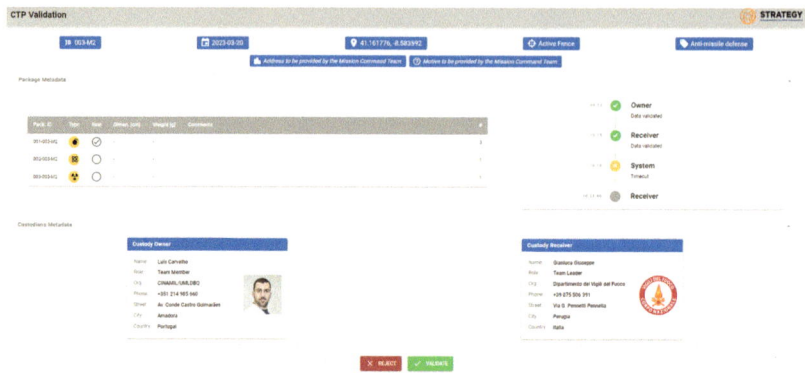

Fig. 1.4   Example of the data structure for a specific CTP

Table 1.2   Data inconsistency levels

Category	Severity levels
Severe: the integrity of the evidence package is compromised	Level 1: Evidence package inconsistency
Moderate: the integrity of the evidence package is not compromised	Level 2: Custodianship inconsistency
	Level 3: Mission inconsistency
	Level 4: Timeout event

- At the bottom, the information of the intervenient stakeholders (Custody owner and custody receiver) is presented.

The user can select a package row to obtain additional details about the package and its sample(s). If any suspicious situation arises, such as a package with a broken seal or serious cracks that could jeopardise sample's integrity (severity level L1), the user should press the reject button to report data inconsistency to the Mission Command Team. A similar procedure is followed for moderate severity level issues where the problem does not compromise the samples (see Table 1.2). In such cases, the user can capture an image of the package's exterior, add it to the image gallery and provide comments on the package's status for additional information. Access to this layout requires proper credentials, and a timeout event is triggered if no user action is detected within a specific timeframe. Default values and timeout events are configured for each process step to ensure timely completion.

The metadata that describes the transportation conditions of the package, along with the attributes that enable stakeholders to verify and validate the reported information for the CTP, holds significant importance for all parties involved [13]. Only after both stakeholders acknowledge the data, can the receiver accept custody of the physical package, thereby completing the CTP and marking it as successfully executed. As shown in Fig. 1.5, additional complementary data such as weight and package dimensions are key for the Carrier Team to handle the package's transportation effectively.

Stakeholders authorised to access samples in secure conditions, typically in intermediate or final destination laboratories, must also verify and validate the related information. This data should be accessible in the system, enabling stakeholders to ensure the digital data aligns with the physical samples inside the package. Different samples may follow various paths,

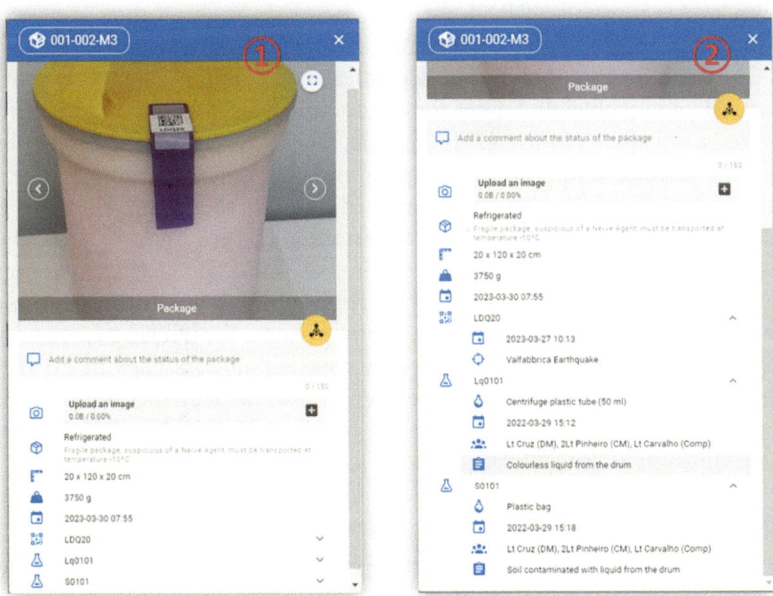

Fig. 1.5 Example of the package metadata with two sample bags

requiring specific transportation conditions, which the Carrier Teams must comply with. Since a package can contain multiple samples, standardised metadata is essential for consistent data sample descriptions.

Key information, including the Sample Collection Form, along with supporting data like pictures and videos, completed by the Sample Team at the incident location as the initial custodian, forms an essential part of this validation process within the CTP dendrogram.

Conclusions

The paper highlights the necessity for standardised recommendations, encompassing harmonised terminology and a data model, to enhance system interoperability. These guidelines empower stakeholders to adhere to applicable laws and regulations while facilitating the seamless integration of new and existing products, services and processes. Given the sensitivity and criticality of the data, preserving the dCoC becomes paramount, providing essential authenticity hallmarks required by the court. Implementing a standardised dCoC process is of utmost importance to ensure the traceability and security of CBRNE evidence items throughout the entire procedure, ensuring evidence consistency and admissibility in court.

Guidance for the dCoC process is offered, providing stakeholders with practical and reliable guidelines for conducting a DCM audit and identifying key participants in the CTP lifecycle. These guidelines establish rules for CTP actions to uphold the integrity and chain of evidence in administrative, disciplinary and judicial proceedings. Furthermore, emphasis is placed on the significance of a consistent data governance workflow, ensuring a traceable digital fingerprint of the digital evidence item throughout the dCoC process. The proposed approach introduces two new concepts, CTP and DCM, as key components for tracking custodianship within the dCoC process. It also seeks to address gaps in existing standards concerning the chain of evidence and the necessary metadata to characterise the chain of responsibility for digital evidence items as they move through each CTP within the dCoC process.

The involvement of various stakeholders highlights the need to thoroughly document all actions in the chain of custodianship, enhancing transparency and accountability. Additionally, the dCoC process should address challenges related to metadata standardisation and promptly trigger alert messages to stakeholders when monitoring the custodianship of digital CBRNE evidence items. Data protection requirements are of utmost importance in executing secure digital transfers and identifying stakeholders as contributors to the evidentiary materials at each stage of the dCoC process.

While the specific technology for implementing DCM's non-repudiation is not discussed, the technology-agnostic approach of the framework emphasises focusing on process description rather than specific implementation details. However, encryption and blockchain (or similar secure solutions) should be carefully considered to safeguard data privacy and integrity. Additionally, since a package can contain multiple samples, standardised metadata is essential for consistent data sample descriptions, and the DCM ensures this consistency by establishing a standardised data format structure, allowing for result comparability across the CTP dendrogram.

Acknowledgements The research presented in this paper was developed within the scope of the STRATEGY project, which has received funding from the European Union's Horizon 2020 research and innovation programme under grant agreement No. 883250. The information presented in the paper was classified as public. It reflects a summary of the main achievements from the author's viewpoint.

REFERENCES

1. Cosic, J., & Cosic, Z. (2012). Chain of custody and life cycle of digital evidence. *Computer Technology and Application, 3*, 126–129. Accessed: May 13, 2021. [Online]. Available: https://www.researchgate.net/publication/279175015_Chain_of_custody_and_life_cycle_of_digital_evidence

2. Prayudi, Y., & Azhari, S. N. (2015). Digital chain of custody: State of the art. *International Journal of Computers and Applications, 114*(5), 1–9. https://doi.org/10.5120/19971-1856

3. AEP-66. (2015). *Nato handbook for sampling and identification of biological, chemical and radiological agents.* Nato Standardization Agency (NSA).

4. CEN. (2020). BS EN 17173:2020 European CBRNE glossary, Brussels.

5. EN ISO/IEC. (2016). *EN ISO/IEC 27042 Information technology – Security techniques – Guidelines for the analysis and interpretation of digital evidence* (pp. 1–14). International Organization for Standardization. [Online]. Available: www.iso.org

6. ISO. (2020). *ISO/FDIS 22095: 2020 (E) Chain of custody—General terminology and models* (pp. 1–34). International Organization for Standardization. [Online]. Available: https://www.iso.org/

7. ISO/IEC. (2019). *ISO/IEC 27050-1 Information technology—Electronic discovery—Part 1: Overview and concepts* (pp. 1–20). International Organization for Standardization. [Online]. Available: www.iso.org

8. ISO/IEC. (2018). *ISO/IEC 27050-2 Information technology—Electronic discovery—Part 2: Guidance for governance and management of electronic discovery* (pp. 1–9). International Organization for Standardization. [Online]. Available: www.iso.org

9. Pestana, G. F., Carvalho, L. M., Gouveia-Carvalho, J., & Antunes, W. (2022). Digital chain of custody for CBRNE events: Custody transfer governance. In A. Rocha, H. Adeli, G. Dzemyda, & F. Moreira (Eds.), *Information systems and technologies* (pp. 304–314). Springer International Publishing.

10. Bonomi, S., M. Engineering, Ruberti, A., Casini, M., & Ciccotelli, C. (2019). B-CoC: A blockchain-based chain of custody for evidences management in digital forensics. In *International conference on blockchain economics, security and protocols* (Vol. 12, pp. 12:1–12:15). https://doi.org/10.4230/OASIcs. Tokenomics.2019.12

11. Caianiello, M., & Camon, A. (2021). *Digital forensic evidence towards common European standards in antifraud administrative and criminal investigations – OLAF, European Anti-Fraud Office.* Wolters Kluwer.

12. Ermine, J.-L. (2013). A knowledge value chain for knowledge management. *Journal of Knowledge & Communication Management, 3*(2), 85. https://doi.org/10.5958/j.2277-7946.3.2.008

13. Koleoso, R. A. (2018, July). *A digital forensics investigation model with digital chain of custody for confidentiality, integrity and authenticity,* Conimsconference.Com.Ng

Open Access This chapter is licensed under the terms of the Creative Commons Attribution 4.0 International License (http://creativecommons.org/licenses/by/4.0/), which permits use, sharing, adaptation, distribution and reproduction in any medium or format, as long as you give appropriate credit to the original author(s) and the source, provide a link to the Creative Commons license and indicate if changes were made.

The images or other third party material in this chapter are included in the chapter's Creative Commons license, unless indicated otherwise in a credit line to the material. If material is not included in the chapter's Creative Commons license and your intended use is not permitted by statutory regulation or exceeds the permitted use, you will need to obtain permission directly from the copyright holder.

CHAPTER 2

Implementation and Evaluation of Standardisation Activities in Crisis Management Through a Full-Scale Exercise

Spyridon C. Athanasiadis, Panagiotis Michalis,
Vangelis Tsougiannis, Eleftherios Ouzounoglou,
Lazaros Karagiannidis, and Angelos Amditis

INTRODUCTION

Technical and organisational interoperability in a fully transboundary configuration is considered of key importance considering that relevant authorities are required to manage a range of natural and human-made hazards that significantly impact the cascading effects on critical infrastructure and societies [1, 2]. Training exercises are employed by crisis management authorities in order to enhance preparation to effectively manage and respond to hazardous events. However, the majority of crisis

S. C. Athanasiadis (✉) • P. Michalis • V. Tsougiannis • E. Ouzounoglou •
L. Karagiannidis • A. Amditis
Institute of Communication and Computer (ICCS), Zografou, Greece
e-mail: spyros.athanasiadis@iccs.gr; p.michalis@iccs.gr; vangelis.tsougiannis@iccs.
gr; eleftherios.ouzounoglou@iccs.gr; lkaragiannidis@iccs.gr; a.amditis@iccs.gr

© The Author(s) 2025   15
I. Gkotsis et al. (eds.), *Paradigms on Technology Development*
for Security Practitioners, Security Informatics and Law
Enforcement, https://doi.org/10.1007/978-3-031-62083-6_2

management exercises follow traditional approaches such as paper-based scenarios [3, 4]. The development of a common data model for crisis management exercises is therefore considered of key importance and is the main focus of the first standardisation item which is entitled CWA 18019:2023 'Specifications for Digital Scenarios for Crisis Management Exercises' [5]. The Trial Management Tool (TMT) [6] execution manager was used for the validation of the CWA and was adapted to the specifications of the scenarios incorporating the derived information from the Full-scale Exercise (FSX) in the form of injects.

During a crisis incident, first responders are also faced with a number of competing challenges and responsibilities to take control of the situation and deal effectively and efficiently with the wounded and deceased victims. As healthcare resources are limited due to the number of injured individuals, digital victim tracking systems are implemented to offer the greatest good to the greatest amount of people [7, 8]. The goal is to determine as quickly as possible the priority of victims' treatment based on the severity of their condition, and to move patients away from the incident towards resources that offer comprehensive care. The article also presents the implementation of an existing solution entitled IN-SIDER which was modified according to the CWA18004:2023 'Requirements for acquiring digital information from victims during Search and Rescue operations' [9] providing specifications for digital victim tracking systems and part of the solution is the triage of the victims after a mass casualty incident during a Search and Rescue operation. This will assist first responders to prioritise the rescue mission and boost awareness of first responders during emergency operations.

IMPLEMENTATION OF PRE-STANDARDISATION ACTIVITIES

The following two standardisation activities aimed at enhancing interoperability in crisis management area were validated through extended FSX carried out as part of STRATEGY project on the 30th March 2023 in Gualdo Tantino (Italy). This involved more than 200 participants including 65 victims (actors), 70 FRs from different agencies and more than 70 operational stakeholders and experts who participated in the command centre. The target group of the participants included STRATEGY partners and external experts such as:

- Crisis and disaster management practitioners
- First Responders
- Standardisation experts
- Civil protection agencies' representatives
- Technical experts
- Industry representatives

Scenarios (i.e. virtual scenarios that are used for training purposes) used for the FSX comprised of Scenario A which focused on an earthquake hazard occurring in the area causing the partial collapse of a dam element and Scenario B focused on a wildland-urban interface fire. The FSX involved parallel events derived from the two scenarios that were occurring in Italy territory. The event was presented in inject-type format with a total of 82 injects which were incorporated into the TMT execution manager.

Digital Scenario Specifications for Crisis Management Exercises

The standardisation item aims at specifying a digital process for the planning of crisis management exercises. This will involve providing recommendations on the type of digital information exchanged for scenarios in the crisis management area and the use of a common data model for exchanging scenario characteristics. Main target groups are organisations responsible for planning and implementation of exercises, and particularly local, regional, national and international emergency management authorities, exercise planning teams, trainers and first responders. The FSX activities initially involved a roundtable with the participation of various stakeholders from different organisations. A presentation of the proposed data model and a demonstration of the TMT execution manager was carried out, followed by a discussion focused on the evaluation of the CWA18019:2023 content for providing feedback. An example with respect to scenario planning as well as a short demonstration of the TMT execution manager was also provided to participants, offering a better understanding of the proposed pre-standardisation activity and leading to fruitful discussions and collection of feedback by participants. The actual FSX then took place during which the content of the proposed CWA18019:2023 was put into practice incorporating two FSX scenarios which involved a large number of actors, events and injects (see Fig. 2.1). The execution of the virtual scenario was carried out by the TMT while

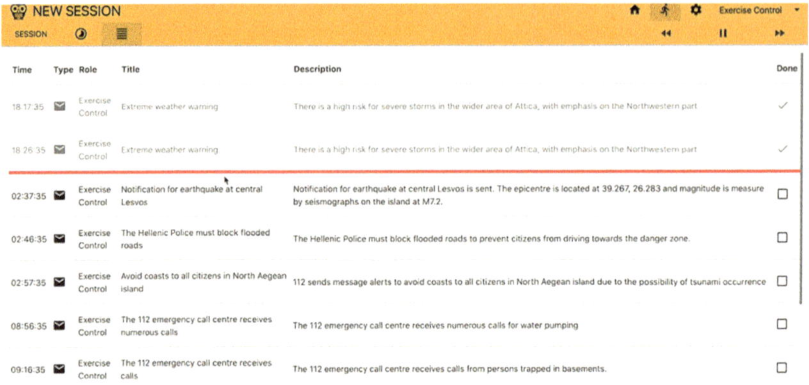

Fig. 2.1 Scenario injects and events generated during the TTX using the TMT application for the two FSX scenarios demonstrating the applicability of the proposed data model

certain injects that contained alarm messages were published to the message broker provided by the Driver+ framework [6] and then received by a Command and Control (C2) system (see Fig. 2.2).

Digital Victim Tracking System for Mass Casualty Incidents

The proposed CWA18004:2023 was tested in the FSX in the scenario of mass casualty incident where 65 victims and first responders from different agencies were involved. In total, 70 triage tags were used (i.e. 20 digital victim tracking tags and 50 paper-based with a QR code) and 370 transactions conducted between the tags and the mobile applications. All the tags are scanned by the same mobile application by the first responders in three phases (i.e. primary triage, secondary triage and before dispatching to hospitals or after a major update identified). The progress of the operation was available on the main platform IN-SIDER, shown in Fig. 2.3, in real time shared with a C2 system and covered the whole part of the Scenario A. Figure 2.4 also presents actual First Responders using the mobile application to triage and track victims during the FSX.

Fig. 2.2 View of the FSX control room with TMT generating injects containing alarm messages (bottom screen), published to the message broker provided by the D+ framework and then received and presented by the C2 system (top left screen)

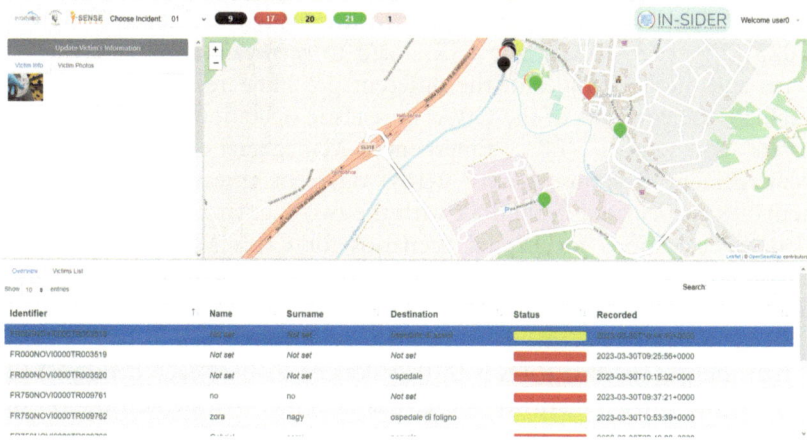

Fig. 2.3 IN-SIDER platform with the overall information after the end of the mission

Fig. 2.4   First responders in actions during the FSX evaluation

EVALUATION OF PRE-STANDARDISATION ACTIVITIES

The evaluation for both standardisation activities conducted for Search and Rescue and Training is mainly based on three pillars (TTX evaluation, FSX evaluation and comments received through the public commenting period). The evaluation performed during the TTX is dedicated to Search and Rescue conducted 23rd of June 2022 and for the Training conducted 22nd of June 2022. For the commenting period, in total received 36 comments which improved the final versions of the pre-standardisation activity. The final evaluation was performed during the FSX week in March 2023 (27-31/3/2023). The FSX aimed to showcase all integrated systems and methodologies of the different pre-standardisation items, and evaluate the concepts of the CWAs in realistic operational conditions and assets and actors (FRs, Emergency Management Centres, Crisis Management Response teams) across different threat scenarios. Both activities participated in the FSX with the two standardisation items entitled 'Specifications for Digital Scenarios for Crisis Managements' and 'Requirements for acquiring digital information from victims during Search and Rescue operations'.

Both CWAs involvement in the FSX week took place in two phases. Firstly, during the first dates with two slots of 4 hours each dedicated to present the current status of the CWAs documents, a demonstration of

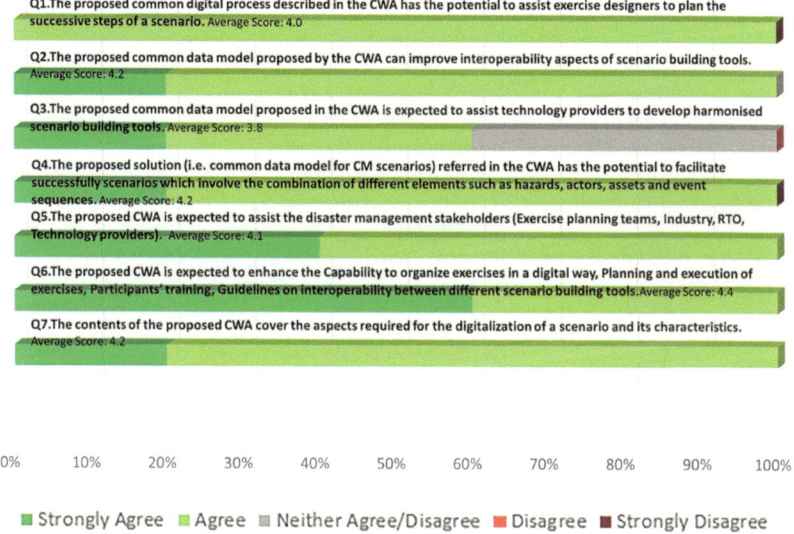

Fig. 2.5 Scores' distribution obtained for all the statements of CWA: 'Specifications for Digital Scenarios for Crisis Management Exercises'

potential solutions adapted to the propositions of the CWAs and an open discussion based on the end-user questions. In the end of the discussions, a questionnaire was shared among the participants to provide input. The following Figs. 2.5 and 2.6 present the feedback collected by crisis management stakeholders for both CWAs respectively.

The obtained results for CWA 18019:2023 and CWA 18004:2023 indicate an overall score ranging from 3.8 to 4.4 (out of 5) and 4.1 to 4.7 (out of 5) with average scores of 4.1 and 4.2 respectively, which highlights that the participants evaluated satisfactorily the CWA activities and rated them as promising and well-structured.

Secondly on 30/3/2023 leading the first part and large part of the 2nd scenario of the exercise providing a solution to triage and track victims after a mass casualty incident and a digital scenario builder. Indicatively Figs. 2.7 and 2.8 are available for large-scale acceptance from the operational stakeholder participating either as players or as observers.

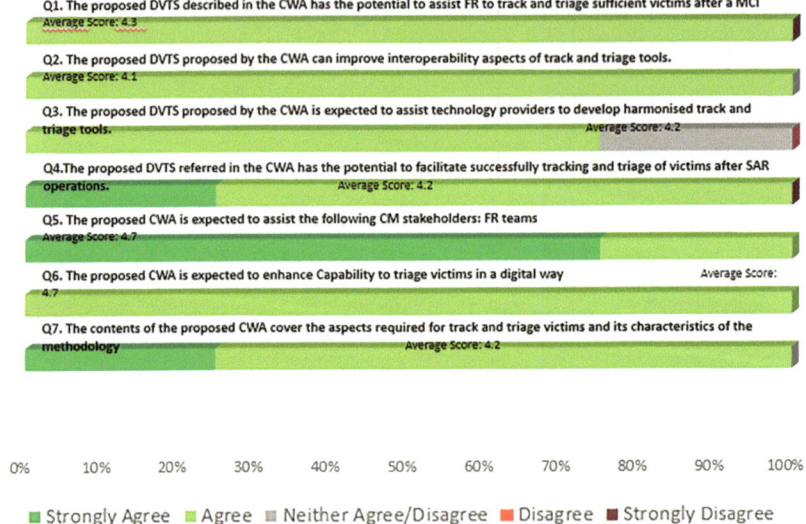

Q1. The proposed DVTS described in the CWA has the potential to assist FR to track and triage sufficient victims after a MCI
Average Score: 4.3

Q2. The proposed DVTS proposed by the CWA can improve interoperability aspects of track and triage tools.
Average Score: 4.1

Q3. The proposed DVTS proposed by the CWA is expected to assist technology providers to develop harmonised track and triage tools.
Average Score: 4.2

Q4.The proposed DVTS referred in the CWA has the potential to facilitate successfully tracking and triage of victims after SAR operations.
Average Score: 4.2

Q5. The proposed CWA is expected to assist the following CM stakeholders: FR teams
Average Score: 4.7

Q6. The proposed CWA is expected to enhance Capability to triage victims in a digital way
Average Score: 4.7

Q7. The contents of the proposed CWA cover the aspects required for track and triage victims and its characteristics of the methodology
Average Score: 4.2

0% 10% 20% 30% 40% 50% 60% 70% 80% 90% 100%

■ Strongly Agree ■ Agree ■ Neither Agree/Disagree ■ Disagree ■ Strongly Disagree

Fig. 2.6 Scores' distribution obtained for all the statements of CWA: 'Requirements for acquiring digital information from victims during Search and Rescue operations'

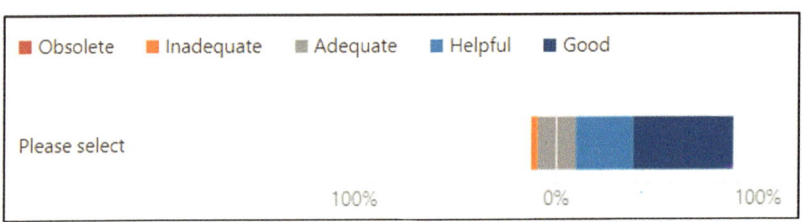

■ Obsolete ■ Inadequate ■ Adequate ■ Helpful ■ Good

Please select

100%   0%   100%

Fig. 2.7 FSX participant responses on 'How do you rate the methodology adopted for running the FSX, using an IT infrastructure for supporting the exercise and the interaction among the participants?'

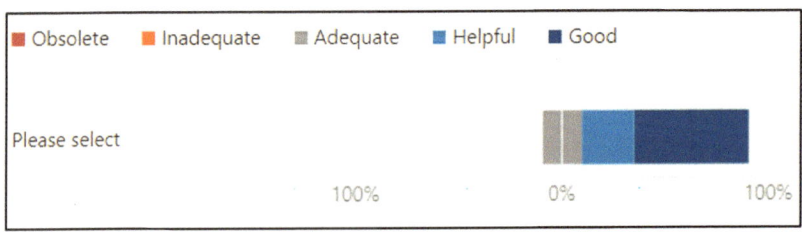

■ Obsolete ■ Inadequate ■ Adequate ■ Helpful ■ Good

Please select

100%   0%   100%

Fig. 2.8 FSX participant responses on 'How do you rate the scenarios used during the FSX? Do you think it helped to achieve the objective of the exercise?'

Conclusions

The technical and validation activities of STRATEGY project were concluded with the FSX which provided the basis to demonstrate how the proposed pre-standardisation activities of Stream 1 'Search and Rescue' and Stream 7 'Training' could be integrated and implemented within specific technical tools, demonstrating their practical application and opening up future opportunities for exploitation. Meanwhile, the development of the CWAs was enriched by the experiences and feedback of all involved external participants and experienced partners and submitted for publication on May 2023 at CEN/CENELEC and both accepted while receiving the unique numbers below. CWA 18019:2023 'Specifications for Digital Scenarios for Crisis Management Exercises' and CWA18004:2023 'Requirements for acquiring digital information from victims during Search and Rescue operations'.

Acknowledgements This project has received funding from the European Union's Horizon 2020 research and innovation programme under grant agreement No. 883520, project STRATEGY. This article reflects only the authors' views, and the Research Executive Agency and the European Commission are not responsible for any use that may be made of the information it contains.

References

1. Michalis, P., & Vintzileou, E. (2022). The growing infrastructure crisis: The challenge of scour risk assessment and the development of a new sensing system. *Infrastructures, 7*, 68. https://doi.org/10.3390/infrastructures7050068
2. Michalis, P., & Sentenac, P. (2021). Subsurface condition assessment of critical dam infrastructure with noninvasive geophysical sensing. *Environmental Earth Sciences, 80*, 556. https://doi.org/10.1007/s12665-021-09841-x
3. Amditis, A. J., Ouzounoglou, E., Michalis, P., Misichroni, F., Perlepes, L., & Sdongos, E. (2023). Interoperability and standardisation supporting preparedness and response to disasters. In C. Fonio, A. Widera, & T. Zwęgliński (Eds.), *Innovation in crisis management* (1st ed.). Routledge. https://doi.org/10.4324/9781003256977. ISBN: 9781003256977.
4. Athanasiadis S. C., Michalis P., Ouzounglou E, Karagiannidis L., & Amditis A. (2002). Interoperability in crisis management: Specifications for digital scenarios to enhance acquisition of triage information during search and rescue operations. In *SafeThessaloniki, 2002, proceedings*. ISSN 2265-1823.

5. CEN workshop Agreement 18019:2023. *Specifications for digital scenarios for crisis management exercises.* https://www.cencenelec.eu/news-and-events/news/2023/workshop/2023-03-23-digital-scenarios-for-crisis-management/
6. Trial Management Tool (TMT). Available at https://github.com/DRIVER-EU/scenario-manager
7. Misichroni, F., Stamou, A., Kuqo, P., Tousert, N., Rigos, A., Sdongos, E., & Amditis, A. (2021). A novel, reliable and real-time solution for triage and unique identification of victims of mass casualty incidents. *Engineering Proceedings, 2021*(6), 72. https://doi.org/10.3390/I3S2021Dresden-10180
8. Rigos, A., Sdongos, E., Misihroni, F., Stamou, A., Kuqo, P., Latsa, E., & Amditis, A. (2019). A novel triage system using Bluetooth devices in support of incident management. In *13th International Conf. on Sensing Technology (ICST).* https://doi.org/10.1109/ICST46873.2019.9047744
9. CEN workshop Agreement 18004:2023. *Requirements for acquiring digital information from victims during Search and Rescue operations.* https://www.cencenelec.eu/news-and-events/news/2023/workshop/2023-01-19-digital-information-search-and-rescue-operations/

Open Access This chapter is licensed under the terms of the Creative Commons Attribution 4.0 International License (http://creativecommons.org/licenses/by/4.0/), which permits use, sharing, adaptation, distribution and reproduction in any medium or format, as long as you give appropriate credit to the original author(s) and the source, provide a link to the Creative Commons license and indicate if changes were made.

The images or other third party material in this chapter are included in the chapter's Creative Commons license, unless indicated otherwise in a credit line to the material. If material is not included in the chapter's Creative Commons license and your intended use is not permitted by statutory regulation or exceeds the permitted use, you will need to obtain permission directly from the copyright holder.

CHAPTER 3

Proposed Actions Toward Streamlining Cyprus Wide Prevention Strategies in Wildfire Management

Pooja Pandey, George Boustras, Miriam Arenas Coneio, and Guillermo Rein

INTRODUCTION

The wildfire issue in Europe is undergoing significant changes, primarily driven by warmer temperatures, drier conditions, and several other factors such as land abandonment and increased fuel load [10, 26]. These factors make wildfires harder to contain, more expensive to suppress, and a great threat to firefighters and the public [19, 43, 44]. The rising risk of wildfires in a warmer and drier environment indicates an ongoing menace [1]. In 2022, the highest number of fires since 2006 occurred, foreshadowing

P. Pandey (✉) • G. Boustras
Centre of Excellence in Innovation and Technology, European University Cyprus, Engomi, Cyprus

M. A. Coneio
Estudis de Psicologia i Ciències de l'Educació, Universitat Oberta de Catalunya, Barcelona, Spain

G. Rein
Department of Mechanical Engineering, Imperial College London, London, UK

© The Author(s) 2025   25
I. Gkotsis et al. (eds.), *Paradigms on Technology Development for Security Practitioners*, Security Informatics and Law Enforcement, https://doi.org/10.1007/978-3-031-62083-6_3

the worst drought in Europe in 500 years [34]. Managing this increasing concern has become difficult due to the high suppression costs, damage to resources, and disproportionate risks to lives [5, 32].

To reduce extreme wildfire damage, it is crucial to implement Integrated Fire Management (IFM) approach. This requires addressing a contradiction researchers have contemplated for a decade [10, 14, 16, 24, 33, 40]: relying solely on firefighting personnel and tactics is not viable [12, 17]. Rather proactive prevention strategies at the EU and the national levels can effectively limit extreme fire danger.

To initiate this transformative process, existing practices may need modification: objectives must be redefined, and the "fear trap" (the negative stress imposed on response agencies as they continuously strive to enhance efficiency and safety against increasingly rapid, intense, and overwhelming wildfires through resource and technological advancements) described by [8] should be overcome. Key aspects to analyze this situation include understanding the rationale for prioritizing IFM (*WHAT*), assessing the effectiveness of current approaches (*WHY*), identifying impactful measures (*HOW*), and determining the optimal timing for action (*WHEN*).

However, possessing knowledge alone doesn't guarantee pragmatic policy decisions. Factors such as politics, economics, and societal norms can influence policymaking and hinder evidence-based strategies [6, 22, 31, 37]. Therefore, contextual factors that impact policy implementation should be considered alongside scientific research [2]. This comprehensive approach ensures policies are based on knowledge and account for sociopolitical intricacies, leading to effective and pragmatic solutions [28, 35].

This paper explores the sociopolitical challenges to wildfire prevention strategies using a case study approach. Consequently, providing recommendations to overcome these challenges and enhance wildfire management practices.

METHODOLOGY

This paper stems from the interviews conducted as part of a PhD thesis, focusing on wildfire prevention strategies within the IFM framework in Cyprus. The research methodology involved conducting ten semi-structured interviews with key stakeholders involved in wildfire management in Cyprus. In addition, the sociopolitical context in Cyprus was examined to gain insights into how it impacts wildfire management strategies. Factors influencing the implementation of prevention measures were also identified. Drawing from the insights gathered during these interviews, recommendations aimed at enhancing the coherence and effectiveness of wildfire prevention strategies in Cyprus were developed. These recommendations underscore the critical role of education in promoting public awareness and engagement. Implementing these recommendations has the potential to bolster Cyprus's resilience in the face of escalating wildfire threats.

Case Study—Cyprus

The Sociopolitical Situation in Cyprus

The sociopolitical landscape in Cyprus poses significant challenges to the effective implementation of wildfire prevention strategies. The island's divided nature has resulted in limited collaboration among authorities responsible for environmental and natural resource management. This political division often leads to disputes over land ownership and management, which, in turn, hinder the successful execution of wildfire prevention strategies. Moreover, Cyprus's reliance on tourism creates conflicting interests, with the need to safeguard natural resources sometimes conflicting with the imperative to sustain the economy. This can potentially result in less stringent regulations and an increased risk of wildfires in the wildland-urban interface. It is worth noting that this paper primarily focuses on the Republic of Cyprus due to the author's geographical location and the specific political context that restricts cooperation and data sharing.

Wildfire Situation and Related Framework in the Republic of Cyprus

Wildfire management in the Republic of Cyprus primarily focuses on suppression [4, 29]. Despite multiple organizations participating in fire management efforts, a notable lack of coordination among them poses a significant constraint [18]. However, it is widely recognized that prevention is the most crucial strategy for managing wildfires [21, 23]. Various initiatives, including policymaking, forest management, education, and training, aim to reduce the likelihood of fires and control their spread and intensity [11, 15, 42]. Therefore, understanding the existing strategies is vital for identifying areas that require improvement. The institutional framework for wildfire management in the Republic of Cyprus is defined by The Forest Law and the Firefighting Action Plan in Rural Areas. The Department of Forests, operating under the Ministry of Agriculture, Rural

Development, and Environment, holds legal, administrative, and technical responsibility for the effective control and extinguishing of forest fires. The Forest Law governs forest management, with a particular emphasis on the prevention and suppression of wildfires, including penalties for those responsible for starting or allowing fires in forested areas.

The Forest Department places significant emphasis on prevention, preparedness, response, and recovery. Regarding prevention strategies, several measures have been implemented in Cyprus. Law Enforcement strictly prohibits lighting fires within 1 km of state forest boundaries, resulting in penalties of imprisonment or fines. The department occasionally conducts awareness campaigns to engage the public and mitigate human-caused fires. Management of picnic and camping sites concentrates visitors in designated areas with fireplaces and amenities to reduce fire risk.

Despite the Department of Forest's efforts to enhance wildfire prevention, several factors still hinder the effective implementation of prevention strategies in Cyprus. These challenges were identified through an examination of barriers to interagency exchange and collaboration in wildfire management, supported by the Assessment Report by the Audit Office of the Republic of Cyprus and interview transcripts.

Factors Influencing Wildfire Prevention in the Republic of Cyprus
- Availability of resources—For an organization, it is most important that resources are available and accessible to effectively manage wildfire programs [42]. Several key resources are critical for effective wildfire prevention efforts. These resources, as identified in the case study, include *personnel, equipment such as fire trucks and firefighting tools, adequate funding for prevention measures, and advanced technology like early warning systems.* However, a lack of resources, including data and funding, has made it difficult to keep up with the increasing wildfire risks.
- Employee training and situation awareness—Employee training and situation awareness are crucial factors in wildfire prevention in Cyprus. Effective training programs can improve the knowledge and skills of employees, allowing them to be aware of their surroundings and any potential fire hazards, as well as any changes in weather conditions that may increase the risk of wildfires. This can also help them take proactive measures to prevent wildfires and respond quickly and effectively if a fire does occur.
- Organizational collaboration—Effective organizational collaboration is crucial for several reasons in Cyprus. Firstly, it can help to

ensure that all relevant stakeholders are involved in the prevention efforts, which can improve the effectiveness of those efforts. Secondly, organizational collaboration can help to ensure that resources are used efficiently and effectively. By working together, organizations can share information about the areas most at risk of wildfires and coordinate the allocation of resources such as firefighting equipment and personnel. Thirdly, organizational collaboration can help to build trust and foster a sense of shared responsibility among stakeholders. When different organizations work together toward a common goal, they are more likely to feel a sense of ownership and commitment to the success of the prevention efforts.

- Government law and policies—Government policy plays an essential role by setting regulations and guidelines for land management, fire prevention, and suppression activities [7, 20, 25, 39]. Effective policies encourage public engagement in wildfire programs and ensure alignment between organizational and governmental efforts [9]. Cyprus' Forest Law (N. 25(I)/2012) designates the responsibility of the Director of the Department of Forests for preventing and suppressing forest fires within state forests and two kilometers from their borders. However, the law's limitations hinder prevention measures in the zone mostly comprising private plots and forest communities protected by constitutional rights Articles 16 and 23. Thus, by neglecting to enforce laws aimed at minimizing the probability of fires happening beyond their purview, there is ample space for mistakes to occur.
- Community engagement—In the wildfire literature, much attention is given to community engagement, highlighting the sensitive issues of *who* should participate in decision-making [13]. It involves active participation and collaboration between local communities, firefighting agencies, and local governments. Through community engagement, awareness can be raised, education provided, and responsible behavior promoted to reduce the risk of wildfires.

Wildfire management strategies are influenced by various factors, including the involvement of individuals, administrative and resource contexts, and the specific geographic location under consideration [3, 30, 38, 41]. While wildfire prevention activities do not normally reduce the number of natural ignitions [27], education can be effective in mitigating the losses in areas, in addition to the historical data of natural ignitions [11, 15, 36]. An example of successful wildfire preparedness through education and awareness campaign can be seen in Firewise communities of Italy. In addition, some potential strategies for effective wildfire prevention in Cyprus are provided in the following section. The potential strategies discussed here are the result of the interviews conducted by the corresponding author in Cyprus.

PROPOSED ACTIONS TOWARD STREAMLINING CYPRUS WIDE PREVENTION STRATEGIES IN WILDFIRE MANAGEMENT

- Collaborative Approach—Foster collaboration and partnership between EU countries, local communities, and stakeholders, including forest owners, land managers, fire officials, and firefighters. Encourage the sharing of knowledge, resources, and expertise in preventing, detecting, and managing wildfires.

  For example, the "National Operational Program for Forest Fire Control" in Greece is a collaborative program that aims to prevent and manage wildfire through coordinated actions by various agencies, including the Ministry of Environment and Energy, the Hellenic Fire Service, and the Forest Service.

- Enhance Preparedness—Increase preparedness by conducting regular risk assessments, developing evacuation plans, and improving communication system. Utilize new technologies such as fire detection systems, drones, and satellite imagery for early detection and monitoring of wildfires.

  For example, in Portugal, the Integrated Rural Fire Management (SGIFR) program, established in 2018, promotes a coordinated and integrated approach to rural fire management. The program includes measures such as fuel reduction and fire prevention education.

- Promoting Prevention Measures and Land Use Planning— Encourage and support forest management practices that reduce the risk of wildfires, such as thinning and grazing. Implement strict regulations and enforcement mechanisms to prevent illegal activities that can cause wildfires. Prioritize land use planning to reduce the risk of extreme wildfires, particularly in high-risk areas such as forests, grasslands, and shrublands. This can be achieved through measures such as zoning regulations.

  For example, Italy has developed a National Forest Fire Prevention Plan that aims to reduce the risk of wildfires through better land management practices.

- Increasing Public Awareness—Raise awareness among the general public about the dangers of wildfires, the role they can play in preventing them, and the need for responsible behavior. Use age-appropriate education campaigns, social media, and community engagement to spread awareness and encourage public participation.

  For example, Italy has implemented a National Fire Prevention Week, which includes public awareness campaigns and educational programs on fire prevention and safety measures.

- Improve Coordination—Improve coordination among different agencies and stakeholders involved in wildfire management, including national and local authorities, emergency services, and civil protection agencies. Establish clear lines of communication and collaboration to ensure a timely and effective response to wildfires.

  For example, Spain, the National Plan for the Prevention and Control of Forest Fires (INFOMA), established in 2006, coordinates the response to wildfires. The plan includes representatives from various agencies, including the fire service, police, civil defense, volunteers, and private companies. Similarly, in Portugal, the National Civil Protection Authority (ANPC) is responsible for coordinating the response to wildfires and includes representatives from various agencies.

- Research and Innovation—Invest in research and innovation to develop new technologies, practices, and strategies to prevent and manage wildfires. Develop a knowledge base of best practices and share it across the neighboring states and with the EU to enhance prevention efforts. For example, in Finland, the Finnish Meteorological Institute (FMI) conducts research on wildfire prevention and management. The FMI develops new technologies and strategies for predicting and monitoring wildfires, as well as providing training and education on wildfire management.

  Similarly, in Spain, the Center for Forest Fire Research (CIFOR), established in 1987, conducts research on wildfire prevention and management. The center collaborates with various stakeholders, including universities, government agencies, and private companies.

These measures target the primary causes of wildfires in Cyprus based on research interviews. However, they can be applied in other fire-prone countries. Additionally, education plays a critical role in preventing wildfires because many wildfires in Cyprus are caused by human activity, such as careless use of fire or littering, and can be prevented by changing human behavior. Table 3.1 provides a comprehensive list of education measures for wildfire prevention in Cyprus.

Table 3.1 Proposed educational wildfire prevention strategies with evidence from effective initiatives from other countries

Wildfire prevention strategy

Initiatives that use public awareness and education

Community	Media	Signage and high-visibility public contact	Public education
Deliver homeowner fire safety resources (FireWise) Deliver fire safety programs in conjunction with community association (forest intervention zone) Organizing and getting involved in neighborhood town meetings (building fire resilient communities) Organize collaborative activities through yearly and bi-yearly campaigns (spring cleanup, neighborhood watch, weed abatement program)	Prepare public service announcement. Coordinate the local stations' participation in the national fire prevention program. Airing recurring radio and television messages about benefits of fire prevention activities and community programs. Release timely news to local printed media. Training media personnel on effectively communicating about wildfire (national day radio)	Develop and annually update the fire prevention sign plans such as location signs (highway, road, etc.), type of message (seasonally) Maintain and repair current signs. Create engine patrol routes and use them in need such as on vacations, weekends. Creating interagency education programs to maintain high visibility of fire prevention effort.	Take part in educational initiatives like fairs, exhibits, and service groups to keep preventative efforts visible (school campaign in Cyprus) Create and lead a fire prevention month or week (wildfire awareness month) Run outdoor fire safety campaigns. Create a safety equipment education campaign. Create and conduct character appearances programs as per standard guidelines (respect the flame). Create bilingual and multicultural fire prevention such as team teaching, special use permits, fire regulations

Source: Barriers to Interagency Exchange and Collaboration [45], Interviews

CONCLUSIONS

Wildfires are a recurring and significant threat in Cyprus, fuelled by a combination of high temperatures, low humidity, and strong winds. These conditions create ideal circumstances for the rapid ignition and spread of wildfires. In recent years, wildfires in Cyprus, such as the one that occurred in the Arakapas region in 2021, have caused significant damage to forests, vegetation, and infrastructure, endangering human lives and property.

Currently, wildfire management in Cyprus primarily focuses on suppression efforts. However, adopting an integrated approach could be beneficial in reducing the likelihood of fire incidents and controlling their spread and intensity. The insights presented in this paper will be particularly valuable to individuals and organizations involved directly or indirectly in planning, implementing, and enhancing wildfire prevention measures. This includes wildfire managers, policymakers, scientists, and others committed to combating these fires effectively. Furthermore, the media can use the insights from this paper to refine their strategies for wildfire prevention and to communicate targeted and impactful messages to the public. Additionally, evaluating and implementing the suggested measures on a Europe-wide scale could also be beneficial.

Acknowledgments This project has received funding from the European Union's Horizon 2020 research and innovation program MSCA-ITN-2019—Innovative Training Networks under grant agreement No. 860787 (Pyrolife). This article reflects only the authors' views, and the Research Executive Agency and the European Commission are not responsible for any use that may be made of the information it contains.

REFERENCES

1. Barriers to Interagency Exchange and Collaboration. PhD Thesis. (2023). Interviewed by Pandey, P., Cyprus.
2. Bednar-Friedl, B., Biesbroek, R., & Schmidt, D. N. (2022). IPCC Sixth Assessment Report (AR6): Climate change 2022-Impacts, adaptation and vulnerability: Regional factsheet Europe.
3. Beierle, T. C., & Cayford, J. (2002). *Democracy in practice: Public participation in environmental decisions. Resources for the future.* Routledge.
4. Blanchard, B., & Ryan, R. L. (2003, April). Community perceptions of wildland fire risk and fire hazard reduction strategies at the wildland-urban inter-

face in the northeastern United States. In *Proceedings of the 2003 northeastern recreation research symposium, 317* (Vol. 317, pp. 285–294).

5. Boustras, G., & Boukas, N. (2013). Sci-Hub | Forest fires' impact on tourism development: A comparative .... *Management of Environmental Quality: An International Journal, 24*(4). https://doi.org/10.1108/MEQ-09-2012-0058

6. Bowman, D. M., Kolden, C. A., Abatzoglou, J. T., Johnston, F. H., van der Werf, G. R., & Flannigan, M. (2020). Vegetation fires in the Anthropocene. *Nature Reviews Earth & Environment, 1*(10), 500–515.

7. Brownson, R. C., Baker, E. A., Deshpande, A. D., & Gillespie, K. N. (2017). *Evidence-based public health.* Oxford University Press.

8. Carreiras, M., Ferreira, A. J. D., Valente, S., Fleskens, L., Gonzales-Pelayo, Ó., Rubio, J. L., Stoof, C. R., Coelho, C. O. A., Ferreira, C. S. S., & Ritsema, C. J. (2014). Comparative analysis of policies to deal with wildfire risk. *Land Degradation & Development, 25*(1), 92–103.

9. Castellnou, M., Prat-Guitart, N., Arilla, E., Larrañaga, A., Nebot, E., Castellarnau, X., Vendrell, J., Pallàs, J., Herrera, J., Monturiol, M., & Cespedes, J. (2019). Empowering strategic decision-making for wildfire management: Avoiding the fear trap and creating a resilient landscape. *Fire Ecology, 15*, 1–17.

10. Colavito, M. (2021). The human dimensions of spatial, pre-wildfire planning decision support systems: A review of barriers, facilitators, and recommendations. *Forests, 12*(4), 483.

11. Costa, P., Larrañaga, A., Castellnou, M., Miralles, M., & Kraus, D. (2011). *Prevention of large wildfires using the fire types of concept.* European Forest Institute.

12. Department of Forests, Cyprus. http://www.moa.gov.cy/moa/fd/fd.nsf/fd51_en/fd51_en?OpenDocument

13. Dube, O. P. (2013). Emissions mitigation opportunities for savanna countries from ... – Nature. *Weather and Climate Extremes, 1*, 26–41.

14. Eckerberg, K., & Buizer, M. (2017). Promises and dilemmas in forest fire management decision-making: Exploring conditions for community engagement in Australia and Sweden. *Forest Policy and Economics, 80*, 133–140.

15. Europe, F. (2010, May). Assessment of forest fire risks and innovative strategies for fire prevention. In *Proceedings of the workshop on the assessment of forest fire risks and innovative strategies for fire prevention*, Rhodes, Greece, pp. 4–6.

16. European Commission. (2021). https://op.europa.eu/en/publication-detail/-/publication/4e6cc1f1-8b8a-11eb-b85c-01aa75ed71a1

17. FAO. (2010). https://www.fao.org/news/story/en/item/44808/icode/

18. Heikkilä, T. V., Grönnqvist, R., & Jurvélius, M. (2007). *Wildland fire management. Handbook for trainers.* Ministry for Foreign Affairs of Finland.

19. Herrero, G., Lázaro, A., & Montiel, C. (2010). *A comparative assessment of the European forest policies and their influence in wildfire management.* DIANE Publishing.

20. Jenkins, M. J., Page, W. G., Hebertson, E. G., & Alexander, M. E. (2012). Fuels and fire behavior dynamics in bark beetle-attacked forests in Western North America and implications for fire management. *Forest Ecology and Management, 275*, 23–34.
21. Kocher, S. D., & Butsic, V. (2017). Governance of land use planning to reduce fire risk to homes Mediterranean France and California. *Land, 6*(2), 24.
22. Lasanta, T., Khorchani, M., Pérez-Cabello, F., Errea, P., Sáenz-Blanco, R., & Nadal-Romero, E. (2018). Clearing shrubland and extensive livestock farming: Active prevention to control wildfires in the Mediterranean mountains. *Journal of Environmental Management, 227*, 256–266.
23. Liverani, M., Hawkins, B., & Parkhurst, J. O. (2013). Political and institutional influences on the use of evidence in public health policy. A systematic review. *PLoS One, 8*(10), e774.
24. Marino, E., Hernando, C., Planelles, R., Madrigal, J., Guijarro, M., & Sebastián, A. (2014). Forest fuel management for wildfire prevention in Spain: A quantitative SWOT analysis. *International Journal of Wildland Fire, 23*(3), 373–384.
25. Mateus, P., & Fernandes, P. M. (2014). Forest fires in Portugal: Dynamics, causes and policies. In *Forest context and policies in portugal: Present and future challenges* (pp. 97–115). Springer.
26. Mockrin, M. H., Fishler, H. K., & Stewart, S. I. (2018). Does wildfire open a policy window? Local government and ... – PubMed. *Environmental Management, 62*, 210–228.
27. Mohammadi, Z., Lohmander, P., Kašpar, J., Berčák, R., Holuša, J., & Marušák, R. (2022). The effect of climate factors on the size of forest wildfires (case study: Prague-East district, Czech Republic). *Journal of Forestry Research, 33*, 1–10.
28. National Wildfire Coordinating Group. https://www.nifc.gov/sites/default/files/pio/documents/3%20Fire%20Management.PDF
29. Nikolakis, W., & Roberts, E. (2022). Wildfire governance in a changing world: Insights for policy learning and policy transfer. *Risk, Hazards & Crisis in Public Policy, 13*(2), 144–164.
30. Papageorgiou, K., & Papageorgiou, G. (2012). Management of Forest Fires, Prevention, Detection, Control and Restoration; the Case of Cyprus. In 1st International Conference on Safety and Crisis Management in the Construction, Tourism and SMEs Sectors (1st CoSaCM), Nicosia, Cyprus.
31. Paton, D. (2013). Disaster resilient communities: Developing and testing an all-hazards theory. *IDRiM Journal, 3*(1), 1–17.
32. Pivello, V. R., Vieira, I., Christianini, A. V., Ribeiro, D. B., da Silva Menezes, L., Berlinck, C. N., Melo, F. P., Marengo, J. A., Tornquist, C. G., Tomas, W. M., & Overbeck, G. E. (2021). Understanding Brazil's catastrophic fires: Causes, consequences and policy needed to prevent future tragedies. *Perspectives in Ecology and Conservation, 19*(3), 233–255.

33. Plucinski, M. P. (2019). Fighting flames and forging firelines: Wildfire suppression effectiveness at the fire edge. *Current Forestry Reports, 5*, 1–19.
34. Rego, F., Alexandrian, D., Fernandes, P., & Rigolot, E. (2007, May). Fire paradox: An innovative approach of integrated wildland fire ... In *Proceeding of 4th international wildland fire conference* (pp. 13–17).
35. San-Miguel-Ayanz, J., Durrant, T., Boca, R., Maianti, P., Libertá, G., Oom, D., Branco, A., de Rigo, D., Ferrari, D., Roglia, E., & Scionti, N. (2023). *Advance report on forest fires in Europe, Middle East and North Africa 2022.* Publications Office of the European Union. https://doi.org/10.2760/091540, JRC133215.
36. Sarkies, M. N., Bowles, K. A., Skinner, E. H., Haas, R., Lane, H., & Haines, T. P. (2017). The effectiveness of research implementation strategies for promoting evidence-informed policy and management decisions in healthcare: A systematic review. *Implementation Science, 12*, 1–20.
37. Swedish Services Rescue Agency. https://ec.europa.eu/echo/files/civil_protection/civil/prote/pdfdocs/fire_prevention.pdf
38. Swinburn, B., Gill, T., & Kumanyika, S. (2005). Obesity prevention: A proposed framework for translating evidence into action. *Obesity Reviews, 6*(1), 23–33.
39. Syphard, A. D., & Keeley, J. E. (2015). Location, timing, and extent of wildfire vary by cause of ignition. *International Journal of Wildland Fire, 24*(1), 37–47.
40. Tedim, F., Xanthopoulos, G., & Leone, V. (2015). Forest fires in Europe: Facts and challenges. In *Wildfire hazards, risks, and disasters* (pp. 77–99). Elsevier.
41. Tedim, F., Leone, V., & Xanthopoulos, G. (2016). A wildfire risk management concept based on a social-ecological approach in the European Union: Fire smart territory. *International Journal of Disaster Risk Reduction, 18*, 138–153.
42. Tedim, F., McCaffrey, S., Leone, V., Delogu, G. M., Castelnou, M., McGee, T. K., & Aranha, J. (2020). What can we do differently about the extreme wildfire problem: An overview. In *Extreme wildfire events and disasters* (pp. 233–263). Elsevier.
43. Tymstra, C., Stocks, B. J., Cai, X., & Flannigan, M. D. (2020). Wildfire management in Canada: Review, challenges, and opportunities. *Progress in Disaster Science, 5*, 100045.
44. Wollstein, K., O'Connor, C., Gear, J., & Hoagland, R. (2022). Minimize the bad days: Wildland fire response and suppression success. *Rangelands, 44*(3), 187–193.
45. Xanthopoulos, G., Caballero, D., Galante, M., Alexandrian, D., Rigolot, E., & Marzano, R. (2006). Forest fuels management in Europe. In *Fuels management-how to measure success: Conference proceedings.* USDA Forest Service.

Open Access   This chapter is licensed under the terms of the Creative Commons Attribution 4.0 International License (http://creativecommons.org/licenses/by/4.0/), which permits use, sharing, adaptation, distribution and reproduction in any medium or format, as long as you give appropriate credit to the original author(s) and the source, provide a link to the Creative Commons license and indicate if changes were made.

The images or other third party material in this chapter are included in the chapter's Creative Commons license, unless indicated otherwise in a credit line to the material. If material is not included in the chapter's Creative Commons license and your intended use is not permitted by statutory regulation or exceeds the permitted use, you will need to obtain permission directly from the copyright holder.

CHAPTER 4

SAFERS: Structured Approaches for Forest Fire Emergencies in Resilient Societies

Edoardo Arnaudo, Luca Bruno, Federico Oldani,
Marko Laine, Conrad Bielski, Alberto Croci,
Andrea Trucchia, Panagiota Masa, and Claudio Rossi

INTRODUCTION

Forest fires have become an escalating concern due to the intensifying frequency and severity of extreme weather conditions, exacerbated by the effects of climate change. This global issue not only results in the loss of human lives and habitats but also contributes to the release of millions of tons of CO^2 and other pollutants [1]. The development of effective forest

E. Arnaudo (✉)
AI, Data & Space, LINKS Foundation, Torino, Italy

Dipartimento di Automatica e Informatica (DAUIN), Politecnico di Torino, Torino, Italy
e-mail: edoardo.arnaudo@linksfoundation.com; edoardo.arnaudo@polito.it

L. Bruno • F. Oldani • C. Rossi
AI, Data & Space, LINKS Foundation, Torino, Italy
e-mail: luca.bruno@linksfoundation.com; federico.oldani@linksfoundation.com; claudio.rossi@linksfoundation.com

© The Author(s) 2025   39
I. Gkotsis et al. (eds.), *Paradigms on Technology Development for Security Practitioners*, Security Informatics and Law Enforcement, https://doi.org/10.1007/978-3-031-62083-6_4

fire emergency management systems becomes therefore crucial in mini-mizing the impacts of future fire events.

The SAFERS project (Structured Approaches for Forest fire Emergencies in Resilient Societies) addresses this pressing challenge by proposing a comprehensive Emergency Management System (EMS) designed to man-age forest fires across all critical phases of the emergency management cycle. Leveraging a service-oriented approach, SAFERS integrates a set of Intelligent Services (ISs), independent modules exploiting Artificial Intelligence (AI), and other advanced capabilities to provide more refined and informative outputs. These services utilize and produce diverse data sources, including Earth Observation from the EU Copernicus program, meteorological forecasts, hazards and risk forecasts, propagation models, data from social media and chatbot applications, as well as real-time data from in situ camera sensors. The outputs generated by the SAFERS IS are stored within a geospatial data lake, harmonized, and made accessible to end users through an interactive web-based dashboard. This integration enables informed decision-making throughout the entire emergency man-agement cycle, enhancing preparedness, response, and recovery efforts.

Compared to other solutions, the SAFERS platform provides two key strengths. First, it encompasses all key phases of the emergency manage-ment cycle, ensuring a comprehensive management of the emergencies.

M. Laine
Finnish Meteorological Institute (FMI), Helsinki, Finland
e-mail: marko.laine@fmi.fi

C. Bielski
Riscognition GmbH, Geltendorf, Germany
e-mail: conrad.bielski@riscognition.com

A. Croci
WaterView s.r.l., Torino, Italy
e-mail: alberto.croci@waterview.it

A. Trucchia
CIMA Foundation, Savona, Italy
e-mail: andrea.trucchia@cimafoundation.org

P. Masa
Centre for Research and Technology Hellas (CERTH), Marousi, Greece
e-mail: gmasa@iti.gr

Second, the system embraces a modular and service-oriented architecture, allowing each IS to operate independently with minimal dependencies. This modularity is further enhanced by asynchronous communication through a message broker, allowing for more flexibility. To further maximize interoperability, the platform leverages several software standards, including REST API communication for web-based components, INSPIRE metadata, OGC (Open Geospatial Consortium)-compliant services for GIS data management. This standardized approach allows for easier integration with existing systems and applications when necessary, fostering collaboration among various stakeholders involved in forest fire emergency management.

This work aims to provide an overview of the SAFERS ISs and platform, focusing on its architecture and technical aspects. Through this comprehensive analysis, we aim to highlight the innovative features and capabilities of the overall platform, with particular attention to the innovations brought to the domain of forest fire emergency management.

The remainder of this document is organized as follows. Section "The SAFERS Architecture" will delve into the SAFERS architecture, followed by an examination of the ISs and their operational and on-demand functionalities in section "The SAFERS Intelligent Services". Additionally, the data layer, which plays a critical role in integrating diverse data sources, will be explored in section "The Data Layer". Section "The SAFERS Dashboard" provides an overview of the SAFERS dashboard, which is a user-friendly web-based interface designed to enhance decision-making and situational awareness during forest fire emergencies. Finally, section "Conclusions" draws the final conclusions, highlighting future directions for further enhancements in this field.

The SAFERS Architecture

In the SAFERS platform, the architectural design shown in Fig. 4.1 is crucial in facilitating efficient communication, data management, and visualization of heterogeneous information.

The main functionalities of the SAFERS platform are provided by a set of services independent of one another. The ISs are grouped into three main categories, namely *operational services*, *on-demand services*, and *crowdsourcing solutions*.

First, the operational services represent modules with a simpler communication paradigm: they autonomously provide periodical outputs for

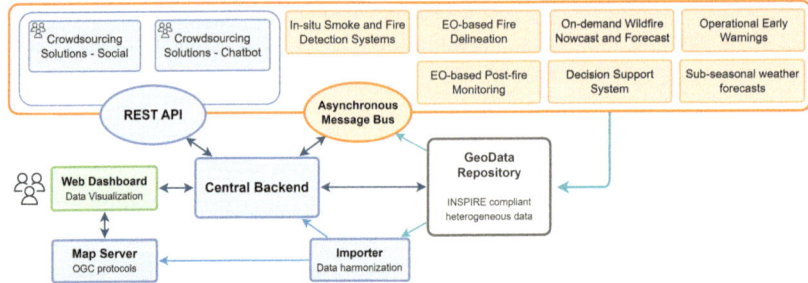

Fig. 4.1 Overview of the SAFERS architecture. The dashboard represents the main entry point, visualizing the data produced by the ISs (top pane). These ISs provide information either via REST or async API. Geospatial data is handled by the GeoData repository (GDR), then processed by the importer and map server for visualization

specific data. These include Subseasonal Weather Forecasts, Operational Early Warnings, In situ Smoke and Fire Detection Systems, and the Decision Support System.

Second, on-demand services, namely the EO-based Fire Delineation, Postfire Monitoring, and On-demand Wildfire Forecast services, use instead a more complex communication mechanism: they are triggered by a specific message type named *map request*. These services act on demand, retrieving the necessary data based on user-defined parameters, and producing one or more outputs, depending on the request content. Once the results become available, the other services are notified through a message bus, allowing for the required follow-up actions to take place.

Last, SAFERS also incorporates *crowdsourcing solutions*, including a chatbot and a social media module, to gather information directly or indirectly from citizens. The chatbot enables structured geolocated and multimedia data collection, while the social module gathers and classifies real-time Twitter posts, extracting relevant emergency events and estimating their impacts.

To support efficient data management, the platform includes a data layer, which serves as a central storage system for heterogeneous data. This module adheres to INSPIRE metadata standards,[1] ensuring proper organization and accessibility of diverse datasets. When applicable, the raw data

---

[1] https://inspire.ec.europa.eu

stored in the GeoData Repository (GDR) is processed by the Importer and Mapper modules. These components allow the creation and management of OGC-compliant[2] layers, enabling the integration and presentation of spatial data within the system.

The main communication point between the user dashboard and the services is the central backend, which facilitates the exchange of information, allowing users to interact with various platform components seamlessly. This backend ensures the smooth flow of data and requests throughout the system, while a web-based dashboard is provided to users for intuitive data visualization. The dashboard supports the display of geospatial information through Web Map Service (WMS) and Web Map Tile Service (WMTS) layers. Additionally, it incorporates data from crowdsourcing solutions and real-time information from in situ cameras, enabling users to gain valuable insights and make informed decisions.

THE SAFERS INTELLIGENT SERVICES

The ISs have the purpose of providing heterogeneous data in a timely manner to the SAFERS Dashboard. These services are carefully designed to address specific aspects of the emergency management cycle, enhancing preparedness, response, and mitigation efforts.

Operational Services These play a critical role in providing high-quality information relevant to forest fires. Data is essential for preventing, preparing, and responding to potentially dangerous wildfire hazard weather situations. To achieve this, an automated processing chain has been established, allowing for the accessibility of necessary resources within the SAFERS platform, provisioned by four services. First, the Subseasonal Weather Forecasts service provides forecast data collected from the ECMWF,[3] focusing on relevant variables for forest fire hazard and risk prediction. The data includes high-resolution deterministic forecasts with hourly updates up to 72 h, as well as medium-range probabilistic predictions for lead times up to 15 days (360 h). Additionally, extended range/subseasonal probabilistic predictions with lead times up to 46 days (1104 h) are provided twice a week. All weather data is postprocessed, automatically uploaded to the SAFERS GeoData Repository, and made available for visualization and sharing with other partner services. Second,

[2] https://www.ogc.org
[3] https://www.ecmwf.int

the Operational Early Warnings service augments operational wildfire hazard and risk mapping to improve the dissemination, uptake, and updating of early warnings. By integrating open datasets from complementing sources and utilizing novel risk models, this service delivers valuable information to stakeholders, enabling daily assessment and preparation in case of a wildfire. The datasets used include the European Forest Fire Information System (EFFIS),[4] Fire Weather Index (FWI), European weather forecasts from FMI, and real-time information about active fires. The wildfire risk modeling module integrates current and forecast weather information, and it is expected to incorporate lightning forecasts.

Third, the Smoke and Fire Detection System (smoCAM) utilizes standard surveillance cameras to monitor large areas for wildfire and smoke, automatically detecting fire and smoke plumes in various outdoor conditions. This AI-based service is capable of accurately detecting smoke and fire, even in challenging environmental conditions such as clouds and fog. SmoCAM is deployed in six test sites, each equipped with a PTZ (Pan-Tilt-Zoom) camera and an LTE router to enable remote connections. The camera patrols different points of view to widen the monitored area, and relevant information describing the detected smoke and fire is transmitted as JSON messages to the SAFERS bus. The messages are then integrated and displayed on the dashboard, providing real-time updates on detected wildfire events.

Last, the Decision Support System (DSS) analyses data from various sources and provides valuable recommendations to stakeholders. It utilizes semantic reasoning on top of the SAFERS domain ontology to derive new insights from preexisting knowledge. The DSS has been designed to meet user requirements and address specific scenarios, such as FWI forecasts, Fine Fuel Moisture, duff moisture, and drought anomalies. The system's functionalities have been implemented to accommodate the SAFERS ontology, derived from previous works [2]. The DSS plays a crucial role in aiding decision-makers during forest fire emergencies, contributing to the resilience and effectiveness of the overall system.

On-Demand Services SAFERS also offers services that can be activated on demand to provide rapid mapping, severity assessment, and wildfire forecast functionalities, as shown in Fig. 4.2. First, the EO-based delineation

---

[4] https://effis.jrc.ec.europa.eu

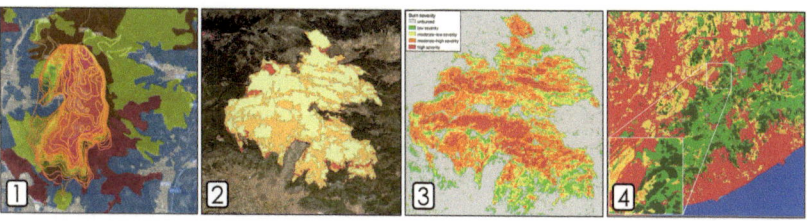

Fig. 4.2 Example outputs derived from the on-demand services: (1) wildfire forecast simulation, (2) severity estimation, (3) postfire dNBR analysis, and (4) fuel map delineation

service leverages Sentinel-2 satellite imagery to generate thematic maps for burned area delineation [3], severity estimation, fire front detection, and land cover maps [4] through machine learning models. By employing state-of-the-art segmentation networks, this service generates output maps for a given area and time interval accurately, and in a short amount of time. The service is deployed using a multimodule containerized architecture, ensuring fault-tolerance and distribution of components, and asynchronous communication through the SAFERS message bus. Similarly, the Postfire Monitoring service generates long-term assessments during the recovery period, providing dNBR and other layers to better observe and understand the regrowth in the affected areas.

The Wildfire Forecast service is instead achieved through the implementation of PROPAGATOR [5], an operational tool based on cellular automata. It rapidly simulates the potential spread of wildfires, considering various factors such as ignition point, topography, fuel cover, and meteorological data. The resulting probabilistic fire fronts stem from averaging several realizations of the stochastic core, and the service provides useful accessory data, including maximum and mean Rate of Spread (RoS) and Fireline Intensity (FI) of the affected area. The simulation also allows for the incorporation of firefighting actions, such as Canadair and helicopter drops, and waterlines. The IS is deployed with async communication using the message broker, enabling efficient orchestration of simulation requests. The results are uploaded to the data repository, and the dashboard receives timely updates from both services.

Crowdsourcing Solutions Currently, operational procedures primarily involve one-way communication from authorities and monitoring agencies to citizens, lacking active citizen engagement. This deficiency leads to inadequate risk awareness and preparedness among citizens, hindering disaster impact reduction in the case of natural hazards. Valuable information shared by citizens on social media during emergencies often gets lost in the vast sea of data. To address these challenges, SAFERS proposes implementing two essential crowdsourcing solutions: the Social Module and the Chatbot.spiepr Par26 The Social Module is designed to fetch and analyze real-time content posted on Twitter,[5] with a particular focus on wildfires and other natural disasters. By combining text and image analysis with location information, the module utilizes natural language and image processing to extract meaningful data related to emergency situations, early warning signals, and damage assessment. First rule-based filtering excludes erroneous or misleading data, retaining only informative content relevant to emergency scenarios. The module then classifies social media posts using text mining and deep learning techniques, mapping them into relevant categories [6], and groups them by meaningful events [7, 8]. Implemented and deployed in a cloud platform, the Social Module seamlessly integrates with the SAFERS message bus through secured web-based (HTTPS) REST APIs, ensuring efficient data sharing and enhancing the overall emergency response capabilities.

The Chatbot [8] facilitates instead structured geolocated and multimedia data collection from first responders and volunteers, serving as an innovative and user-centric tool by enhancing communication and collaboration during disaster management efforts. It enables two-way communication between citizens, field forces, and control centers, crucial for effective disaster risk reduction. Built on a robust technology stack, the Chatbot ensures scalability and seamless integration with the Telegram API, enabling efficient real-time communication and personalized functionalities for different user roles. Two primary user roles are supported: professionals with specialized training and citizens. For professionals, the chatbot offers advanced features like real-time location sharing, reporting activities, detailed measurements, damage assessments, and instructions exchange with the control room, facilitating efficient emergency coordination. Citizens, with limited emergency response training, can actively

[5] https://twitter.com/

contribute by reporting events, sharing real-time information, receiving geolocated broadcast messages, and accessing safety guidelines through the Chatbot.

Both the Social Module and the Chatbot integrate with the project-level API Gateway and OAuth2[6] authentication module to ensure secure data management and user authentication, safeguarding user accounts and associated data.

THE DATA LAYER

Every service introduced so far generates heterogeneous outputs that need to be stored and processed to be visualized and exploited by expert users. This is the task of the data layer, which consists of two main components: the Geodata Repository (GDR), and the Importer-Mapper coupling, displayed in Fig. 4.3.

The GDR consists of a management component that facilitates data ingestion and retrieval by serving as a central repository for metadata. It enables efficient data discovery and exploration within the data repository. This component is built on CKAN,[7] an open-source data management system that has been customized with plugins and extensions to meet the specific requirements of the GDR. CKAN provides a user-friendly web-based GUI and API for users to upload, edit, delete, and search data. The large amount of data in various formats is securely stored in a cloud big data storage system, using a data lake approach and employing the HDFS

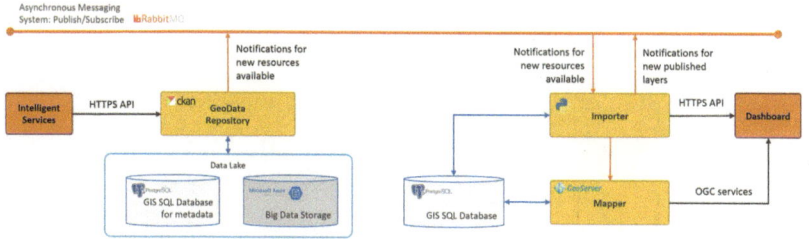

Fig. 4.3 Data layer components and their interaction with the other modules

[6] https://oauth.net/2/
[7] https://ckan.org

file system to preserve data in its original form within Hadoop clusters [9]. To ensure effective communication and collaboration between microservices, the GDR exploits the message bus to notify other services about new data availability and changes to existing datasets.

The Importer-Mapper consists instead of two interconnected elements. The Mapper represents a map server that exposes geospatial data through OGC-compliant services, i.e., WMS, WMTS, WFS, or WCS, implemented through GeoServer.[8] The Importer routine handles changes in the raw data stored in the GeoData Repository and updates the content available through the Mapper. It can handle data in various formats, transforming and publishing it as layers. The Importer component provides additional APIs to enhance its functionality. These APIs allow users to retrieve the list of available layers, delete layers from both GeoServer and the GDR, obtain metadata for specific layers, download raw files associated with layers, and access the temporal series of values for specific coordinates from a list of layers. By supporting multiple data formats and offering standardized services, the Importer-Mapper components ensure seamless integration and retrieval of geospatial data from different sources.

THE SAFERS DASHBOARD

The SAFERS dashboard, illustrated in Fig. 4.4, stands as a pivotal interface for managing and visualizing critical data throughout various phases of forest fire emergencies effectively. Developed on React JS framework and integrating essential libraries, this dashboard ensures a seamless and intuitive user experience. The dashboard handles the complexities of

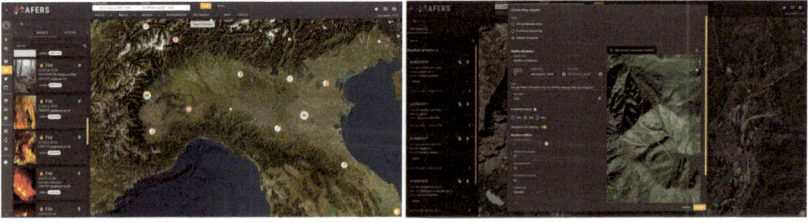

Fig. 4.4 Screenshots of the SAFERS dashboard, displaying the main screen (left), and an example form to create a new map request

[8] https://geoserver.org/

spatiotemporal data representation, with the central map serving as a geographical anchor, and the top-level date and time selectors as global temporal reference for the whole system.

The visualization methods employed within the dashboard offer multiple approaches to better understand the data. Dynamic filters in the top bar enhance data exploration, enabling users to refine their focus and select increasingly specific information. This allows for a more precise analysis, crucial in making informed decisions during emergency situations. The central map not only provides essential geographical context but also allows users to interact directly with the data.

A user-friendly list pane on the left side offers a streamlined approach to navigation, ensuring that users can effortlessly access the necessary information. The dashboard integrates and handles OGC-compliant layers, complete with their respective legends and metadata. This feature enables users to grasp spatial relationships and patterns. Moreover, the use of INSPIRE metadata standards ensures seamless data sharing and interoperability among European stakeholders. To further enhance the temporal analysis, the inclusion of a dedicated time widget grants users the ability to explore data across different timeframes, including the map layers. Additionally, the dashboard allows retrieving data at specific points, offering the tools for the analysis on the time series, displaying the evolution of a given metric in a specific location (e.g., precipitation, temperature).

Lastly, the dashboard also provides a dedicated summary page, offering a comprehensive overview of ongoing emergency situations. This page consolidates critical information, providing stakeholders with a comprehensive snapshot, and allowing users to grasp the current situation at a glance.

CONCLUSIONS

The SAFERS project presents a comprehensive and modular forest fire management system, addressing the challenges posed by increasing forest fires and extreme weather events. The SAFERS ISs form a powerful backbone, offering essential functionalities across all emergency management phases. Operational services provide high-quality weather forecasts for early warnings and risk prediction, while on-demand services facilitate rapid mapping and severity assessment using satellite imagery. Crowdsourcing solutions actively engage citizens in data collection and real-time communication. Overall, the SAFERS platform leverages intelligent technologies like AI and EO, empowering informed decision-making and fostering resilience in forest fire emergency management, contributing to more resilient societies and mitigating future fire impacts.

Acknowledgments This work was carried out in the context of the Horizon 2020 project SAFERS (Grant Agreement n.869353).

REFERENCES

1. Knorr, W., Jiang, L., & Arneth, A. (2016). Climate, CO2 and human population impacts on global wildfire emissions. *Biogeosciences, 13*, 267–282.
2. Maltoni, C., Rossi, C., & Sánchez, G. (2017). Improving resilience to emergencies through advanced cyber technologies: The I-REACT project. *Geomedia, 21*, 18–22.
3. Arnaudo, E., Barco, L., Merlo, M., & Rossi, C. (2023). Robust burned area delineation through multitask learning. In *Conference on machine learning and principles and practice of knowledge discovery in databases.* Springer. https://2023.ecmlpkdd.org/program/proceedings/
4. Galatola, M., Arnaudo, E., Barco, L., Rossi, C., & Dominici, F. (2023). Land cover segmentation with sparse annotations from Sentinel-2 imagery. In *IEEE international geoscience and remote sensing symposium (IGARSS 2023).* IEEE.
5. Trucchia, A., D'Andrea, M., Baghino, F., Fiorucci, P., Ferraris, L., Negro, D., Gollini, A., & Severino, M. (2020). PROPAGATOR: An operational cellular-automata based wildfire simulator. *Fire, 3*, 26.
6. Piscitelli, S., Arnaudo, E., & Rossi, C. (2021). Multilingual text classification from twitter during emergencies. In *2021 IEEE international conference on consumer electronics (ICCE).* IEEE.
7. Salza, D., Arnaudo, E., Blanco, G., & Rossi, C. (2022). A 'glocal' approach for real-time emergency event detection in twitter. In *ISCRAM 2022 Conference proceedings-19th international conference on information systems for crisis response and management.*
8. Konstantoudakis, K., Christaki, K., Sainidis, D., Babic, I., Kogias, D. G., Inglese, G., Bruno, L., Giunta, G., Patrikakis, C. Z., Balet, O., et al. (2022). Common operational picture and interconnected tools for disaster response: The FASTER toolkit. In *Disaster management and information technology: Professional response and recovery management in the age of disasters* (pp. 83–110). Springer.
9. Borthakur, D., et al. (2008). HDFS architecture guide. *Hadoop Apache Project, 53*, 2.

Open Access This chapter is licensed under the terms of the Creative Commons Attribution 4.0 International License (http://creativecommons.org/licenses/by/4.0/), which permits use, sharing, adaptation, distribution and reproduction in any medium or format, as long as you give appropriate credit to the original author(s) and the source, provide a link to the Creative Commons license and indicate if changes were made.

The images or other third party material in this chapter are included in the chapter's Creative Commons license, unless indicated otherwise in a credit line to the material. If material is not included in the chapter's Creative Commons license and your intended use is not permitted by statutory regulation or exceeds the permitted use, you will need to obtain permission directly from the copyright holder.

CHAPTER 5

Measuring Forest Resilience Against Wildfires and Climate Change: Methods and Technical Approaches

Konstantinos Demestichas, Dimitrios Sykas,
Dimitrios Zografakis, Spyridon Kaloudis,
Nikolaos Kalapodis, Georgios Sakkas,
Miltiadis Athanasiou, and Constantina Costopoulou

INTRODUCTION

Forest ecosystems are complex, dynamic, and resilient. They are home to a vast array of biodiversity, act as significant carbon sinks, and provide essential ecosystem services such as water regulation and soil conservation.

K. Demestichas (✉) • D. Sykas • D. Zografakis • C. Costopoulou
Department of Agricultural Economics and Rural Development, Agricultural University of Athens, Athens, Greece
e-mail: cdemest@aua.gr; dsykas@aua.gr; tina@aua.gr

S. Kaloudis
Department of Forestry and Natural Environment Management, Agricultural University of Athens, Karpenisi, Greece
e-mail: kaloudis@aua.gr

© The Author(s) 2025   53
I. Gkotsis et al. (eds.), *Paradigms on Technology Development for Security Practitioners*, Security Informatics and Law Enforcement, https://doi.org/10.1007/978-3-031-62083-6_5

However, these ecosystems are increasingly under threat from a range of disturbances, most notably wildfires and climate change. The capacity of a forest to withstand, adapt to, and recover from these disturbances is referred to as forest resilience [1].

Wildfires, while a natural and essential part of many forest ecosystems, have been increasing in frequency and severity due to changing climatic conditions. Droughts, heatwaves, and shifting precipitation patterns exacerbate the risk and impact of these fires [2]. Climate change, on the other hand, poses a more insidious threat. It alters the very conditions under which these ecosystems have evolved, leading to shifts in species distributions, changes in community composition, and potential disruptions to ecological functions [3].

Understanding and measuring forest resilience against wildfires and climate change is crucial for forest management and conservation. It allows us to assess the health of our forests, predict their response to future disturbances, and develop strategies to enhance their resilience. This is a complex task that requires a combination of quantitative and qualitative methodologies, each providing unique insights into different aspects of forest resilience [4].

This chapter aims to provide a comprehensive overview of the methods and technical approaches used to measure forest resilience against wildfires and climate change. We will explore both quantitative methodologies, which focus on measuring specific ecosystem parameters, and qualitative methodologies, which aim to understand the social and ecological factors contributing to forest resilience. We will also discuss the role of Earth Observation (EO) as a critical tool for monitoring changes in forest health, identifying areas of risk, and informing management strategies [5].

The importance of this topic cannot be overstated. As wildfires become more frequent and climate change continues to alter our environment, ensuring the resilience of our forests is not just an ecological imperative, but a matter of global concern. By understanding how to measure and

N. Kalapodis • G. Sakkas • M. Athanasiou
Center for Security Studies (KEMEA), Ministry of Citizen Protection,
Athens, Greece
e-mail: n.kalapodis@kemea-research.gr; g.sakkas@kemea-research.gr;
m.athanasiou@kemea-research.gr

enhance forest resilience, we can better equip ourselves to face these challenges and ensure the continued survival and prosperity of our forest ecosystems.

UNDERSTANDING FOREST RESILIENCE

Resilience, as a concept, has been interpreted in diverse ways across various scientific fields. In the context of ecology, resilience is often understood as the capacity of an ecosystem to withstand disturbances and recover to its original state [6]. This capacity is not static but varies over time and in response to different types of disturbances. In the context of forest ecosystems, resilience is the ability of a forest to withstand, adapt to, and recover from disturbances such as wildfires and climate change [1].

Forest resilience is a multifaceted concept that encompasses several key aspects. Firstly, it involves the ability of a forest to resist disturbances, such as wildfires, and maintain its ecological functions and services. This includes carbon sequestration, biodiversity conservation, and water regulation, even under changing conditions [1]. Forest resilience involves the capacity of a forest to recover effectively from disturbances. This recovery can be seen in the regeneration of forests, the support of postfire ecosystem processes, and the reduction of the risk of future fires. The rate and extent of this recovery can serve as a measure of the forest's resilience [6]. Also, forest resilience includes the ability of a forest to adapt to changing environmental conditions. This includes shifting temperature and precipitation patterns, changes in the distribution of species and communities, and changes in the frequency and severity of disturbances such as wildfires [1].

Understanding forest resilience is not just about understanding the forest's response to disturbances. It also involves understanding the social and ecological factors that contribute to a forest's resilience. This includes the role of local communities, indigenous peoples, and forest managers in supporting forest resilience, as well as the impact of broader socioeconomic and political factors. Forest resilience is a complex and dynamic concept that reflects the forest's capacity to resist, recover from, and adapt to disturbances. Understanding this concept is crucial for managing and conserving our forest ecosystems in the face of increasing threats from wildfires and climate change.

QUANTITATIVE METHODOLOGIES FOR MEASURING FOREST RESILIENCE

Introduction to Quantitative Methodologies

Quantitative methodologies provide a rigorous and objective means of measuring forest resilience. These methods focus on measuring specific ecosystem parameters that can indicate a forest's ability to resist, recover from, and adapt to disturbances such as wildfires and climate change. Quantitative methodologies often involve the collection of numerical data that can be analyzed statistically to conclude the resilience of a forest. This data can be collected through various means, including field measurements, remote sensing technologies, and laboratory analyses.

The quantitative methods are organized into three main categories, (a) Vegetation monitoring, (b) Soil sampling, (c) Hydrological monitoring. This categorization reflects the domain of focus of the techniques. In each of these categories, multiple approaches exist on how to quantify forest resilience. This section provides an overview of the relevant techniques, while the detailed explanation is out of scope.

Vegetation Monitoring

Vegetation monitoring is a key quantitative methodology for measuring forest resilience. It involves monitoring changes in vegetation, such as plant cover and species composition, which can provide insights into the recovery of the ecosystem after a wildfire. Vegetation monitoring can be conducted using various technical means, depending on the specific objectives of the action. Ground observations can provide important information on plant density, height, and other quantitative measurements that can help assess changes in vegetation after a wildfire. Remote sensing methods, such as drones and satellite imaging, can provide a broader view of changes in vegetation over large areas.

Remote sensing methods have revolutionized the field of vegetation monitoring, allowing for the collection of data over large areas and inaccessible locations, and providing a broader view of changes in vegetation over time. These methods are particularly useful for monitoring changes in forest cover, including the extent of deforestation, forest fragmentation, and the impacts of natural disturbances, such as wildfires and climate change.

One of the most common remote sensing methods used in vegetation monitoring is the use of optical satellite imagery. This type of imagery can be used to monitor changes in forest cover, including the extent of deforestation and forest fragmentation, and the impacts of natural disturbances, such as wildfires and climate change [7].

Another important remote sensing technology is Light Detection and Ranging (LiDAR). LiDAR is a remote sensing technology that can be used to measure forest structure and biomass, providing valuable information on forest health and productivity [8].

Radar data is also used in remote sensing for vegetation monitoring. Radar data can be used to monitor changes in forest cover and structure, including the detection of forest disturbance and the mapping of forest biomass. Radar data can also be used to monitor changes in soil moisture levels, which can affect forest health and resilience [9].

Soil Sampling

Soil sampling is another important quantitative methodology for assessing forest resilience. Sampling soil after a wildfire can help to assess changes in soil fertility, organic matter content, and nutrient cycling, which can affect the ability of vegetation to regrow [10]. Soil sampling typically involves the collection of soil samples from various depths and locations within the forest, which are then analyzed in a laboratory for various chemical and physical properties. This information can provide valuable insights into the health of the soil and its capacity to support vegetation recovery after a disturbance.

Hydrological Monitoring

Hydrological monitoring is a crucial aspect of assessing forest resilience, as it provides insights into the impacts of wildfires on the water cycle and the potential for erosion and landslides [11, 12]. The availability and quality of water in a forest ecosystem can significantly affect its ability to recover from disturbances such as wildfires. For instance, a study by Yuan et al. [11] used machine learning models to predict road flooding risks based on topographic, hydrologic, and temporal precipitation features. This approach could be adapted to predict the risk of flooding in forest areas following wildfires, which could inform management strategies for forest recovery and future fire prevention. The study by Tomelleri and Tonon

[12] linked sap flow measurements with earth observations to model canopy transpiration. This approach could be used to monitor changes in water availability in forest ecosystems following wildfires, providing valuable insights into the recovery of the ecosystem.

The Role of Earth Observation (EO) in Measuring Forest Resilience

Earth Observation (EO) is a powerful tool for measuring and supporting forest resilience. It allows for the monitoring of changes in forest health, the identification of areas at risk, and the informing of management strategies [13]. Optical imagery, LiDAR data, radar data, climate data, and topographic data are all types of EO data that can be used to measure forest resilience. For instance, the study by Tomelleri and Tonon [12] demonstrated the relevance of Sentinel-2 data for the data-driven upscaling of ecosystem fluxes from plot scale sap flow data.

Ge et al. [13] introduced a novel semi-supervised contrastive regression framework for wall-to-wall mapping of continuous forest variables using multisensory satellite data. This approach could be used to monitor changes in forest cover and structure, including the detection of forest disturbance and the mapping of forest biomass. Furthermore, Filatov and Yar [14] proposed a solution for observing the area covered by the forest and water using the U-Net model, an image segmentation model. This approach could be used to monitor changes in forest cover, including the extent of deforestation, forest fragmentation, and the impacts of natural disturbances, such as wildfires and climate change.

LiDAR is a remote sensing technology that can be used to measure forest structure and biomass. LiDAR data can be used to create detailed 3D models of forest ecosystems, providing valuable information on forest health and productivity. For instance, a study by Sun et al. [15] correlated airborne LiDAR data statistics with the Landsat 8 satellite's surface temperature product to understand the cooling effect of urban forests. In another study, Finley et al. [16] proposed a two-stage hierarchical Bayesian model to estimate forest biomass density and total given sparsely sampled LiDAR and georeferenced forest inventory plot measurements. This approach provides a model-based method to estimate forest variables when inventory data are sparse or resources limit the collection of enough data to achieve desired accuracy and precision using design-based methods. Furthermore, Huang et al. [17] quantified the aboveground biomass

(AGB) of a plateau mountainous forest reserve using a system that synergistically combines an unmanned aircraft system (UAS)-based digital aerial camera and LiDAR. This approach leverages the complementarity between digital aerial photogrammetry (DAP) and LiDAR measurements, providing a cost-efficient method for large-scale wall-to-wall AGB mapping.

QUALITATIVE METHODOLOGIES FOR ASSESSING FOREST RESILIENCE

Qualitative methodologies play a crucial role in assessing forest resilience, providing insights into the complex interactions between social, economic, and ecological factors that influence forest ecosystems. These methodologies often involve the use of case studies, interviews, and participatory approaches to understand the experiences and perceptions of local communities, forest managers, and other stakeholders. In contrast with the quantitative methods, qualitative methodologies aim at providing information that cannot be directly materialized using in situ or remote measurements, rather to model the interactions among heterogeneous socio-environmental-economic systems.

For instance, a study by Mukul et al. [18] explored the role of nontimber forest products (NTFPs) in sustaining forest-based rural livelihoods in and around a protected area in Bangladesh. The study used a combination of household surveys and interviews to gather data on the collection, processing, and selling of NTFPs, and their contribution to household income and resilience. The findings revealed that NTFPs played a significant role in providing primary, supplementary, and emergency sources of income for local households, and were particularly important for households located close to the park.

Moreover, Itter et al. [1] developed a hierarchical Bayesian state-space model to investigate the variable effects of climate on forest growth in relation to climate extremes, disturbance, and forest stand dynamics. The study used dendrochronology data from forest stands of varying composition, structure, and development stages in northeastern Minnesota. The results indicated that forest growth was most sensitive to variables describing climatic water deficit, and that forest growth was both resistant and resilient to climate extremes, particularly when these coincided with insect defoliation events.

THE INTERPLAY BETWEEN QUANTITATIVE
AND QUALITATIVE APPROACHES

The interplay between quantitative and qualitative approaches is crucial for a comprehensive understanding of forest resilience. These approaches complement each other, providing a holistic view of the complex dynamics of forest ecosystems.

Tinchev et al. [19] introduced Natural Segmentation and Matching (NSM), an algorithm for reliable localization in both urban and natural environments using laser data. The authors highlighted the challenges of working in structure-poor vegetated areas such as forests, where clutter and perceptual aliasing prevent repeatable extraction of distinctive landmarks. The study demonstrated the interplay between quantitative data (laser measurements) and qualitative understanding (recognition of natural features) in achieving reliable localization.

Lade et al. [20] proposed a framework for conceptualizing and modeling social-ecological transformations based on adaptive networks. The authors emphasized the interplay between the structure of a social-ecological system and the dynamics of individual entities. This approach could be used to understand the complex dynamics of forest ecosystems and their resilience to disturbances.

CONCLUSION

Forest resilience is a complex and multifaceted concept, encompassing the ability of forest ecosystems to withstand, recover from, and adapt to disturbances. The measurement of forest resilience is a challenging task, requiring a combination of quantitative and qualitative methodologies and a deep understanding of the interplay between social, economic, and ecological factors. A key takeaway is the fact that forest resilience is not described by a single metric or parameter, rather multiple viewpoints are needed to assess the resilience of a forest.

The studies reviewed in this chapter highlight the diverse approaches to measuring forest resilience, from hydrological monitoring and earth observation techniques to the use of machine learning models and decision support systems. These methodologies provide valuable insights into the impacts of disturbances such as wildfires, climate change, and deforestation on forest ecosystems and their ability to recover.

However, these studies also highlight the challenges of measuring forest resilience, including the difficulty of dealing with large amounts of heterogeneous data, the complexity of forest ecosystems, and the variety of disturbances they face. Despite these challenges, there are also significant opportunities for improving our understanding and measurement of forest resilience, thanks to advances in technology, data collection, and analysis.

In conclusion, while measuring forest resilience is a complex and challenging task, it is also a crucial one. Forests play a vital role in maintaining global biodiversity, regulating climate, and supporting human livelihoods. Understanding and enhancing their resilience is therefore of paramount importance for both people and the planet. As technology and our understanding of forest ecosystems continue to evolve, so too will our ability to measure and enhance forest resilience.

Acknowledgments This study has been conducted in the framework of the SILVANUS—Integrated Technological and Information Platform for Wildfire Management project and has received funding from the European Union's Horizon 2020 research and innovation program under grant agreement No 101037247. The contents of this publication are the sole responsibility of the authors and can in no way be taken to reflect the views of the European Commission.

Declaration of Competing Interest The authors declare that they have no known competing financial interest or personal relationship that could have appeared to influence the work reported in this paper.

References

1. Itter, M. S., Finley, A. O., D'Amato, A. W., Foster, J. R., & Bradford, J. B. (2016). Variable effects of climate on forest growth in relation to climate extremes, disturbance, and forest stand dynamics. arXiv, 1602.07228. Retrieved from https://arxiv.org/abs/1602.07228v2
2. Tomaselli, L., Jen, C., & Lee, A. B. (2020). Wildfire smoke and air quality: How machine learning can guide forest management. arXiv, 2010.04651. Retrieved from https://arxiv.org/abs/2010.04651v2
3. Ballard, T., Cooper, M., Lowrie, C., & Erinjippurath, G. (2023). Widespread increases in future wildfire risk to global forest carbon offset projects revealed

by explainable AI. arXiv, 2305.02397. Retrieved from https://arxiv.org/abs/2305.02397v1

4. Zhou, W., & Klein, L. (2020). Monitoring the impact of wildfires on tree species with deep learning. arXiv, 2011.02514. Retrieved from https://arxiv.org/abs/2011.02514v2

5. Imteaj, A., Amini, M. H., & Mohammadi, J. (2019). Leveraging decentralized artificial intelligence to enhance resilience of energy networks. arXiv, 1911.07690. Retrieved from https://arxiv.org/abs/1911.07690v1

6. Meyer, K. (2015). A dynamical systems framework for resilience in ecology. arXiv, 1509.08175. Retrieved from https://arxiv.org/abs/1509.08175v1

7. Sannigrahi, S., Bhatt, S., Rahmat, S., Rana, V., & Chakraborti, S. (2018). Effects of forest fire severity on terrestrial carbon emission and ecosystem production in the Himalayan region, India. arXiv, 1805.11680. Retrieved from https://arxiv.org/abs/1805.11680v1.

8. Ballester-Berman, J. D. (2020). Reviewing the role of the extinction coefficient in radar remote sensing. arXiv, 2012.02609. Retrieved from https://arxiv.org/abs/2012.02609v1

9. Baur, M., Jagdhuber, T., Link, M., Piles, M., Entekhabi, D., & Fink, A. (2020). Estimation of vegetation loss coefficients and canopy penetration depths from SMAP radiometer and IceSAT lidar data. arXiv, 2012.03318. Retrieved from https://arxiv.org/abs/2012.03318v1

10. Makarieva, A. M., Gorshkov, V. G., Sheil, D., Nobre, A. D., Bunyard, P., & Li, B.-L. (2013). Why does air passage over forest yield more rain? Examining the coupling between rainfall, pressure and atmospheric moisture content. arXiv, 1301.3083. Retrieved from https://arxiv.org/abs/1301.3083v2

11. Yuan, F., Mobley, W., Farahmand, H., Xu, Y., Blessing, R., Dong, S., et al. (2021). Predicting road flooding risk with machine learning approaches using crowdsourced reports and fine-grained traffic data. arXiv, 2108.13265. Retrieved from https://arxiv.org/abs/2108.13265v2

12. Tomelleri, E., & Tonon, G. (2021). Linking sap flow measurements with earth observations. arXiv, 2108.01290. Retrieved from https://arxiv.org/abs/2108.01290v1

13. Ge, S., Gu, H., Su, W., Lönnqvist, A., & Antropov, O. (2023). A novel semisupervised contrastive regression framework for forest inventory mapping with multisensor satellite data. *IEEE Geoscience and Remote Sensing Letters, 20*, 1–5. https://doi.org/10.1109/LGRS.2023.3281526

14. Filatov, D., & Yar, G. N. A. H. (2022). Forest and water bodies segmentation through satellite images using U-Net. arXiv, 2207.11222. Retrieved from https://arxiv.org/abs/2207.11222v1

15. Sun, W., Sun, Y., Liu, C., & Albrecht, C. M. (2023). DeepLCZChange: A remote sensing deep learning model architecture for urban climate resilience. arXiv, 2306.06269. Retrieved from https://arxiv.org/abs/2306.06269v1

16. Finley, A. O., Andersen, H.-E., Babcock, C., Cook, B. D., Morton, D. C., & Banerjee, S. (2023). Models to support forest inventory and small area estimation using sparsely sampled LiDAR: A case study involving G-LiHT LiDAR in Tanana, Alaska. arXiv, 2302.06410. Retrieved from https://arxiv.org/abs/2302.06410v3.
17. Huang, R., Yao, W., Xu, Z., Cao, L., & Shen, X. (2022). Information fusion approach for biomass estimation in a plateau mountainous forest using a synergistic system comprising UAS-based digital camera and LiDAR. arXiv, 2204.06746. Retrieved from https://arxiv.org/abs/2204.06746v1
18. Mukul, S. A., Rashid, A. Z. M. M., Uddin, M. B., & Khan, N. A. (2015). Role of non-timber forest products in sustaining forest-based livelihoods and rural households' resilience capacity in and around protected area-A Bangladesh study. arXiv, 1508.02056. Retrieved from https://arxiv.org/abs/1508.02056v1
19. Tinchev, G., Nobili, S., & Fallon, M. (2018). Seeing the wood for the trees: Reliable localization in urban and natural environments. arXiv, 1809.02846. Retrieved from https://arxiv.org/abs/1809.02846v2
20. Lade, S. J., Bodin, Ö., Donges, J. F., Kautsky, E. E., Galafassi, D., Olsson, P., & Schlüter, M. (2017). Modelling social-ecological transformations: An adaptive network proposal. arXiv, 1704.06135. Retrieved from https://arxiv.org/abs/1704.06135v1

Open Access This chapter is licensed under the terms of the Creative Commons Attribution 4.0 International License (http://creativecommons.org/licenses/by/4.0/), which permits use, sharing, adaptation, distribution and reproduction in any medium or format, as long as you give appropriate credit to the original author(s) and the source, provide a link to the Creative Commons license and indicate if changes were made.

The images or other third party material in this chapter are included in the chapter's Creative Commons license, unless indicated otherwise in a credit line to the material. If material is not included in the chapter's Creative Commons license and your intended use is not permitted by statutory regulation or exceeds the permitted use, you will need to obtain permission directly from the copyright holder.

CHAPTER 6

EU-Integrated Multifunctional Forest and Fire Management, Policies, and Practices: Challenges Between "As-Is" and "To-Be" State

Nikolaos Kalapodis, Georgios Sakkas, Alexandre Lazarou,
Domenica Casciano, Konstantinos Demestichas,
Miltiadis Athanasiou, Spyridon Kaloudis,
and Dimitrios Sykas

INTRODUCTION

One of the challenges for the EU Forest Strategy 2030 is to address and harmonize economic, social, and environmental aspects of forest management, including integrating fire management practices while climate

N. Kalapodis (✉) • G. Sakkas
Center for Security Studies (KEMEA), Ministry of Citizen Protection,
Athens, Greece
e-mail: n.kalapodis@kemea-research.gr

A. Lazarou • D. Casciano
Zanasi Alessandro SRL (Z&P), Modena, Italy

© The Author(s) 2025
I. Gkotsis et al. (eds.), *Paradigms on Technology Development*
for Security Practitioners, Security Informatics and Law
Enforcement, https://doi.org/10.1007/978-3-031-62083-6_6

change projections over the Mediterranean Basin show that long and occasionally extremely dry periods are more likely to occur in the future [1]. Moreover, climate change brings cascading effects like pest and insect outbreaks, windthrows and biomass accumulation and continuity, and increased flood risk [2]. Healthy forest ecosystems provide essential ecosystem services [3], but climate change is expected to impact them along with their ability to meet multiple demands.

Therefore, forest management, including wildfire management, needs to be integrated and adapt to these challenging conditions in order to protect and preserve healthy and productive forests. Since forest ecosystems are vital both for well-being and the economy, postfire restoration, sustainable management, and resilience improvement are also crucial, responding to threats that come from climate change. Despite their degree of vulnerability, forests are increasingly exposed to climate change impacts which are expected to worsen with Global warming [4, 5].

This paper analyzes the current state of forest and fire management policies through a literature review, identifies needed changes, and highlights challenges and gaps through an As-Is and To-Be analysis.

EU LAWS, POLICIES, INSTITUTIONAL FRAMEWORK, AND STRATEGIES ON FOREST AND WILDFIRE SUSTAINABLE MANAGEMENT

EU forests cover 42% of the total land area of the continent, equivalent to over 182 million hectares. With 0.36 hectares of forest per capita, the EU is committed to protecting this valuable natural asset for the benefit of all

K. Demestichas • S. Kaloudis • D. Sykas
Agricultural University of Athens (AUA), Athens, Greece

M. Athanasiou
Center for Security Studies (KEMEA), Ministry of Citizen Protection, Athens, Greece

Institute of Mediterranean Forest Ecosystems, Hellenic Agricultural Oganisation "Dimitra" (ELGO DIMITRA), Athens, Greece

Europeans.[1] The European Green Deal aims to reduce greenhouse gas emissions by at least 55% by 2030, making Europe the first climate-neutral continent by 2050.[2] The new EU Strategy on Adaptation recognizes the increasing frequency and severity of climate extremes, leading to a rise in disasters and damage over the past two decades.[3] The new EU Strategy promotes a comprehensive policy framework for European forests, focusing on accelerating adaptation and implementing strategies at all governance levels. In addition, for the first time in EU legislation, the LULUCF Regulation includes a binding emission reduction commitment for 2021–2030, accounting for all land uses and wetlands by 2026.[4] The EU Biodiversity Strategy for 2030 acknowledges forests' vital role in biodiversity, climate regulation, food provision, and more, with specific forest-related actions.[5] The EU proposal for Nature Restoration Law aims to improve protected areas, restore diverse nature, reduce pressures on habitats, support nature recovery, limit urban sprawl, tackle pollution, and create jobs. It is a key part of the EU Biodiversity Strategy with binding targets for ecosystem restoration and disaster prevention.[6] The EU must enhance the quantity, quality, and resilience of forests to climate-related threats, preserving biodiversity, ecosystem services, and supporting resilient economies and communities in the circular bioeconomy. The EU Forest Strategy aims to protect, restore, and expand multifunctional forests to combat climate change and biodiversity loss.[7] Regulation 2158/92/EEC of 23 July 1992 established a forest fire protection scheme from 1992 to 2002. The objective of Forest Focus (Regulation (EC) No 2152/2003 of 17 November 2003) aimed to protect EU forests by monitoring pollution and fires. In 1998, the Joint Research Center (JRC) established a research group to develop advanced forest fire risk assessment methods, supporting preparedness and ecological impact

---

[1] Over 40% of the EU covered with forests—Products Eurostat News—Eurostat (europa.eu).

[2] https://commission.europa.eu/strategy-and-policy/priorities-2019-2024/european-green-deal_en

[3] https://climate.ec.europa.eu/eu-action/adaptation-climate-change/eu-adaptation-strategy_en

[4] https://climate.ec.europa.eu/eu-action/land-use-sector_en#eu-rules-on-land-use-land-use-change-and-forestry-lulucf

[5] https://environment.ec.europa.eu/strategy/biodiversity-strategy-2030_en

[6] https://environment.ec.europa.eu/topics/nature-and-biodiversity/nature-restoration-law_en

[7] https://environment.ec.europa.eu/strategy/forest-strategy_en

assessments.[8] In 2006, the European Parliament urged improvements to EFFIS. The Commission formed a working group to propose post-2006 forest fire prevention policy. Furthermore, EU has committed to protecting the world's forests under several international agreements, initiatives, and policies, including UN Sustainable Development Goal 15, the New York Declaration on Forests, the UN Convention on Biological Diversity, and the Paris Agreement on climate change.

CONCEPTUAL FRAMEWORK

Definition of Sustainable Forest Management

The definition of Sustainable Forest Management (SFM) that has been agreed by the Ministerial Conference on the Protection of Forests (Forest Europe) was developed by [6] and is defined as: "*Sustainable management means the stewardship and use of forests and forest lands in such a way, and at a rate, that maintain their biodiversity, productivity, re-generation capacity, vitality, and their potential to fulfill, now and in the future, relevant ecological, economic, and social functions, at local, national, and global levels, and that does not cause damage to other eco-systems.*". The Food and Agriculture Organization (FAO) accepted this definition and defined SFM in [7] as: "*A dynamic and evolving concept, that is intended to maintain and enhance the economic, social, and environmental value of all types of forests, for the benefit of present and future generations.*"

As stated above, SFM has evolved from a single objective to multiple objectives and spatial planning, coexisting with ecosystem services (e.g., [8]).

Criteria and indicators assess forest management sustainability, but global agreement is lacking due to contextual variations [9]. The set of indicators that is mainly used to assess SFM is the Montreal Process, which was formed in 1994 and updated in 2015 [10]. Another set of criteria and indicators were adopted in 1998, reviewed and improved in 2003, and updated in 2015 by Forest Europe [11]. The impact of forest certification is uncertain due to limited scientific evidence [12], but positive effects have been observed in some cases, especially regarding the interest of forest owners in adopting the certification [13].

---

[8] https://effis.jrc.ec.europa.eu/about-effis

The Importance of Integrated Fire Management

Myers [14] defines IFM "*as an approach to address the problems and issues posed by both damaging and beneficial fires within the context of the natural environments and socio-economic systems in which they occur, by evaluating and balancing the relative risks posed by fire with the beneficial or necessary ecological and economic roles that it may play in a given conservation area, landscape or region. Integrated fire management facilitates cost-effective approaches to prevent damaging fires and maintain desired fire regimes. It evaluates fire effects, weighs benefits and risks, and responds based on objectives.*" By integrating science, society, and fire management technology and processes, an integrated fire management approach considers the interaction of biological, environmental, cultural, societal, economic, and political factors. To encourage such an approach for fire management and maintain native species, habitats, and landscape, several critical factors need to be considered [14]:

- Understanding the role of fire in ecosystems and its use: Distinguish between fire-dependent, fire-independent, fire-sensitive, and fire-influenced ecosystems.
- Identifying fire management goals and desired future conditions: Define management objectives and desired ecological outcomes.
- Considering community needs: Recognizing local community perspectives and requirements is crucial for achieving sustainable and culturally sensitive fire management practices.
- Promoting collaborative approaches among decision-makers: Collaboration among government, NGOs, local communities, and stakeholders is crucial for developing integrated fire management plans and policies.
- Tailoring strategies to the ecological and social context: Customized prevention, preparedness, and response strategies should consider ecological characteristics and social dynamics of the region.
- Supporting research efforts: Research on fire ecology, behavior, and social sciences is vital to understand fire's impact on ecosystems and societies.

The Role of Climate Change and Its Impact on SFM in the EU

Forests and climate are interrelated, as changes in forests affect climate and vice versa. The impact varies by region and severity [15]. However, due to the complexity of forest ecosystems and the interactions between various biotic and abiotic factors, determining climate change's specific impact is challenging. Uncertainties make it difficult to guide forest decision-makers in climate change planning [16]. Deforestation reduces carbon sequestration, increases CO_2, and raises temperatures [17]. Climate change can also limit tree regeneration [18]. Therefore, adapting forest and fire management to climate change is crucial to achieving the important objectives of resilient and healthy forests [19].

RESULTS

Considering all the above, the analysis of the major current policy challenges and gaps in the current state analysis for the EU follows.

Achieving Proactive Forest Management

As-Is Forest ecosystems face escalating pressure and threats due to climate change.

To-Be To address this challenge, proactive forest management is crucial for the EU to enhance the quantity, quality, resistance, and resilience of its forests. This includes targeted treatments to mitigate the increasing risks posed by fires, droughts, pests, diseases, and other emerging threats amplified by climate change.

Policy Challenge The "Closer-to-Nature Forest Management" [20] approach seems to be promising in mitigating the main impacts of climate change on forest management, but its implementation requires careful consideration of local conditions and objectives, particularly in various biomes and habitats.

Gaps Recognizing the limitations of a "one size fits all" approach to forest management is crucial due to differing national definitions, diverse forest types, and conflicting interests in resource management. Additionally, the varying intensity of natural disturbances and historical forest utilization necessitate a nuanced understanding of management requirements. To effectively implement proactive strategies, improving EU knowledge and practice is essential, considering these multifaceted factors and tailoring approaches accordingly. This will ensure the feasibility and success of forest management initiatives across all EU countries.

Achieving Integrated Multifunctional Forest Management

As-Is Sustainable yield forestry, while not a distinct and original management concept, can be viewed as a comprehensive framework that encompasses various management models within a shared spatial context.

To-Be Integrated multifunctional forest management is essential in achieving a harmonious balance between the social, economic, and environmental objectives of forest resources within this context.

Policy Challenge To address this policy challenge, there is a need for the development and improved implementation of forest policies that adopt a multifaceted approach, effectively balancing these objectives. This should remain a top priority for the future EU Forest Strategy.

Gaps Gaps in understanding and implementing integrated multifunctional forest management persist, including a lack of practical understanding, monitoring deficiencies, and challenges in addressing diverse stakeholder interests. Addressing these gaps is critical for advancing successful management strategies.

Integrating Adaptation and Mitigation Measures Through Climate-Smart Forestry

As-Is Forests play a crucial role in mitigating the impacts of climate change.

To-Be In order to effectively address climate change, forest management strategies must take into account forest disturbances, and adaptation and mitigation measures should be integrated.

Policy Challenge "Climate-Smart Forestry" [21, 22] is a promising approach that integrates mitigation and adaptation measures to enhance forest resilience and ecosystem services, addressing the needs of a growing population.

Gaps However, several gaps hinder the implementation of best practices in this field including the absence of standardized indicators to monitor the sustainability and climate resilience of forest management practices in the EU. Increased awareness and education among forest managers, policymakers, and citizens is needed to promote the adoption of this approach and its potential benefits.

Promoting Integrated Fire Management

As-Is The increasing wildfire activity across Europe poses a significant threat to heterogeneous forest ecosystems and the vital ecosystem services they provide.

To-Be Improve protection and preparedness through the implementation of climate- and fire-smart landscapes [23], especially "firescapes," with a particular focus on empowering local communities. To manage fires at the landscape scale, fire-prone vegetation should be manipulated to reduce fuel loads, decrease continuity of distribution, and minimize the spread of homogeneous fuel types, thereby reducing fire hazard, i.e., the ignition, extent, intensity, and severity of fires.

Policy Challenge A key policy challenge lies in promoting resilient landscapes mosaics and communities through IFM, which holds promise for mitigating the risks and impacts of wildfires.

Gaps Collaboration among governments, land managers, policymakers, scientific organizations, and local communities is essential to achieve the goal of effective IFM and the preservation of Europe's valuable ecosystems. In addition, EU legislation on Nature Restoration Law should establish clear targets, timetables, funding mechanisms, and support to preserve Europe's ecosystems through cohesive efforts and coordination.

Discussion

IMFM involves using forests in a multifunctional way that maintains biodiversity, productivity, and ecological, economic, and social functions (triangle of sustainability). The Paris Agreement unites the members of the United Nations Framework Convention on Climate Change (UNFCCC) also in combating climate change and adapting to its effects [24]. IMFM and the adoption of alternative policy measures, such as the combination of mitigation and adaptation measures, are crucial actions in this regard [25]. However, traditional forest policies and management practices often fail to consider the multifunctionality of forests, posing challenges for SFM that aims to achieve socially justice, ecologically sound, and economically viable outcomes (e.g., [26]). Challenges in implementing SFM policies and management strategies (e.g., IMFM) include monitoring, information reporting, government incentives and funding, communication between organizations, identification of the national authorities for forest policy and planning [27], government budgetary cycles, etc. [28].

A holistic approach to IFM, as defined above, is crucial for effective coordination of various stakeholders and levels of fire management, leading to increased resilience, sustainability, and ecosystem preservation. It considers multiple factors and requires collaboration, context-specific strategies, and ongoing research. In this way, the risk of wildfires is reduced and natural ecosystems and communities are protected. IFM also faces challenges, particularly in reintroducing fire into fire-dependent ecosystems. Many policies and laws still view all fires as negative, preventing prescribed fires in numerous countries. This hinders ecosystem maintenance and restoration within protected areas [14]. Certainly, prescribed burning is a useful tool and one of the more effective and cost-efficient means of managing vegetation for multiple purposes, but it will be possible to fine-tune the burns through technology and scientific research. Another challenge is the increased number of values at risk due to climate change in fire-prone areas [29].

Overcoming these challenges requires integrating the needs of local communities, promoting prescribed fire use, preventing undesired ignitions, improving fuel management for biodiversity conservation [30], and sustaining ecosystem services. Long-term solutions could include IMFM and IFM approaches such as: (a) reducing fuels (thinning, prescribed burns, grazing, removal of potentially hazardous dead timber), (b) planting of fire-tolerant tree species in a mixture with an increase in the number of broadleaves, and (c) creating fire breaks to reduce fireline intensities in conditions most likely to cause damages.

In conclusion, implementing IMFM and IFM involves addressing various challenges, such as protecting values at risk due to climate change, reducing wildfire impacts, recognizing the multifunctionality of different landscape mosaic types, minimizing the interaction of external and internal threats to biodiversity, enhancing continuous monitoring and reporting of landscape mosaic types change dynamics, bridging policy gaps on fire reintroduction.

By integrating the needs of local communities and adopting IFM practices, policymakers can promote sustainable practices that balance social, ecological, and economic considerations.

Acknowledgments This study has been conducted in the framework of the SILVANUS—Integrated Technological and Information Platform for Wildfire Management project and has received funding from the European Union's Horizon 2020 research and innovation program under grant agreement No 101037247. The contents of this publication are the sole responsibility of the authors and can in no way be taken to reflect the views of the European Commission.

Author Statement The idea for this manuscript was developed by Nikolaos Kalapodis (first author, conceptualization). All authors collectively developed the outline and structure. Data contributions were made by different coauthors. The writing process was led by the first author, with revisions and input provided by all authors.

Declaration of Competing Interest The authors declare that they have no known competing financial interest or personal relationship that could have appeared to influence the work reported in this paper.

REFERENCES

1. Korená Hillayová, M., Holécy, J., Korísteková, K., Bakšová, M., Ostrihoň, M., & Škvarenina, J. (2023). Ongoing climatic change increases the risk of wildfires. Case study: Carpathian spruce forests. *Journal of Environmental Management,* *337,* 117620. https://doi.org/10.1016/j.jenv-man.2023.117620. Epub 2023 Mar 17. PMID: 36934505.
2. Khabarov, N., Krasovskiy, A., Obersteiner, M., Swart, R., Dosio, A., San-Miguel-Ayanz, J., Durrant, T., Camia, A., & Migliavacca, M. (2016). Forest

fires and adaptation options in Europe. *Regional Environmental Change, 16*, 21–30. https://doi.org/10.1007/s10113-014-0621-0

3. Jenkins, M., & Schaap, B. (2018). Background analytical study 1: Forest ecosystem services. Global Forests Goals; United Nations Forum on Forests. URL   https://www.un.org/esa/forests/wp-content/uploads/2018/05/ UNFF13_BkgdStudy_ForestsEcoServices.pdf. Accessed July 21, 2023.

4. Seidl, R., Thom, D., Kautz, M., Martin-Benito, D., Peltoniemi, M., Vacchiano, G., Wild, J., Ascoli, A., Petr, M., Honkaniemi, J., Lexer, M. J., Trotsiuk, V., Mairota, P., Svoboda, M., Fabrika, M., Nagel, T. A., & Reyer, C. P. O. (2017). Forest disturbances under climate change. *Nature Climate Change, 7*, 395–402. https://doi.org/10.1038/NCLIMATE3303

5. Gauthier, S., Bernier, P., Burton, P. J., Edwards, J., Isaac, K., Isabel, N., Jayen, K., Le Goff, H., & Nelson, E. A. (2014). Climate change vulnerability and adaptation in the managed Canadian boreal forest. *Environmental Reviews, 22*(3), 256–285. https://doi.org/10.1139/er-2013-0064

6. Helsinki: Ministry of Agriculture and Forestry. (1993). Ministerial conference on the protection of forests in Europe, Helsinki, 16–17 June 1993: Conference proceedings. In *Ministerial conference on the protection of forests in Europe*, 2nd, Helsinki, FI, 16–17 June 1993, Helsinki.

7. FAO. (2018). *Global forest resource assessment 2020 – Terms and definition (FRA 2020)*, Rome.

8. Pülzl, H., Wydra, D., & Hogl, K. (2018). Piecemeal integration: Explaining and understanding 60 years of European Union forest policy-making. *Forests, 9*(11), 719.

9. Castañeda, F. (2000). *Criteria and indicators for sustainable forest management: International processes, current status and the way ahead*. FAO. https:// www.fao.org/3/x8080e/x8080e06.htm. Last accessed July 21, 2023.

10. The Montréal Process. (2015). Montréal Process: Criteria and indicators for the conservation and sustainable management of temperate and boreal forests.

11. FOREST EUROPE. (2015). Madrid 2015. In *FORES EUROPE 7th ministerial conference*, Madrid.

12. Di Girolami, E., Kampen, J., & Arts, B. (2023). Two systematic literature reviews of scientific research on the environmental impacts of forest certifications and community forest management at a global scale. *Forest Policy and Economics, 146*, 102864.

13. Lombardo, E. (2023). Analysis of the propensity of Italian and German forest owners towards forest certification for ecosystem services. *Journal of Forest Science, 69*, 266–276.

14. Myers, R. L. (2006). Living with fire—Sustaining ecosystems & livelihoods through integrated fire management. *The Nature Conservancy*.

15. Venäläinen, A., Lehtonen, I., Laapas, M., Ruosteenoja, K., Tikkanen, O. P., Viiri, H., Ikonen, V. P., & Peltola, H. (2020). Climate change induces multi-

ple risks to boreal forests and forestry in Finland: A literature review. *Global Change Biology, 26*(8), 4178–4196.

16. Lindner, M., Fitzgerald, J. B., Zimmermman, N. E., et al. (2014). Climate change and European forests: What do we know, what are the uncertainties, and what are the implications for forest management? *Journal of Environmental Management, 145,* 69–83.

17. Li, Y., Brando, P. M., Morton, D. C., Lawrence, D. M., Yang, H., & Randerson, J. T. (2022). Deforestation-induced climate change reduces carbon storage in remaining tropical forests. *Nature Communications, 13,* 1964.

18. Mok, H.-F., Arndt, S. K., & Nitschke, C. R. (2012). Modelling the potential impact of climate variability and change on species regeneration potential in the temperate forests of South-Eastern Australia. *Global Change Biology, 18*(3), 1053–1072.

19. FOREST EUROPE 2020. (2020). *Adaptation to climate change in sustainable forest management in Europe.* FOREST EUROPE Liaison Unit Bratislava.

20. Larsen, J. B., Angelstam, P., Bauhus, J., Carvalho, J. F., Diaci, J., Dobrowolska, D., Gazda, A., Gustafsson, L., Krumm, F., Knoke, T., Konczal, A., Kuuluvainen, T., Mason, B., Motta, R., Pötzelsberger, E., Rigling, A., & Schuck, A. (2022). *Closer-to-nature forest management. From science to Policy 12.* European Forest Institute. https://doi.org/10.36333/fs12

21. Hetemäki, L., & Verkerk, H. (2022). Climate-smart forestry approach. In L. Hetemäki, J. Kangas, & H. Peltola (Eds.), *Forest bioeconomy and climate change. Managing forest ecosystems* (Vol. 42). Springer. https://doi. org/10.1007/978-3-030-99206-4_9

22. Verkerk, P. J., Costanza, R., Hetemäki, L., Kubiszewski, I., Leskinen, P., Nabuurs, G. J., Potočnik, J., & Palahí, M. (2020). Climate-smart forestry: The missing link. *Forest Policy and Economics, 115.* https://doi.org/10.1016/j. forpol.2020.102164. ISSN 1389-9341.

23. Iglesias, M. C., Hermoso, V., Campos, J. C., Carvalho-Santos, C., Fernandes, P. M., Freitas, T. R., Honrado, J. P., Santos, J. A., Sil, Â., Regos, A., & Azevedo, J. C. (2022). Climate- and fire-smart landscape scenarios call for redesigning protection regimes to achieve multiple management goals. *Journal of Environmental Management, 322,* 116045. https://doi.org/10.1016/j. jenvman.2022.116045. ISSN 0301-4797.

24. UNFCCC. (2015). The Paris Agreement – Publication, Paris.

25. Bratu, I. A. (2019). Open-source solutions to improve the quality of sustainable forest management. *MATEC Web of Conferences, 290,* 11003 . EDP Sciences..

26. Primdahl, J., Arler, F., Angelstam, P., Christensen, A. A., & Elbakidze, M. (2018). Rural landscape governance and expertise: On landscape agents and democracy. In *Defining landscape democracy: A path to spatial justice* (pp. 153–164). Edward Elgar Publishing.

27. Lazdinis, H., Angelstam, P., & Pülzl, H. (2019). Towards sustainable forest management in the European Union through polycentric forest governance and an integrated landscape approach. *Landscape Ecology, 34,* 1737–1749.
28. Hickey, G. M. (2008). Evaluating sustainable forest management. *Ecological Indicators, 8*(2), 109–114.
29. Schultz, C. A., Thompson, M. P., & McCaffrey, S. M. (2019). Forest Service fire management and the elusiveness of change. *Fire Ecology, 15,* 13.
30. Bowman, D. M. J. S., Kolden, C. A., Abatzoglou, J. T., Johnston, F. H., van der Werf, G. R., & Flannigan, M. (2020). Vegetation fires in the Anthropocene. *Nature Reviews Earth & Environment, 1,* 500–515.

Open Access This chapter is licensed under the terms of the Creative Commons Attribution 4.0 International License (http://creativecommons.org/licenses/by/4.0/), which permits use, sharing, adaptation, distribution and reproduction in any medium or format, as long as you give appropriate credit to the original author(s) and the source, provide a link to the Creative Commons license and indicate if changes were made.

The images or other third party material in this chapter are included in the chapter's Creative Commons license, unless indicated otherwise in a credit line to the material. If material is not included in the chapter's Creative Commons license and your intended use is not permitted by statutory regulation or exceeds the permitted use, you will need to obtain permission directly from the copyright holder.

CHAPTER 7

The Search and Rescue Project: Emerging Technologies for Early Location of Entrapped Victims Under Collapsed Structures and Advanced Wearables for Risk Assessment and First Responder Safety in SAR Operations

Sofia Karma and Christos Ntanos

INTRODUCTION

Extreme weather events currently observed on a global scale have been correlated with the current "climate crisis," which, combined with population growth and urbanization, has exacerbated the consequences in terms of disaster casualties, property losses, and environmental impacts. The Search and Rescue project, among other related initiatives, has implemented part of the Disaster Risk and Resilience priorities outlined in the UN's 2030 Sustainable Development Agenda. Its primary goal was to establish and promote a comprehensive framework encompassing system

S. Karma (✉) • C. Ntanos
National Technical University of Athens, Athens, Greece
e-mail: skarma@epu.ntua.gr

© The Author(s) 2025
I. Gkotsis et al. (eds.), *Paradigms on Technology Development for Security Practitioners*, Security Informatics and Law Enforcement, https://doi.org/10.1007/978-3-031-62083-6_7

and equipment interoperability, training, and awareness by providing cutting-edge technologies and innovative tools for the first responders. The technologies, tools, and techniques developed within the Search and Rescue project were successfully tested and validated through seven field exercises based on various use-case scenarios. Within this framework, a promising field technology—Membrane Inlet Mass Spectrometry (MIMS)—was adapted to meet the needs of first responders and tested for the first time in search and rescue operations with the "RESCUE-MIMS" prototype (TRL 6). Specifically, the device was used in relevant exercises and tested (a) as an early warning system on-board a ground robotic platform for first responders' safety, and (b) as an "artificial sniffer" for detecting compounds related to human presence "human signs" according to literature, complementing the rescue dogs. The human chemical signs can be considered the total of the chemical compounds evolved by the human body and skin like axilla and sweat, as well as and the ones emitted via human breath [1, 2]. According to literature, the human body emits different types of VOCs like alcohols, aldehydes, amines, aromatics, ketones, etc. [3] while the majority of VOCs in breath may have concentrations ranging from ppb to ppt levels [4, 5].

In Table 7.1, the most representative VOCs and gases that have been correlated with human presence in literature are summarized that can be

Table 7.1 Representative VOCs and gases that have been correlated with "human presence" in literature and their characteristic mass fragments for monitoring with mass spectrometry-based technologies [6]

Compound	Type	Characteristic mass fragments
Water vapors of the exhaled air	Gas phase	17, 18
Ammonia	Inorganic gas	16, 17
Carbon Dioxide	Inorganic gas	44
Oxygen	Inorganic gas	28, 12
Acetone	VOC	43, 58
Isoprene	VOC	53, 67, 68
Hexane	VOC	57, 86
Pentane	VOC	41, 42, 57, 72
1-Pentene	VOC	55, 70
Lactic acid	VOC	45, 90
Ethanol	VOC	27, 29, 30, 31, 45, 46
Acetaldehyde	VOC	29
Limonene	VOC	68, 121

used for field analysis [6]; the characteristic mass fragments of those compounds are also provided for monitoring in Single Ion Monitoring (SIM) mode by using the RESCUE-MIMS prototype.

According to the literature, the above compounds may behave differently as they permeate the ruins of buildings, interacting with brick, concrete, wood, furnishings, and other materials [8].

Another important issue for the rescuers that operate in the disaster environment, as well as the canines, is the safety and security concerns arise. For example, in case of collapsed structures there is a possibility of toxic or explosive gases release due to destruction of pipelines or explosion of gas cylinders under pressure that might be under the ruins, or even fire spots that can be generated due to short circuits.

As a result of the above, it seems crucial to have field technologies capable of monitoring on-site and online such hazardous environments and at a safe distance from the source, in order to protect the firefighters or the first responders and their canines in general. However, chemical hazards' release creates a strong chemical background under the ruins that may burden search and rescue operations of entrapped victims.

Methodology

As previously mentioned, mimicking the rescue dogs in the disaster scene is a complex issue since hundreds of chemical compounds with different origins can be present. In Fig. 7.1, testing of the RESCUE-MIMS in the

Fig. 7.1 Testing of the RESCUE-MIMS in the field under the scenario "People Trapped Under the Rubbles" that took place in Limoges, France in cooperation with the PUI team

field under the scenario "People Trapped Under the Rubbles" that took place in Limoges, France in cooperation with the PUI team (Pompiers de l' Urgence Internationale), is shown.

Based on the scenario, the RESCUE-MIMS device was deployed in the field for measuring the compounds inside the voids of the collapsed structure and for providing a possible "alarm of a human" under the rubbles.

For the needs of the SnR project, a number of key compounds were selected for monitoring the exhaled air of individuals during preparatory lab-experiments, as well as for the field exercises, namely the aforementioned pilot "People trapped under rubbles." These key compounds were monitored via their representative masses, based on the Table 7.1. More specifically, the selected masses for monitoring with the RESCUE-MIMS prototype were: *mass 43* (acetone), *mass 44* (carbon dioxide), *mass 45* (ethanol), *mass 58* (acetone).

The RESCUE-MIMS field technology has also been tested under the SnR project framework for remote sensing on-board robotic platforms under a simulated industrial fire scenario; forest fire expanded threatening an industrial zone.

In that scope, a number of chemical hazards for the safety of the first responders were selected at first place for initial testing and validation of the RESCUE-MIMS prototype in lab-scale. Specifically, the selected masses that were finally monitored with the RESCUE-MIMS prototype were: *mass 17* (Ammonia), *mass 28* (Carbon Monoxide), *mass 30* (Nitrogen Oxide) *mass 44* (Carbon Dioxide), *mass 46* (Nitrogen Dioxide), *mass 78* (Benzene) [9, 10]. The chemical data were transferred online and displayed as alarm messages on the SnR platform, informing the first responders at the command and control center.

As shown in Fig. 7.2, the robot was able to pass through the intense smoke and measure online the various airborne components, utilized as a screening tool of the toxicity of the area. Though, the thermal resistance of the materials of the whole equipment when using it in fire incidents needs to be considered to avoid any damage. Moreover, possible contamination of the detector due to smoke or tars should be taken into account; filters in front of the sampling probe can be used to protect it.

Fig. 7.2 The RESCUE-MIMS on-board the SeekurJr robotic platform by DFKI was used as a remote early warning system of a hazardous smoky environment for the first responders

RESULTS

The RESCUE-MIMS was able to successfully detect online the increased intensities of masses that are correlated with compounds relevant to human exhaled air according to the literature, like Carbon Dioxide and Acetone, compared to the background measurements recorded inside the voids (See Figs. 7.1 and 7.3).

The RESCUE-MIMS was also used in another use case inside the Search and Rescue project (forest fire expanded and threat an industrial zone), and tested as an early warning system on-board a ground robotic platform for the first responders' safety; hazardous compounds, such as ammonia (NH3), carbon monoxide (CO), Benzene, were successfully monitored online and the respective alarms were sent to the SnR platform at the command and control center.

Specific Key Performance Indicators (KPIs) were used to evaluate the RESCUE-MIMS prototype in the aforementioned use cases; portability; robustness; easiness to operate; easiness to deploy; friendliness to the user; fast response times; high sensitivity; minimum false positives/negatives. Also, it has to be mentioned that in both use cases, usability testing of the Rescue MIMS prototype took place with the assistance of the end users. An important consideration when measuring online in the field is the

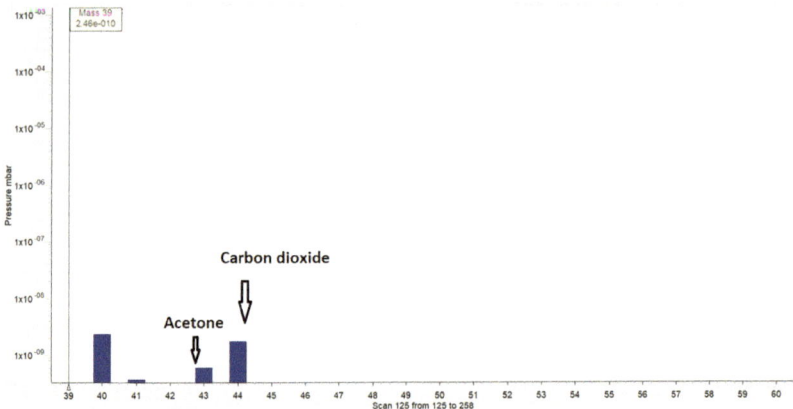

Fig. 7.3 An indicative screenshot of the RESCUE-MIMS display while measuring inside the voids of the collapsed structure where the victim was entrapped (peaks attributed to acetone and carbon dioxide)

potential background interferences that may create false positives or negatives; for this reason, background measurements were recorded by the RESCUE-MIMS in the respective pilots.

CONCLUSIONS

Based on the first responders' feedback, it became apparent that it was possible to use a highly sensitive analytical technique in terms of ultra-low detection capabilities with minimum false alarms alongside sophisticated analytical equipment in the field. It is a promising field technology that can be used as a complementary tool to the classical detection options for location of entrapped victims under collapsed structures, in order to help the first responders in search and rescue operations, e.g., to cope with limitations when using rescue dogs in terms of availability, fatigue or incomplete/insufficient training. The RESCUE-MIMS technology has proven its effectiveness, but there is still room for improvement in terms of weight, ease of use, resistance to environmental, field conditions, and operating autonomy. Also, special training is needed by the users, e.g., one or two members from the firefighters team should be trained every 2–3 months, as already do for similar technologies, like the scanner or the drones.

Acknowledgments The Search and Rescue project has received funding from the European Union's Horizon 2020 research and innovation program under grant agreement No. 882897. This article reflects only the authors' views, and the Research Executive Agency and the European Commission are not responsible for any use that may be made of the information it contains.

REFERENCES

1. Callagher, M., Wysocki, C. J., Leyden, J. J., Spielman, A. I., Sun, X., & Preti, G. (2008). Analyses of volatile organic compounds from human skin. *British Journal of Dermatology, 159*, 780–791.
2. Curran, A. M., Rabin, S. I., & Furton, K. G. (2005). Analysis of the uniqueness and persistence of human scent. *Forensic Science Communications, 7*, 114.
3. Ratcliffe, N. (2020). A mechanistic study and review of volatile products from peroxidation of unsaturated fatty acids: An aid to understanding the origins of volatile organic compounds from the human body. *Journal of Breath Research, 14*, 034001.

4. Wang, C., & Sahay, P. (2009). Breath analysis using laser spectroscopic techniques: Breath biomarkers, spectral fingerprints, and detection limits. *Sensors, 9*, 8230–8262.

5. Statheropoulos, M., Sianos, E., Agapiou, A., Georgiadou, A., Pappa, A., Tzamtzis, N., Giotaki, H., Papageorgiou, C., & Kolostoumbis, D. (2005). Preliminary investigation of using volatile organic compounds from human expired air, blood and urine for locating entrapped people in earthquakes. *Journal of Chromatography B: Analytical Technologies in the Biomedical and Life Sciences, 822*, 112–117.

6. Giannoukos, S. (2015). *Portable mass spectrometry for artificial sniffing*, Thesis submitted in accordance with the requirements of the University of Liverpool for the degree of Doctor in Philosophy, Department of Electrical Engineering and Electronics, The University of Liverpool.

7. Statheropoulos, M., Agapiou, A., Pallis, G. C., Mikedi, K., Karma, S., Vamvakari, J., Dandoulaki, M., Andritsos, F., & Thomas, C. L. P. (2015). Factors that affect rescue time in urban search and rescue (USAR) operations. *Natural Hazards, 75*, 57–69. https://doi.org/10.1007/s11069-014-1304-3

8. Down, S. (2012). *Buried biomarkers: Rubble affects urinary volatiles in earthquake wreckage*, 04 06 2012. [Online]. Available: https://analyticalscience. wiley.com/do/10.1002/sepspec.137bddb94f8/full/. Accessed August 27, 2023.

9. Karma, S. (2018). Tools for analyzing risks from human exposure to chemical environments: The case of exposure to smoke components during forest or other field fires. In *Novel approaches in risk, crisis and disaster management*. Nova Science Publishers, p. Chapter 9.

10. Karma, S. (2019). Challenges and lessons learned from past major environmental. Disasters due to technological or wildland urban interface fire incidents. Contributing Paper to Global Assessment Report on Disaster Risk Reduction (GAR). Available at: https://www.preventionweb.net/publications/view/66718

Open Access This chapter is licensed under the terms of the Creative Commons Attribution 4.0 International License (http://creativecommons.org/licenses/by/4.0/), which permits use, sharing, adaptation, distribution and reproduction in any medium or format, as long as you give appropriate credit to the original author(s) and the source, provide a link to the Creative Commons license and indicate if changes were made.

The images or other third party material in this chapter are included in the chapter's Creative Commons license, unless indicated otherwise in a credit line to the material. If material is not included in the chapter's Creative Commons license and your intended use is not permitted by statutory regulation or exceeds the permitted use, you will need to obtain permission directly from the copyright holder.

CO-PROTECT: Greek Cluster for Cooperative and Interoperable Crisis and Disaster Management Solutions

Ilias Gkotsis, Dimitrios Diagourtas, Nikolaos Zafeiropoulos, Victoria Katsarou, Dimitris Katsaros, Theodora Galani, Dimitris Bliziotis, and Eleftherios Ouzounoglou

INTRODUCTION

CO-PROTECT, endorsed by the Greek General Secretariat of Research and Innovation, is a collaborative cluster project involving various technological enterprises and SMEs within Greece. Its mission is to foster innovation by integrating and enhancing high TRL products and services dedicated to natural disaster management, environmental crises, and civil

I. Gkotsis (✉) • D. Diagourtas
Satways Ltd, Neo Irakleio, Greece
e-mail: i.gkotsis@satways.net

N. Zafeiropoulos • V. Katsarou
Space Hellas SA, Cholargos, Greece

D. Katsaros • T. Galani
EXUS Software Single Member Limited Liability Company, Athens, Greece

D. Bliziotis • E. Ouzounoglou
Institute of Communications and Computer Systems, Athens, Greece

© The Author(s) 2025   89
I. Gkotsis et al. (eds.), *Paradigms on Technology Development for Security Practitioners*, Security Informatics and Law Enforcement, https://doi.org/10.1007/978-3-031-62083-6_8

protection. This pioneering initiative, aligned with the Greek Disaster Resilience Innovation Cluster, Defkalion (DRIC), seeks to develop globally applicable interoperable products. The consortium, comprising SMEs and top Greek research institutions, works to bridge technological gaps and deliver competitive solutions. CO-PROTECT's offerings include interoperable tools for public and private entities such as Civil Protection Organizations, Emergency Management Services, and Law Enforcement Agencies. These tools enhance the efficiency of operational tasks and missions. Emphasizing international collaboration, the project's strategy aims to elevate Greek SMEs and extend the reach of safety and security products to global markets. CO-PROTECT is uniquely designed to provide components that work together seamlessly, culminating in a unified platform that supports disaster and crisis management. The anticipated result is a significant enhancement in disaster and crisis management support, reflecting Greece's leadership in this vital field.

Specifically, CO-PROTECT's results encompass the following key areas:

Interoperability Framework: Creation of a standardized framework at the data models and services level. This framework facilitates communication and sharing between current products, services, methods, and solutions within CO-PROTECT's platform, and can be extended to future developments.

Technological Solutions for Earthquakes: Provision of early warning systems for earthquakes and rapid damage assessment, leveraging risk assessment reasoning, fragility curves, and networks of specialized sensors and digital equipment.

Data Modeling for Flood and Weather Events: Development of data modeling solutions aimed at flood events and extreme weather forecasting, enhancing preparedness and informed decision-making.

Wildfire Management Solutions: Implementation of comprehensive solutions for managing wildfires, including risk mapping, fire spot detection and location, fire spread simulation, and support for fire control operations.

Critical Infrastructure Security: Introduction of solutions that assist in the surveillance and monitoring of security issues within critical infrastructures, ensuring robust protection.

These innovative outcomes represent a significant stride in disaster and crisis management, integrating multiple domains and technologies to provide a holistic approach to safety and security.

CO-PROTECT Common Interoperability Framework

As delineated above, the CO-PROTECT product repository encompasses various tools and solutions designed to assist with disaster and crisis management tasks. The repository is meticulously organized according to risk categories, including extreme weather, floods, wildfires, earthquakes, and the protection of critical infrastructures. Within this structure, the relevant components are crafted to be compatible with platforms, ensuring seamless interoperability of data and services.

Key to the success of these solutions is the ability to share common information and model data exchange appropriately to enable interoperable services. This is realized through implementing CO-PROTECT's Common Interoperability Framework (CIF), a guideline that all integrated systems and applications must adhere to within the project's platforms. The CIF consists of two primary models—data and service models—designed to facilitate maximum cooperation and exchange among disparate applications that are incorporated into the platform.

In the ensuing sections of this publication, readers will find a concise overview of the ontology representation and high-level architecture, supplemented by an in-depth exploration of the common data and service models. These models are instrumental in ensuring interoperability and cohesive functionality among the solutions provided within the CO-PROTECT initiative. The comprehensive integration approach sets the stage for an advanced, cooperative disaster and crisis management framework.

CO-PROTECT Architecture

Methodological Approach
Collaborative interoperability in CO-PROTECT requires harmonizing communication across various entities, including organizations, applications, command systems, and technological equipment, especially in civil protection crises and emergency management. CO-PROTECT's

framework fosters real-time data exchange, interagency communication, and operational synchronization during crises.

Figure 8.1 illustrates CO-PROTECT's general data model, integrating diverse information sources with user requirements to formulate responses for different civil protection crises. Precise data acquisition is paramount, and robust systems ensure accuracy. Data models furnish essential definitions and formats, playing a pivotal albeit modest role in data systems development, yet with likely overarching influence on the outcomes.

Postuser definition and requirement elicitation, the project recognized fragmented solutions within the Greek market by consortium organizations. Their unique characteristics were scrutinized, and the optimal data models for each crisis scenario were adopted. This selection focused on the prerequisites for subsequent integration into a seamless interoperable framework solution, highlighting the methodical approach in unifying disparate technologies and methodologies for enhanced crisis response.

This integrative strategy is central to CO-PROTECT's innovative approach to disaster management.

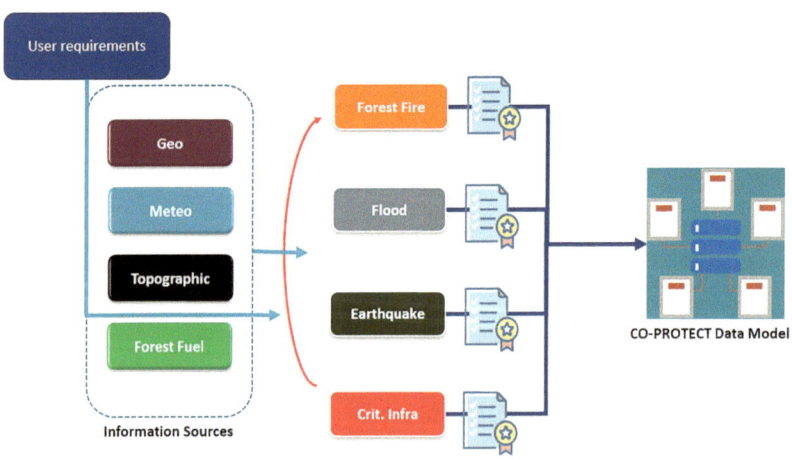

Fig. 8.1   CO-PROTECT data model

CO-PROTECT Reference Architecture

Figure 8.2 illustrates the reference architecture of CO-PROTECT, delineating the core architecture (right) and the governance framework for interoperability (left).

This architecture ensures an interoperable framework regardless of the specific natural disaster addressed, encompassing Legal, Organizational, Semantic, and Technical dimensions.

Legal Interoperability aligns cross-border information exchange laws; Organizational Interoperability molds cross-organizational goals and processes; Semantic Interoperability ensures shared understanding through standards like semantic dictionaries; Technical Interoperability manages system-level connections, such as data integration.

Collectively, these frameworks orchestrate public or private environmental incident management, culminating in a holistic solution that meets user needs and offers unified public services. Such coordination is vital for managing crises involving multiple entities, steered by authorized planning and implementation.

Within CO-PROTECT's architecture, stored event data in designated Databases (DBs) fosters intraorganizational communication, with

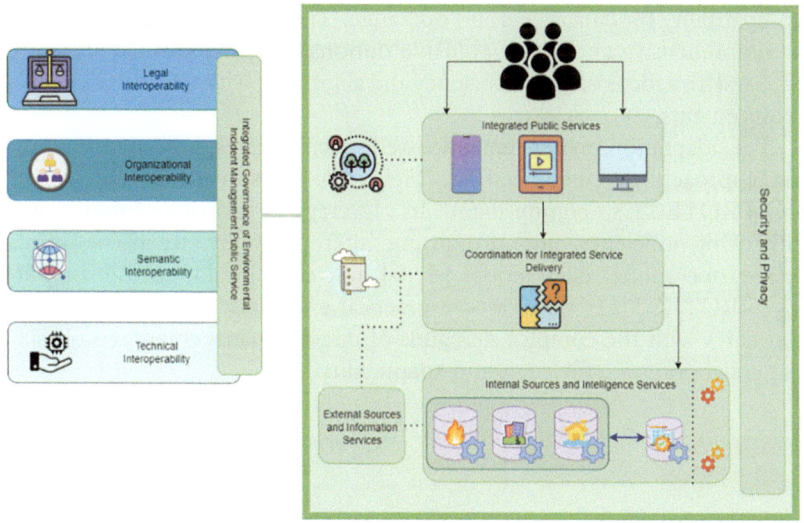

Fig. 8.2 CO-PROTECT reference architecture

integration extending to external sources, enhancing service capabilities. This data amalgamation forms crisis management mechanisms. Accurate interpretation relies on dictionary standards from interoperability repositories. Interoperable data, accessed through specialized mechanisms, is presented to users via various control devices, ensuring timely and competent service provision. This comprehensive approach encapsulates CO-PROTECT's innovative strategy, leveraging interoperability to create a cohesive, real-time response system for effective crisis management.

Common Data Model

This chapter outlines CO-PROTECT's methodology for determining the data model that governs communication across key areas such as earthquakes, wildfires, floods, extreme weather events, and critical infrastructure protection. Recognizing the complexity of disaster management, CO-PROTECT opted to adapt the existing Copernicus EMS service data model (CDM-EMS) [1] rather than create a new model. This choice assures interoperability with existing systems and is economically efficient in terms of time and resources.

The CDM-EMS, already widely adopted and validated for disaster management, was tailored to CO-PROTECT's specific needs, ensuring compatibility for data access and exchange. This decision was grounded in a comprehensive analysis of available options and international literature [2, 3], with a focus on time efficiency and alignment with CO-PROTECT's requirements.

The adaptation process encompassed specific stages, reflecting a strategic approach to customization (Fig. 8.3). This method underscores CO-PROTECT's commitment to leveraging existing innovations, enhancing efficiency, and promoting interoperability. By harnessing a tested operational data model [4, 5] and strategically customizing it, CO-PROTECT has showcased a practical and innovative approach that resonates with the complex demands of disaster management, emphasizing collaboration, efficiency, and adaptability.

Fig. 8.3 Methodological approach for adapting and specializing the EMS data model to CO-PROTECT needs

The envisaged chapter underscores the criticality of standardization and interoperability, aiming for consistency and accuracy in data collection and management within the realm of natural disaster response and critical event handling. The proposed specialized data model CDM-COPROTECT seeks to enhance the efficiency and efficacy of data management and analysis.

CDM-COPROTECT extends the existing CDM-EMS, inheriting its broad applicability and alignment with Copernicus EMS services. Simultaneously, it adds the requisite depth and specialization to fulfill CO-PROTECT's unique requirements.

The structured data model fosters a unified framework streamlining data management, information analysis, and decision-making. A primary challenge was correlating four distinct datasets (Earthquake, Flood, Fire, Critical Infrastructure) using common characteristics to identify disaster types. The solution dynamically accesses the relevant dataset, based on the characterized disaster type, ensuring smooth data management and analysis. Separate datasets were crafted for the four different pillars, with a distinct level (Observed Event) created to establish a stable one-to-many cardinality relationship, signifying the disaster type and corresponding data (Fig. 8.4).

CO-PROTECT's proposed data model embodies a comprehensive framework that facilitates robust management and analysis of information related to various natural disasters and critical infrastructure protection. Designed for seamless communication and interoperability among software and hardware entities within the project's scope, it epitomizes a strategic approach toward integrated and responsive disaster management. This model represents a significant advancement in the field, leveraging standardized practices to drive effective response mechanisms.

The CDM-EMS's Observed_Event level [6] was strategically leveraged and expanded to serve the specific requirements of CO-PROTECT's four principal pillars, encompassing the application and technology interdependencies. Building on the existing structural elements of CDM-EMS (Group 1, Group 2, and Group 3), additional structures and tables were introduced, tailored to the unique requirements of each disaster category (Group 4: EARTHQUAKE, GROUP 5: FLOOD, GROUP 6: FIRE, and GROUP 7: CRITICAL_INFRASTRUCTURE).

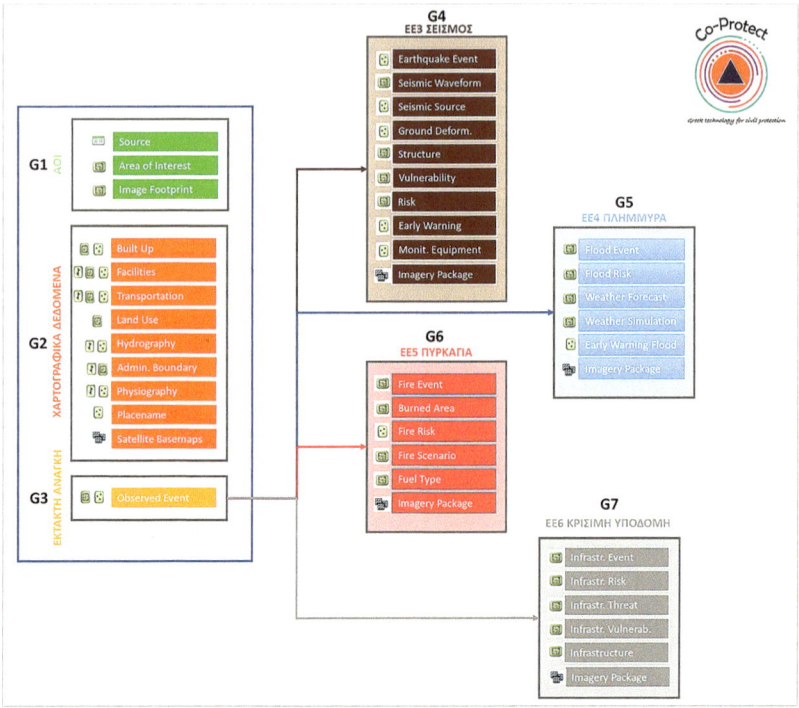

Fig. 8.4   CO-PROTECT common data model

Common Service Model

Interoperability holds paramount significance for civil protection services, enabling seamless communication and information exchange during all phases of crisis management, including preparedness, response, mitigation, and recovery. Employing a common service model to manage information flow in emergencies is instrumental in assuring the interoperability of integrated technology solutions.

CO-PROTECT's logical service model encapsulates logical groupings of tools and services, reflecting abstract representations of the system's main components and articulating their interconnections. These groupings are categorized into the following four levels, each corresponding to the specific functions and processes they support:

This structured approach underscores CO-PROTECT's dedication to fostering coherent and efficient communication within the multifaceted landscape of disaster management, leveraging logical design to enhance functional interplay and responsiveness.

Data Layer

CO-PROTECT's Data Layer focuses on collecting, storing, and organizing primary data. It is where the data from various sensors, including satellites and weather stations, are cataloged and managed. These sensors are chosen based on the technology and environmental risk, detecting specific characteristics for effective intervention strategies. The Data Layer includes systems like satellites and stations that collect raw data, channeling them to higher levels of the platform. This collection of environmental data plays a crucial role in understanding and monitoring natural phenomena, translating raw observations into actionable insights. The meticulous design of the Data Layer emphasizes CO-PROTECT's commitment to precise data handling, employing cutting-edge technology to foster a robust understanding of environmental dynamics and serving as the foundation for effective disaster management.

Data Analysis Layer

The Data Analysis Layer in CO-PROTECT's architecture serves as the processing hub where data captured in the Data Layer is scrutinized, analyzed, and transformed. This critical layer metamorphoses raw data into actionable information through a series of intricate techniques.

Within this layer, a comprehensive suite of procedures and tools is deployed for data cleaning, filtering, normalization, and transformation, priming the data for more nuanced analysis. The analysis engine, a core component of this layer, is equipped with a myriad of algorithms and models specifically designed to distill information and discern patterns from the sensor recordings.

Hosting various analytical models and algorithms, this segment of the platform is tasked with extracting meaningful insights from the preprocessed sensor data. Depending on the specific use case and the complexity of the data, the analysis may encompass a broad spectrum of techniques. This includes but is not limited to statistical analysis, machine learning, artificial intelligence, and other advanced analytical methodologies.

Data Visualization Layer
The Data Visualization Layer within CO-PROTECT's framework focuses on translating analytical results into visually accessible forms. This layer renders complex data into formats that facilitate easy understanding by incorporating visual tools such as graphs, charts, and interactive visualizations. The goal is to make intricate analysis outcomes accessible and actionable without losing complex insights. A distinguishing feature of this layer is its adaptability to user preferences, allowing for customized dashboards that align with individual needs. The Data Visualization Layer reflects CO-PROTECT's commitment to user-centric design, bridging the gap between technical analysis and practical user comprehension. By transforming data into visual narratives, it enhances the platform's usability, fostering informed decision-making through intuitive design. It embodies CO-PROTECT's dedication to making complex environmental data accessible to diverse stakeholders.

Management Layer
The Management Layer, the culmination of CO-PROTECT's process, emphasizes holistic system management, tools, services, and data management. Designed to bolster decision-making and trigger relevant processes, it incorporates various tools and services that empower users to make informed decisions tailored to specific incidents, responsibilities, and operational needs.

As depicted in Fig. 8.5, a seamless correlation exists between successive layers. Information cascades from initial field measurements through data analysis and culminates in user presentation. This integrated flow furnishes users and administrators with a comprehensive understanding of the prevailing situation, arming them with the requisite information to make judicious decisions and execute corresponding actions. This Management

Fig. 8.5 Logical groups of tools/services (layers)

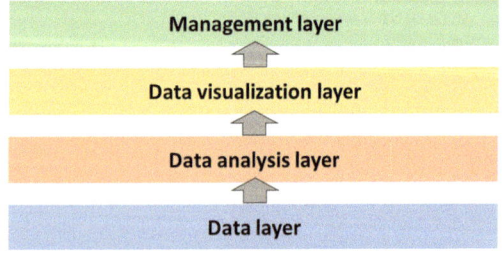

Layer epitomizes CO-PROTECT's strategic alignment of technology and governance, orchestrating a coordinated response to dynamic challenges.

Interoperability Tools
CO-PROTECT utilizes specialized technologies for reliable, immediate, and secure information transmission. These include:

- *Geoserver* [7]: An open-source software enabling users to share and publish geospatial data and maps online, fostering spatial awareness and collaboration.
- *Apache Kafka* [8]: An open-source messaging platform tailored for high-performance, reliable, and scalable real-time data transmission between discrete applications and systems, ensuring seamless integration.
- *REST APIs (Representational State Transfer Application Programming Interfaces)* [9]: A widely adopted architecture instrumental in designing and developing interfaces that facilitate interaction between various applications and systems in the digital landscape.
- *FTP (File Transfer Protocol)* [10] *and SFTP (SSH File Transfer Protocol)*: Distinct protocols utilized for transferring files between computers over the Internet, ensuring secure and efficient data movement.

Together, these technologies form CO-PROTECT's technological backbone, each serving a distinct function within the system. Their combined use ensures seamless integration and robust data flow within the platform, catering to the multifaceted needs of disaster management. CO-PROTECT's choice of these tools reflects a commitment to innovation, interoperability, and security, reinforcing its stance as a responsive and advanced system in the field of civil protection.

Conclusions

CO-PROTECT combines innovative solutions and collaboration with EU entities to enhance civil protection and support Greek SMEs. In summary, the proposed solution's value is substantial and multifaceted. It addresses emergencies at various levels by fostering technological, organizational, and business interoperability. Integrating diverse tools creates a platform for efficient management of environmental crises, reflecting the project's commitment to innovation and cooperation.

The efficacy of this approach is rooted in key foundational elements:

- *Common Data Model*: Ensuring a uniform language across tools and services.
- *Common Service Model*: Facilitating seamless interaction through a unified framework.
- *Connection to Specific Use Cases*: Tailoring responses to unique emergency needs.
- *Internal Structure & Interoperability*: A meticulous understanding of interaction dynamics.

Building upon these principles, CO-PROTECT integrates tools and services, forming a cohesive platform that enhances civil protection efficiency across multiple levels. Its innovative approach sets a precedent for collaboration in disaster management, guiding future endeavors in this vital field.

Acknowledgments This project (code: ΓΓ2CL-0363842) is co-financed by the European Regional Development Fund of the European Union and Greek national funds through the Operational Program Competitiveness, Entrepreneurship, and Innovation.

References

1. https://land.copernicus.eu/eagle/files/documents-and-reports/t13-data-model-uml-documentation/view
2. Eurostat.   https://ec.europa.eu/eurostat/ramon/index.cfm?TargetUrl=DSP_PUB_WELC.
3. https://www.emdat.be/index.php
4. https://www.eea.europa.eu/publications/COR0-landcover
5. Sun, K., Zhu, Y., Pan, P., et al. (2019). Geospatial data ontology: The semantic foundation of geospatial data integration and sharing. *Big Earth Data, 3*(4), 1–28.
6. https://emergency.copernicus.eu/mapping/ems/observed-event
7. http://geoserver.org/
8. https://kafka.apache.org/
9. https://en.wikipedia.org/wiki/Representational_state_transfer
10. https://en.wikipedia.org/wiki/SSH_File_Transfer_Protocol

Open Access This chapter is licensed under the terms of the Creative Commons Attribution 4.0 International License (http://creativecommons.org/licenses/by/4.0/), which permits use, sharing, adaptation, distribution and reproduction in any medium or format, as long as you give appropriate credit to the original author(s) and the source, provide a link to the Creative Commons license and indicate if changes were made.

The images or other third party material in this chapter are included in the chapter's Creative Commons license, unless indicated otherwise in a credit line to the material. If material is not included in the chapter's Creative Commons license and your intended use is not permitted by statutory regulation or exceeds the permitted use, you will need to obtain permission directly from the copyright holder.

CHAPTER 9

Enhancing Pathogen Contamination Incident Management Through Advanced Operational Picture and Collaboration Among Incident Commanders and First Responders

Katerina Valouma, Leonidas Perlepes,
Lefteris Voumvourakis, Antonis Kostaridis,
Iosif Vourvachis, Dimitrios Iliadis, Francesca Grossi,
Dimitra Ioannidou, Arlind Xhelili, Livia El-Khawad,
and Luca E. Sander

INTRODUCTION

Pathogens are a determining factor in emergency response due to their life-threatening nature, both for the public as well as for the First Responder (FR) safety. Waterborne pathogen contamination events can occur anywhere, and may be caused by various reasons, i.e., natural events,

K. Valouma • L. Perlepes • L. Voumvourakis (✉) • A. Kostaridis
Satways Ltd, Neo Irakleio, Greece
e-mail: l.voumvourakis@satways.net

© The Author(s) 2025
I. Gkotsis et al. (eds.), *Paradigms on Technology Development for Security Practitioners*, Security Informatics and Law Enforcement, https://doi.org/10.1007/978-3-031-62083-6_9

accidents or malicious attacks, illegal activities, and cascading effects. To manage such incidents, FRs have ex-pressed the need for a number of new technologies and tools, including but not limited to, sensors for rapid detection of pathogen contaminations, technologies for safe water sampling, decision support tools, tools to isolate the contaminant source and to conduct forensic investigation, restoration guidelines, etc.

To ensure the design and deployment of better and more effective products and services, the PathoCERT project relies on participatory and cocreative approaches. Specifically, the PathoCERT's multistakeholder engagement approach builds upon several interlinked activities, among others the engagement of stakeholders via the establishment of six Communities of Practice (CoPs). A CoP can be defined as a structure that brings together a group of actors who share a common interest in a topic and come together to fulfill both individual and group goals. To identify relevant actors and analyze to be engaged in the CoPs and to better understand their interconnections, relationships and interest within as well as to form an understanding of the current situation or status quo of a system, PathoCERT follows the stakeholder mapping methodology [1] and baseline requirement analysis [2, 3].

The above analysis was conducted by local consortium partners of each region under study (i.e., Granada, Spain; Amsterdam, The Netherlands; Limassol, Cyprus; Thessaloniki, Greece; Sofia, Bulgaria; and Seoul, South Korea) and which will host the validation tests and demonstration exercises. Moreover, it developed a good understanding of the current emergency response and disaster management systems in each region, including applied technologies, and main challenges and opportunities for improvement within. The examination and analysis of requirements, needs, challenges, and opportunities are important to ensure that the project develops and tests appropriate solutions that contribute to improving and advancing the emergency and disaster management system.

I. Vourvachis • D. Iliadis
Hellenic Rescue Team, Thessaloniki, Greece

F. Grossi • D. Ioannidou • A. Xhelili • L. El-Khawad • L. E. Sander
Collaborating Centre for Sustainable Consumption and Production, Wuppertal, Germany

More specifically, the emergency and disaster management system in Greece, is advanced and keeps track of the most recent developments, but lacks coordination among operational actors. Moreover, operational entities effectively follow and implement emergency civil protection plans, nonetheless, factors such as scale or timing of disaster, reduced human resources, frequent staff changes, improper definition of each actor's role and responsibilities, as well as bureaucracy, can diminish this effectiveness. Accordingly, providing relevant guidelines and clarification to each actor involved, on their roles and responsibilities in emergency management would support in effectively addressing such a challenge [4].

PathoCERT aims to strengthen the coordination capability of the FRs during such incidents, allowing the rapid and accurate detection of pathogens, improving their situational awareness and their ability to control and mitigate emergency situations involving waterborne pathogens. Overall, the new solutions will span from individual responders to the Command-and-Control (C2) Center in the field and the Coordination Center in Headquarters. PathoCERT incorporates the PathoIMS incident management system to monitor and coordinate FR activities.

PATHOCERT SOLUTION

To identify all the modules required in PathoCERT architecture, it was necessary to identify the actors involved (Fig. 9.1) in such incidents and identify their needs. These are the FR on the field, that can be from

Fig. 9.1 PathoCERT main actors

different organizations and have different ways of operating. They communicate with the FR form C2 Center which is on the field (usually a vehicle). The C2 Center, communicates with Crisis Management Headquarters (HQ) in which specialists are working collecting, exchanging, and processing information. They communicate with water quality specialists and other water authorities. Each one provides information to the HQ that gives information back to the C2 Center and FR on the fields.

After identifying the actors, it was important to understand how they manage water pathogen contamination events and understand their needs. During several meetings, user stories have been identified and shared to facilitate the collection of requirements. Technically, user stories consist of three parts, the person/role that the solution is developed for, what they are trying to achieve, and what is the overall benefit and problem that needs to be solved. Based on the User Stories, then, the software modules of PathoCERT architecture (Fig. 9.2) and the ways in which these modules interact with each other were identified.

The PathoCERT reference architecture [5] is composed of the following layers:

- Data Collection Layer: Responsible to collect information from heterogeneous sources such as the PathoSENSE IoT gateway, the PathoDRONE, the PathoSAT, the PathoTWEET, the PathoCAM, as well as the SCADA systems of Water Utilities and relevant open data sources. The layer consists of:
  - IoT agents that enable sensors and actuators to exchange data with a Context Data Broker using their own native protocols.
  - IoT Device Manager that allows the centralized management of IoT devices, providing the possibility for a decision-maker to increase situational awareness about the territory and the sensors available on the field.
  - Data Connectors used to get data from non-IoT device (i.e., PathoTWEET, OpenData, PathoSAT, PathoCAM, and PathoDRONE images).

- Interoperability and Data Harmonization Layer: Responsible for the harmonization of the data according to the adopted FiWARE smart data models and ontology used in PathoCERT.
- Data Management Layer: Responsible for the data storage and dispatch to the other components of PathoCERT. It consists of the:

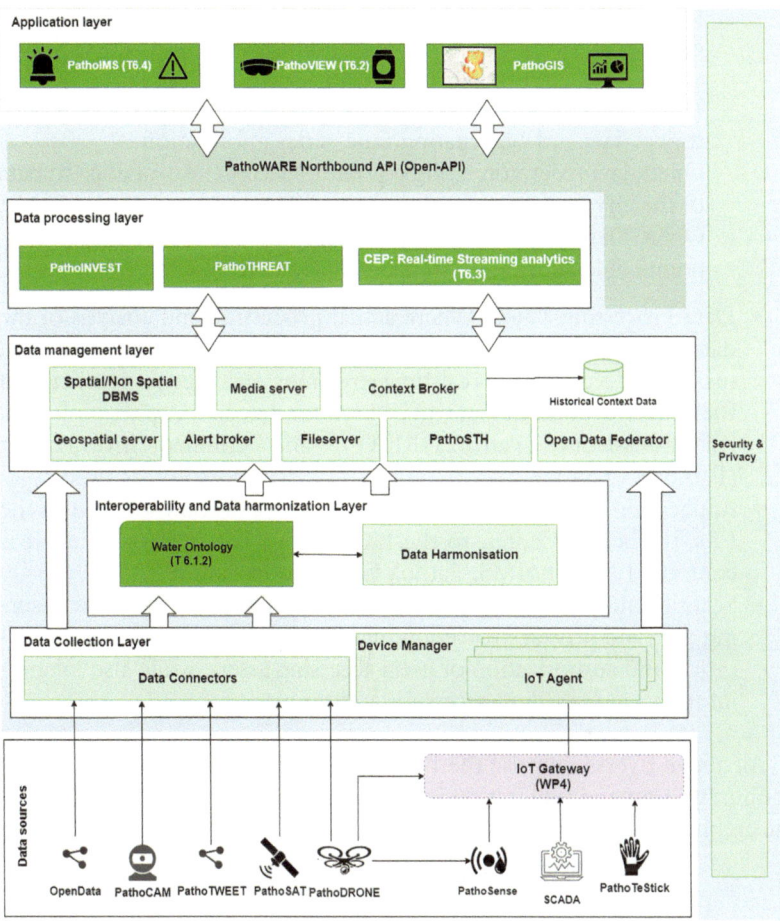

Fig. 9.2 PathoCERT Architecture Overview

- Context Broker (CB) which enables discovering gathering and publishing of context information through NGSI-based APIs.
- Alert Broker, dedicated to generating real-time alerts to PathoVIEW.
- Media Server, responsible for sharing and processing of media.
- Fileserver, which is responsible for storing large files (e.g., GeoTIFF provided by PathoSAT).

- Geospatial server, used to share, process, and edit geospatial data.
- Spatial and nonspatial DBMS, responsible for the storage of geospatial data and nongeospatial data.
- PathoSTH, in charge of managing (storing and retrieving) historical data and aggregating time-series information.
- Open data federator, a single point of access (i.e., catalog) to data of the city/territory, coming from different sources: for example, Open Data portals related to demographic information, endpoints of local legacy systems such as SCADA, etc.

- Data Processing Layer: Where actual processing and analysis of the data is performed in order to generate added value for the decision makers. It leverages the real-time streaming analytics component and the software modules that will be used for detection, epidemiological risk assessment (PathoTHREAT), and criminal investigation (PathoINVEST).
- Application Layer: Contains the services/applications offered to the FRs, to the C2 Center, to the HQ, and to Water Authorities. It is composed of PathoIMS, PathoVIEW, and PathoGIS.
- Security and Privacy Layer: Contains a set of modules aimed at ensuring data and privacy protection, management of identities, authentication and authorization of users accessing assets, while also logging information regarding access for audit purposes.

All above layers comprise the PathoWARE Platform, that provides the communication and integration infrastructure to all the other modules of the architecture.

THE PATHOCERT INCIDENT MANAGEMENT SYSTEM

According to the architecture (Fig. 9.2), the components of the PathoCERT framework are grouped into two main categories: the components that are related to data management and processing and the applications that support the actual interface of the framework with the end users.

PathoIMS is part of the application layer that receives data from the data layers, presenting a situation overview to the commanders and first responders and providing the required incident management capabilities to them. PathoIMS is planned to be deployed in the HQ (desktop version) and on mobile control centers (mobile/tablet version) (Fig. 9.3),

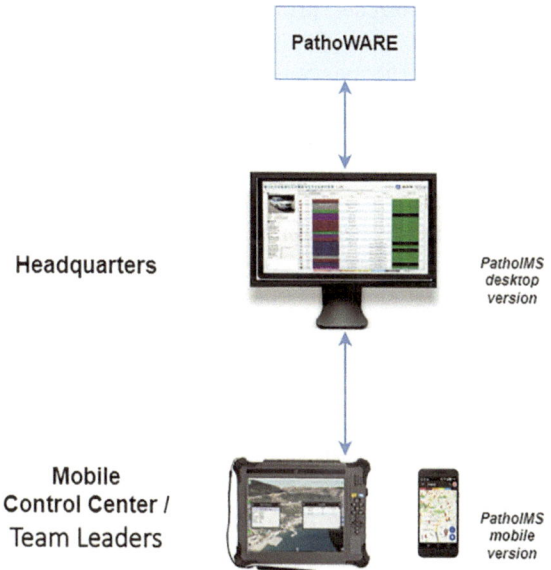

Fig. 9.3 PathoIMS deployment and communication with the framework

supporting the operations of commanders in the HQ and the team leaders on the field, and the sharing of operational information among them. Both types of PathoIMS applications will be synchronized, presenting a common picture of the situation to all users (considering their responsibilities and access rights). The sharing of data among the data and application layers is provided by the PathoWARE component.

In more details, PathoIMS is based on the ENGAGE IMS/CAD system,[1] enhanced in order to support the response activities of the waterborne pathogen contamination events. The design of the PathoIMS capabilities has been based on the initial project objectives, as well as the feedback provided by the stakeholders during the CoP events, considering the following:

- Flexible and intuitive Graphical User Interface (GUI), requiring the minimum intervention of commanders in order to use functionalities

[1] https://satways.net/products-sw/engage-ims-cad/

and take decisions, but also presenting information to users in a user-friendly way.

- Maximizing the performance of the solution to support operations without delay.
- The system is designed to be open and interoperable with third-party and agencies' legacy systems, through using interoperability standards.
- Multiuser collaborative capabilities, enabling situation monitoring and communication among FRs, to support multidisciplinary users that have to be simultaneously active and execute response missions [6].

PathoIMS N-tier architecture comprises a diverse set of Servers and Services suitably interconnected to scale and support clients' needs. More specifically the ecosystem consists of the following layers:

- *Database Layer*: Providing the geospatially enabled relational database for data persistence and supporting the storing of all data managed by PathoIMS.
- *Messaging Layer*: It enables event-driven communication with the internal and external systems (incidents, alerts, and warnings).
- *Communication Layer*: Consists of diverse services for bidirectional communication with the devices on the field, tracking devices, Messaging Layers, and other data providers, allowing open and flexible data sharing from/to third parties.
- *Application Layer*: The back-end service that implements the whole business logic that is required by the PathoIMS clients.
- *Clients Layer*: Contains two types of PathoIMS clients, the desktop and the mobile/tablet version.
- *OS/Virtualization Layer*: The system supports all editions of Windows Server and many editions of Linux Operating Systems. In addition, containerization via Docker technologies is also supported.
- *Load Balancing Layer*: Implemented either via software or via hardware enabling the scaling and support of thousands of users.
- *Security Layer*: Advanced security and authentication techniques are used. Additionally, the access to the information by the users is controlled by the system, considering their hierarchy, access rights, and responsibilities (Fig. 9.4).

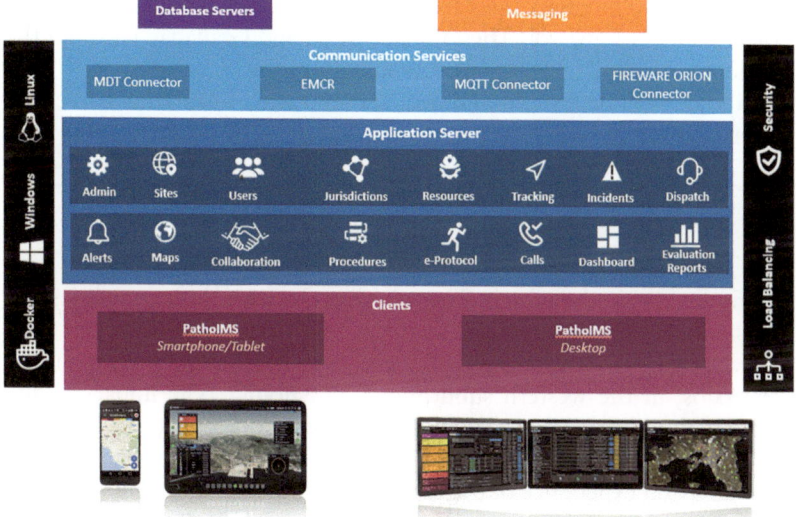

Fig. 9.4 PathoIMS internal design

All PathoIMS functionalities are provided to users through a rich and user-friendly GUI. The desktop version supports an advanced set of incident management functionalities. On the other side, the functionalities and the graphical user interface of the mobile/tablet are optimized to the small screen size of the devices and to the capabilities that are required by the commanders on the field/mobile centers. The main list of functionalities that are supported are:

- (i) Creation of the Common Operational Picture (COP);
- (ii) Incident Management;
- (iii) Alarm Management;
- (iv) Resource Tracking;
- (v) Interconnection with external resource tracking systems;
- (vi) Support of Standard Operational Procedures;
- (vii) Visual presentation of operational information (map, tabular views);
- (viii) Reporting;
- (ix) User management;
- (x) Logging & Audit; and
- (xi) System Administration.

All the above functionalities are demonstrated and validated during the project pilots in Limassol, Thessaloniki, Sofia, and Granada. Through the live demonstration and exercise-based pilots, the developed tools and technologies will be tested, validated, and evaluated by internal and external stakeholders.

PathoCERT Testing and Evaluation: The Greek Pilot

Specifically, within the pilot in Thessaloniki (Greece), IMS capabilities will be tested as part of an exercise scenario related to water pathogen contamination due to extreme rainfall phenomena that cause flooding on the river Axios, in the western suburbs of Thessaloniki. The impact of the flood involves both Water Supply & Sewage Company (EYATH) and Hellenic Rescue Team (HRT), the main actors of this exercise, who will get a hands-on experience of the capabilities offered and the impact of the tools with regard to the management of such an incident.

More specifically two scenarios will take place, one regarding the river overflow due to flooding, with possible pollution of the crossing open channel that transports water to the Thessaloniki Drinking Water Treatment Plant (area A in Fig. 9.5), while the second scenario (area B in Fig. 9.5) is linked with Search and Rescue (SAR) activities for people carried out by the increased flow of the river [7].

These scenarios will set the stage so that on the one hand the internal stakeholders involved in the project, (HRT and EYATH) to test those tools and on the other, the external stakeholders (General Secretary for Civil Protection, Fire Service, Directorate of Public Health and Social Welfare, etc.) to observe their demonstration that will:

- Increase the coordination among operational actors, by providing clarification to the actors based on their role/responsibilities, reducing response time, enhancing resource allocation, etc.
- Exploit the advantages toward involved Authorities, increasing acceptance toward new technologies, and expanding the financial resources allocated.
- Improve citizens' understanding and engagement in emergency events through capacity-building activities, educational and informative campaigns.

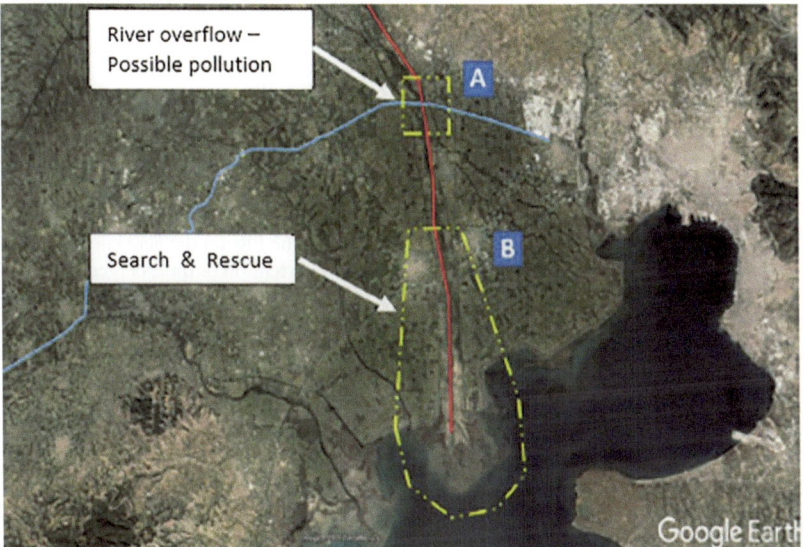

Fig. 9.5 Map of the wider Greek pilot area with the two points of interest

Specifically, the PathoIMS will be validated and evaluated according to the following needs and criteria presented in Table 9.1.

CONCLUSIONS AND NEXT STEPS

In this paper, the authors present the concept and added value of an integrated solution, PathoCERT, which captures the complete spectrum of waterborne pathogen contamination management from detection and situation awareness, to epidemiological, threat risk assessment, and criminal investigation. Overall, the new solutions will address a different part of the FR organization chain, starting from the individual responder who is out in the field, to the mobile Command and Control which coordinates the operations in the field, to the Headquarters that monitor and provide information to the Incident Management System (PathoIMS).

PathoIMS, which will be used by the Incident Commander who is responsible for coordinating the event, is developed to support information exchange and effective coordination on the contamination event

Table 9.1   User needs related to PathoIMS

Type of user	Need	Aim
HQ Commander	See the geo-localized report related to the emergency phenomenon	Organize resources
HQ Commander	Locate exactly the whole area of the emergency	Plan the allocation of resources on the affected region to respond in a timely manner
HQ Commander	Communicate all data collected to and from the FR on the field	The FRs on the field know the extension of the emergency and which precautions to use before approaching the emergency itself
FR on the field	Communicate all data collected from the field to the HQ (e.g., reports and pictures)	The HQ has a wider and updated picture of the emergency and can take the right decisions
FR Commander	Get quick information about the quality of the water a team needs to operate in	Avoiding putting the staff into danger
FR on the field (SAR team)	See on the map the last known position of a missing person, the search areas that need to be or have been covered and important locations	SaR activities can be coordinated efficiently, and missing persons can be found quickly
FR Commander	Be aware of the current status of the emergency situation	Efficiently coordinate the response activities

management between the Command-and-Control Center and the FR Headquarters.

PathoIMS capabilities offered, include but are not limited to, Incident Man-agement, Alarm Management, Resource Tracking, Support of Standard Opera-tional Procedures, Visual presentation of operational information (map, tabular views), Reporting, etc., all of which will be demonstrated and validated during the pilots of the project and specifically in Cyprus, Greece, Bulgaria, and Spain.

Specifically, PathoIMS will be evaluated through the following Key Performance Indicators: (1) Optimized tactical response: 40% reduction in manpower, 40% reduction in response times; (2) Reduced time between victim extraction from hazardous area and proper medical treatment: 50% reduction; (3) Optimized victims/citizens handling, during a hazardous event: >50% increase in worksite declassification time.

Acknowledgments This project has received funding from the European Union's Horizon 2020 research and innovation program under grant agreement No. 883484. This article reflects only the authors' views, and the Research Executive Agency and the European Commission are not responsible for any use that may be made of the information it contains.

REFERENCES

1. AccountAbility. (2015). *AA1000 stakeholder engagement standard.* AccountAbility.
2. PathoCERT Project. (2021). *D3.1 – Stakeholder mapping and requirements analysis.* https://pathocert.eu/wp-content/uploads/2021/12/D3.2.pdf
3. Reed, M. S., Graves, A., Dandy, N., Posthumus, H., Hubacek, K., Morris, J., Prell, C., Quinn, C. H., & Stringer, L. C. (2009). Who's in and why? A typology of stakeholder analysis methods for natural resource management. *Journal of Environmental Management, 90*(2009), 1933–1949.
4. PathoCERT Project. (2021). *D3.2 – Stakeholder engagement plan.* https://pathocert.eu/wp-content/uploads/2021/12/D3.2.pdf
5. PathoCERT Project. (2021). *D6.1 – PathoCERT platform reference architecture.*
6. PathoCERT Project. (2023). *D6.6 – PathoIMS prototype.*
7. PathoCERT Project. (2022). *D8.1 – Horizontal pilot preparation.*

Open Access This chapter is licensed under the terms of the Creative Commons Attribution 4.0 International License (http://creativecommons.org/licenses/by/4.0/), which permits use, sharing, adaptation, distribution and reproduction in any medium or format, as long as you give appropriate credit to the original author(s) and the source, provide a link to the Creative Commons license and indicate if changes were made.

The images or other third party material in this chapter are included in the chapter's Creative Commons license, unless indicated otherwise in a credit line to the material. If material is not included in the chapter's Creative Commons license and your intended use is not permitted by statutory regulation or exceeds the permitted use, you will need to obtain permission directly from the copyright holder.

CHAPTER 10

AI-Powered Microscopy Platform for Airborne Biothreat Detection

János Pálhalmi and Anna Mező

INTRODUCTION

An extensive amount of effort has been put into the development of different sensor solutions to detect, monitor, and identify airborne biological agents. The variety of methods behind the several sensor solutions cannot go unnoticed, but no standard and interoperable EU-wide approach is available to set the threshold for monitoring biothreat either outdoors or within critical infrastructure. Several research and development studies suggest bacteria are capable of surviving the aerosol transport and they can travel much longer distances between the hosts than we previously thought [2, 5, 8–10]. The sporulated form is not the only option for bacteria for airborne dissemination [6], which highlights the importance of using classical optical methods powered by AI-supported solutions for rapid monitoring and detection.

A recent study carried out by the Pentagon revealed several economic barriers of genomic analysis methods to be adopted in daily routine in terms of bioaerosol monitoring and biothreat detection [7] which shows

J. Pálhalmi (✉) • A. Mező
DataSenseLabs Ltd., Budapest, Hungary
e-mail: janos.palhalmi@datasenselabs.net

© The Author(s) 2025

117

I. Gkotsis et al. (eds.), *Paradigms on Technology Development for Security Practitioners*, Security Informatics and Law Enforcement, https://doi.org/10.1007/978-3-031-62083-6_10

that we are unprepared for real time, or at least frequent monitoring of airborne biocontent.

Regarding the currently available solutions for pathogen detection, there is a trade-off between time and accuracy. While the gold standards for genus and strain level identification of bacteria are still the different genomic methods, the classical optical methods like different forms of quantitative phase imaging microscopy powered by AI-supported solutions offer the possibility of rapid and automated detection of suspicious pathogens either in water or air-based samples.

One of the reasons why there is no existing standard, interoperable, and real-time or quasi-real-time, optical sensor-based biothreat monitoring solution is the lack of platforms capable of comparative verification, monitoring, and data archiving for traceable intermethod comparison and cross-validation.

There are several other reasons behind the difficulties of standardization and interoperability of optical sensor-based rapid biothreat monitoring and detection. We do not intend to cover the below-mentioned list in this short publication, but it is important to mention some of these factors to highlight the complexity of the problem:

- The variety of air sampling methods and devices and the lack of knowledge regarding the limit of detection between the several air sampling solutions make it even more difficult to establish protocols for bioaerosol monitoring [1, 3]. The current solution is to focus on use case-specific applications and to comply with the generally well-established, referring NATO (STANAG) and ISO standards regarding the overall sampled volume of the air to avoid statistical down-sampling.
- The lack of widely accepted air quality standards [3] makes it very difficult to determine the exact baseline of the "background noise" within the air [4, 8, 11]—referring to the nonbiological particle components—in order to optimize and finetune the AI-supported solutions for pathogen detection.
- The currently accepted standard pathogen detection and identification protocols are based on culturing following the sample collection which significantly elongates the lengths of the process [3] and also increases the methodological diversity between the existing protocols.

Even if some AI-supported optical sensor-based platform would exist, there are no standard testing and validation protocols to evaluate the application-specific performance metrics. This is the topic we are focusing on in the framework of this publication. We introduce the recent results regarding the performance of the "DataSenseLabs AI-supported biothreat detection platform" using two different approaches: field data (air sample) collection and computer simulation-based testing related to our cross-validation-based development strategy.

Since the disease control authorities are highly vigilant regarding the environmental presence of the most dangerous and unfortunately well-known member of the Bacillus cereus group [4, 12, 13], and it is very easy to access all the necessary components to create a virulent Bacillus anthracis strain, our AI-supported biothreat detection platform is currently being finetuned for the detection of bacillus form objects sampled from the air.

METHODS AND RESULTS

Regarding the process flow of biothreat detection, "DataSenseLabs AI-supported biothreat detection platform" is currently proceeded by three separate components of processes sequentially following each other: (1) air sample collection; (2) sample preparation; (3) light microscopic measurement; and the final step is the AI-supported pathogen detection. In the current phase of the R&D, the optical microscopic measurements were carried out by a digital holographic microscope (DHM-HoloZcan-EPro2) provided by the EU Horizon-supported HoloZcan project, and by a reference differential interference contrast (DIC) light microscope (NIKON Eclipse Ti2).

The performance of the DataSenseLabs AI-platform solution was fine-tuned and tested on three different sample types: laboratory-made mixture of different bacterial strains including Bacillus subtilis (ATCC 6533); environmental field samples collected by the Coriolis Compact air sampling device manufactured by Bertin Technologies; and computer-simulated holographic particle populations in the diameter $(0.5 < d < 2\ \mu m)$ and refractive index $(R = 1.4)$ range compatible with the optical properties of bacteria. In the case of field samples, the air samples were dissolved in 1 ml physiological saline on site and were placed on standard microscopic slides for the DHM and DIC measurements.

Data analysis pipeline developed in MATLAB 2022b and Python consists of the following steps:

1. Use case-specific database building from the captured microscopic image datasets. Currently, the DHM and DIC image data input is supported by the "DataSenseLabs AI-supported biothreat detection platform."

2. Extension of the databases with the metadata of annotated region of interests (ROIs). For supervised deep learning algorithm development, the ROIs of bacterial cells—as the reference or "ground Truth"—were labeled by rectangular bounding boxes in the MATLAB Image Labeler app.

3. Asking the right questions: dividing the database into context-specific and use case-specific training and test sets.

4. Optimization of the convolutional neural network (CNN) model parameters by training and data augmentation.

   (a) Optimizing the training process by finding the optimal number of training iterations based on the mini-batch accuracy and loss function values.

   (b) Image minibatches were created by dividing the original raw images into several smaller images with the pixel dimensions of 300×340. The training process was conducted by applying overlapping and nonoverlapping minibatches as well. During the training process, the saturation and exposure of the input images were modified within a 10% range, to diversify the training set.

5. Testing the trained model, comparing the reference (ground Truth) and predicted values and metrics.

6. Concluding the results and restart the process either from point 3 or 2 or even from point 1.

To support the original interoperability and cross-validation concept of the "DataSenseLabs AI-supported biothreat detection platform," the possibility of multidata input has been developed to compare the results of the AI-supported detection in the case of several light microscopic image databases. In this publication, we show the summary statistics only of the DHM and DIC image datasets.

The two figures below (Fig. 10.1a and b) explain the concept of the cross-reference database-building strategy for the AI-supported platform development:

As referred to in Fig. 10.1a, we followed a custom-designed database-building strategy during the research and development process of the "DataSenseLabs AI-supported biothreat detection platform." We have built seasonal, and geolocation-specific multiimage databases captured from the air samples to be able to estimate the overall diameter distribution of the naturally occurring biological and nonbiological particle components ("background noise"), since these can be crucial carriers of long-distance bacterial dissemination [4, 6, 8, 10, 11]. Three different approaches have been worked out to develop the databases for AI algorithm development and testing:

Cross-validation database building.

Non-stained Bacillus simulant sampled from the air. LM DIC

Gram-stained Bacillus simulant sampled from the air. DHM

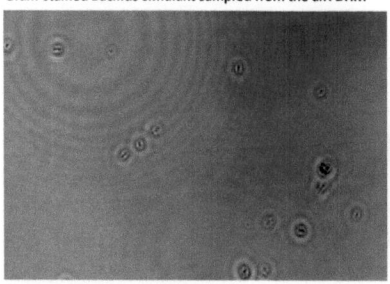

1013 x 682 pixels; 202.6 μm x 136.4 μm (X, Y).

1024 x 1024 pixels; 163.84 x 163.84 μm.

Fig. 10.1 (a) The purpose of the cross-reference database-building approach combines the application of image databases captured from laboratory-made (InLab: on the left) and from field-collected (OnField: on the right) samples. The optimal structure of workflow iteration during the R&D process that is either within the AI development or within the calibration-testing-validation process and between the two different, but interconnected processes is indicated by blue lines. (b) The figure represents a gram-stained DHM image (on the left) and a non-stained DIC image (on the right) of the same "Bacillus simulant" strain that was used during the "field data collection, testing, training and demonstration" events. "DataSenseLabs AI-supported biothreat detection platform" can receive both types of light microscopic images to estimate the number of suspicious bacillus form objects within the air sample

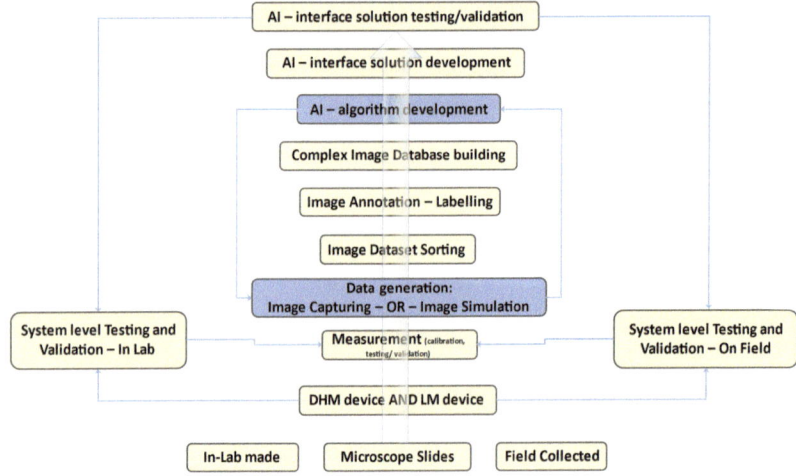

Fig. 10.1 (Continued)

- Calibration-oriented databases including the diameter values (in μm) of the seasonally occurring particles in the air are sampled by a reference air sampling device, the Coriolis Compact. The samples were taken at the same geolocations (forest region, busy road, indoors) in three different seasons (Summer, Autumn, and Spring).
- Digitally mixed databases containing the microscopic images of the environmental (background noise) samples and the samples created under laboratory conditions including specific bacterial strains.
- Directly mixed databases where the environmental fluid samples were directly mixed with the suspensions of specific, known bacterial strains.

The air samples were analyzed by DIC light microscopy to create the reference diameter value databases to verify the presence of the Bacillus simulant applied during the field data collection (air sample collection) and testing events as referenced by the figure below (Fig. 10.2).

Following the calibration-level verification of the presence of the artificially administered Bacillus simulants, the "DataSenseLabs AI-supported biothreat detection platform" was used to detect the presence of suspicious bacillus form objects within the air sample. The performance metrics are summarized below in Table 10.1a and Table 10.1b:

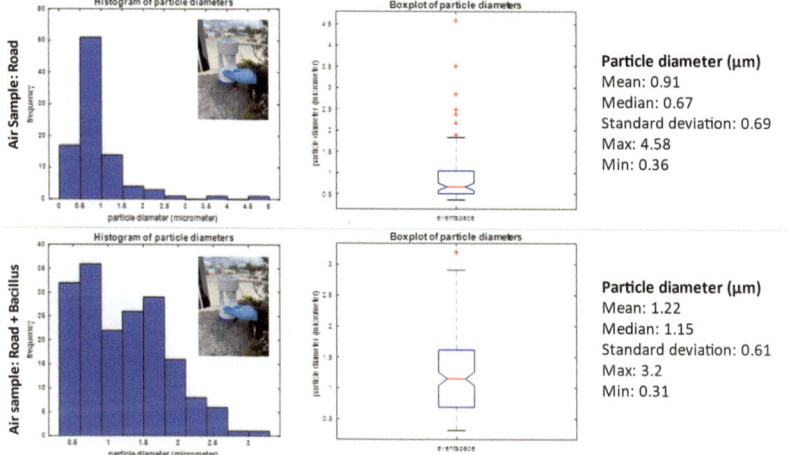

Fig. 10.2   The histogram (on the left), the boxplot (in the middle) of the measured particle diameters, and the basic statistical features of the particle diameters (on the right) are shown. The results on the top demonstrate the diameter distributions of the air samples containing the spontaneously occurring particles at the geolocation "busy road." The results in the bottom demonstrate the diameter distributions of the air samples containing the spontaneously occurring particles + the artificially administered bacillus simulant particles. The distributions are significantly different from each other (non-parametric statistical tests: $p < 0.01$)

Table 10.1 The tables show the results of the Deep-Learning network model-based predictions regarding the presence of suspicious bacillus form objects in the field collected air sample based on the DIC light microscopic measurement (on the left Table 10.1a) and based on the DHM light microscopic measurement (on the right Table 10.1b)

Precision	Recall	F score		Precision	Recall	F score
88.89%	94.12%	0.914		67.71%	98.48%	0.802

Besides the field data collection and testing-based AI algorithm development approach, it is extremely important to test and validate the theoretical limits of performance of an AI-supported algorithm. Simulated holographic particle databases have been created by the Mie method applying the Python-based HoloPy tool [14] to evaluate the performance of the "DataSenseLabs AI-supported biodetection platform" under use case-specific conditions as summarized in Fig. 10.3 and Table 10.2 below.

 **AI algorithm performance testing by simulated databases**

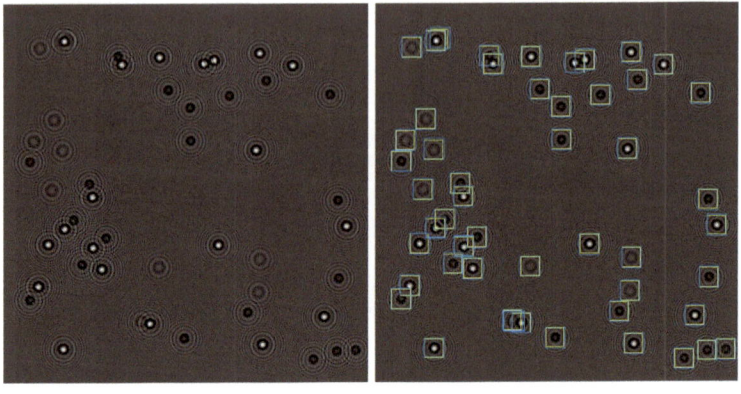

Lens-free Simulation, HoloPy – Mie. d: 0.6; 1.3; 2.1 μm. Lightsource: 520 nm.
Air background. Number of particles: 50. pixel dimension: 512 x 512

Fig. 10.3 The figure shows that the "DataSenseLabs AI-supported biodetection platform" is capable of distinguishing even between spatially extremely close particles in the diameter range of bacterial objects and their spores (0.5–2 μm). On the left: the original simulated image. On the right: the simulated image indicating the reference annotation by yellow, and the predicted AI-based detection by blue bounding boxes

CONCLUSION AND FURTHER STEPS

The "DataSenseLabs AI-supported biodetection platform" supports the quantitative phase imaging light microscopic sensor-based data input (DHM, DIC) and also two different computer-simulated image data inputs for algorithm development, training, testing, and evaluation. The platform's algorithm system can detect and monitor the anomalies in the concentration of bacillus form objects sampled from the air with higher than 80–95% accuracy depending on the study design, sample type, and the light microscopic measurement method.

Evaluation of an AI-supported computer vision platform needs standardized, classic image databases as it is also suggested in the ITU-T F.748.12 (Table 7–3, page 4–5) standard [15], but we also suggest the implementation of a custom-designed, cross-validation-based three step (laboratory-made, field-collected, and simulated data-based) evaluation

Table 10.2 The table shows the influence of the particle density within the sample on the performance metrics of the "DataSenseLabs AI-supported biodetection platform"

AI algorithm performance testing by simulated databases.
Testing the influence of the sample density on the performance metrics.

Comparison of the lens-free and lens-based DHM methods.

Particle diameter: 0.6; 1.3; 2.1 μm

Lens-based simulation Number of Particles	TP	FP	FN	Precision	Recall
10	20	0	0	1.00	1.00
50	100	1	0	0.9901	1.00
100	199	9	1	0.9567	0.9950
Lens-free simulation Number of Particles	TP	FP	FN	Precision	Recall
10	20	0	0	1.00	1.00
50	100	7	0	0.9346	1.00
100	196	7	4	0.97	0.98

Both in the case of simulated lens-based and lens-free DHM configurations, the accuracy of detection and classification of particles in the dimensional range of bacterial objects are not lower than 93.46%, in many cases it is close to 100%. TP: true positive, FP: false positive, FN: false negative detection events.

strategy as we introduced in the case of our AI platform development and evaluation. The simulated databases will be shared in a separate publication according to the "FAIR" data policy directives [16].

Based on the current results, the platform is capable of supporting CBRN and biothreat surveillance-related decision-making and prestandardization processes to establish solid foundations for interoperability in the field of optical sensor-based and light microscopic measurement-based instant biothreat detection. Following the analysis of the final results in the first part of 2024, we will be able to choose to most appropriate and use case-specific sensor type for miniaturization and industrial-level production.

Acknowledgments The project was supported by Horizon 2020 program: HoloZcan (GA: 101021723). (*https://datasenselabs.net/*; *https://datasenselabs. net/horizon2020/*).
Special thanks to the microbiology and CBRN experts of the University of Lodz, Poland for their contribution in the air sample collection and sample preparation.The results summarized in this publication are based on the state of the

R&D process presented at the Research & Innovation Conference for European Security and Defense (https://rise-sd2023.eu/) 29–31 May 2023, Greece.

REFERENCES

1. Zhao, Y., Aarnink, A. J. A., Groot Koerkamp, P. W. G., et al. (2011). Detection of airborne Campylobacter with three bioaerosol samplers for alarming bacteria transmission in broilers. *Biological Engineering, 3*(4), 177–186.
2. Fernandez, M. O., Thomas, R. J., Garton, N. J., et al. (2019). Assessing the airborne survival of bacteria in populations of aerosol droplets with a novel technology. *Journal of the Royal Society Interface, 16*(150), 20180779.
3. Centers for Disease Control and Prevention. (2015). *Background F. Environmental sampling.* https://www.cdc.gov/infectioncontrol/guidelines/environmental/background/sampling.html. Accessed 30 July 2023.
4. Hagiwara, K., Matsumoto, T., Tsedendamba, P., et al. (2020). Distribution of viable bacteria in the dust-generating natural source area of the Gobi region, Mongolia. *Atmosphere, 11*(9), 893.
5. Clark, R. D., Underwood, G., McGenity, T. J., & Dumbrell, A. J. (2021). What drives study-dependent differences in distance–decay relationships of microbial communities? *Global Ecology and Biogeography, 30*(4), 811–825.
6. Hara, K., Zhang, D., Yamada, M., Matsusaki, H., & Arizono, K. (2011). A detection of airborne particles carrying viable bacteria in an urban atmosphere of Japan. *Asian Journal of Atmospheric Environment, 5*(3), 152–156.
7. Srikrishna, D. (2023). The world is completely unprepared for detecting emerging viruses/bacteria in the air. *medRxiv.* https://www.medrxiv.org/content/medrxiv/early/2023/03/29/2022.08.09.22278555.full.pdf. Accessed 30 July 2023.
8. Sanz, S., Olarte, C., Hidalgo-Sanz, R., Ruiz-Ripa, L., Fernández-Fernández, R., García-Vela, S., Martínez-Álvarez, S., & Torres, C. (2021). Airborne dissemination of bacteria (enterococci, staphylococci and Enterobacteriaceae) in a modern broiler farm and its environment. *Animals, 11*(6), 1783.
9. Zhao, J., Jin, L., Wu, D., et al. (2022). Global airborne bacterial community-interactions with Earth's microbiomes and anthropogenic activities. *Proceedings of the National Academy of Sciences of the United States of America, 119*(42), e2204465119.
10. Woo, C., & Yamamoto, N. (2020). Falling bacterial communities from the atmosphere. *Environmental Microbiomes, 15*(1), 22.
11. Hu, W., Murata, K., Fan, C., Huang, S., Matsusaki, H., Fu, P., & Zhang, D. (2020). Abundance and viability of particle-attached and free-floating bacteria in dusty and nondusty air. *Biogeosciences, 17*, 4477–4487.
12. Centers for Disease Control and Prevention. (2020). *Anthrax.* https://www.cdc.gov/anthrax/bioterrorism/threat.html. Accessed 30 July 2023.

13. Johns Hopkins Center for Health Security. (2023). *Factsheet – Bacillus anthracis.* https://centerforhealthsecurity.org/sites/default/files/2023-02/anthrax.pdf. Accessed 30 July 2023.
14. Github – Manoharam Lab Harvard University. HoloPy. (2021). *Hologram processing and light scattering in Python.* https://manoharan.seas.harvard.edu/holographic-microscopy, https://github.com/manoharan-lab/holopy. Accessed 30 July 2023.
15. International Telecommunication Union. (2021). *Deep learning software framework evaluation methodology standard* (ITU-T F.748.12). https://www.itu.int/rec/T-REC-F.748.12/en. Accessed 30 July 2023.
16. Go FAIR. (2020). *FAIR principles.* https://www.go-fair.org/fair-principles/. Accessed 30 July 2023.

Open Access   This chapter is licensed under the terms of the Creative Commons Attribution 4.0 International License (http://creativecommons.org/licenses/by/4.0/), which permits use, sharing, adaptation, distribution and reproduction in any medium or format, as long as you give appropriate credit to the original author(s) and the source, provide a link to the Creative Commons license and indicate if changes were made.

The images or other third party material in this chapter are included in the chapter's Creative Commons license, unless indicated otherwise in a credit line to the material. If material is not included in the chapter's Creative Commons license and your intended use is not permitted by statutory regulation or exceeds the permitted use, you will need to obtain permission directly from the copyright holder.

PART II

Protecting the Infrastructure of Europe and the People Within the Smart City Concept

PLOTO: Improved IWW Resilience Using Predictive Modelling, Environmentally Sustainable and Emerging Digital Technologies and Tools

Dimitris Liparas, Dimitrios Vamvatsikos, Ioannis Drivas,
Anna Zanetti, Alexios Pagkozidis, Didier Bousmar,
Natalia Budescu, Erzsébet Szabó-Aranyi, Alexis Melitsiotis,
Themis Vokali, Vasileios Melissianos, Fotios Barmpas,
Csaba Csiszár, Tomislav Letnik, Gergely Mezo, Joris Hardy,
Romeo Soare, Alina Beatrice Raileanu, Judit Kerényi,
and Antti Hellsten

D. Liparas (✉)
Research and Innovation Development, Netcompany-Intrasoft S.A.,
Brussels, Belgium
e-mail: dimitrios.liparas@netcompany.com

D. Vamvatsikos
School of Civil Engineering, National Technical University of Athens,
Athens, Greece

© The Author(s) 2025
I. Gkotsis et al. (eds.), *Paradigms on Technology Development
for Security Practitioners*, Security Informatics and Law
Enforcement, https://doi.org/10.1007/978-3-031-62083-6_11

131

INTRODUCTION

On 11 December 2019, the European Commission presented the Green Deal [1], intending to provide an initial roadmap of the necessary key policies and measures. The Communication highlights tackling climate and environment-related challenges as 'this generation's challenge'. Achieving zero net greenhouse gas (GHG) emissions by 2050 is arguably

I. Drivas
Diadikasia Business Consulting S.A., Athens, Greece

A. Zanetti
ERTICO – ITS Europe, Brussels, Belgium

A. Pagkozidis
Satways Ltd., Athens, Greece

D. Bousmar
Direction des Recherches Hydrauliques, Service Public de Wallonie, Namur, Belgium

N. Budescu
Asociatia Romanian River Transport Cluster, Galati, Romania

E. Szabó-Aranyi
Freeport of Budapest Logistics Ltd., Budapest, Hungary

A. Melitsiotis
EXUS AI Labs, Athens, Greece

T. Vokali
RISA Sicherheitsanalysen GmbH, Berlin, Germany

V. Melissianos
Societal Resilience and Climate Change Center of Excellence, Brussels, Belgium

F. Barmpas
School of Mechanical Engineering, Aristotle University of Thessaloniki, Thessaloniki, Greece

C. Csiszár
Department of Transport Technology and Economics, Faculty of Transportation Engineering and Vehicle Engineering, Budapest University of Technology and Economics, Budapest, Hungary

T. Letnik
Faculty of Civil Engineering, Transportation Engineering and Architecture, University of Maribor, Maribor, Slovenia

the most central, ambitious and challenging goal set out by the Green Deal in Europe. The latest intermediary target for 2030 is set to at least 55% GHG emission reduction [2].

In accordance, the development of European green ports & multi-modal hubs of the future to 2050 is not only linked to infrastructure but also to smarter approaches, more efficient, effective, innovative technical solutions, sustainable management of goods and freight flows and seamless integration of the port community and inland multimodal terminals & hubs, balancing environmental effects and economic requirements. Every year, 108 trillion tonne-kilometres are transported worldwide, of which the majority is borne by sea (70%), 18% is handled by rail, 9% is carried by road, 2% is shipped on inland waterways and about 0.25% is transported by air [3]. In 2020, road transport accounted for 75.3% of the total inland freight transport and its share has been consistently rising over the last decade. In 2020, more than 10.3 million workers were employed in the freight industry (including warehousing and supporting transport activities), which accounted for 5.3% of the overall EU-27 labour force [4]. Freight transport and logistics are vital to the EU's Single Market and for Europe's prosperity. This industry is an important facilitator in the so-called 'four freedoms' of the EU: it allows the mobility of commodities, resources and people because it links economic players to each other. Additionally, well-performing and dynamic freight and logistics sectors

G. Mezo
National Association of Radio Distress-Signalling and Infocommunications, Budapest, Hungary

J. Hardy
Urban and Environmental Engineering Research Unit, School of Engineering, University of Liege, Liege, Belgium

R. Soare
Lower Danube River Administration, Galati, Romania

A. B. Raileanu
Danubius University of Galati, Galati, Romania

J. Kerényi
Hungarian State Railway Company, Budapest, Hungary

A. Hellsten
Finnish Meteorological Institute, Helsinki, Finland

improve overall productivity and competitiveness through job growth, economies of scale and by generating remarkable economic and social added value [5]. The latest available figures reveal that about EUR 599 billion, or around 5% of the EU's Gross Domestic Product, has been created by the transport and logistics industry, with over EUR 931 billion spent by private household on transport-related goods (13% of total consumption), and more than 1.15 million companies active in transportation and logistics [6]. Even more so, the high quality of Europe's freight and logistics sector is internationally recognised. Indeed, the annual World Bank International Logistics Performance Index ranks no less than 13 European economies in the top 20 global leaders in logistics [7]. Global freight traffic is anticipated to triple for inland modes in the next 30 years [8]. In addition, in the EU, surface freight traffic is expected to rise by 53% by 2050 [9]. In spite of these projections, the growth of the sector is not without complications. The biggest concern is how economic gains from increased demand can be sustained when considering adverse externalities and possible rebound effects.

Transport modes still adhere to silo mentality, despite ports setting moderate goals on transport modal split. Other major challenges include saturated infrastructure, carbon emission goals and energy constraints [10]. Laudable initiatives on alternative energy in ports, remain far too often in a demonstration phase, and do not succeed in spreading the innovation all over the sector. Despite the fact that the ports of Rotterdam, Antwerp and Hamburg claim to be frontrunners in this field, their interaction with 'smaller' ports not belonging to the absolute top, is still below expectations. Smaller ports are in urgent need of a concrete Roadmap to sustainability that has ready-to-use cases, in order not to miss that boat. However, this firm step towards sustainability is not possible without a concrete transition to increased automation, digitalisation, standardisation, interoperability of processes, technologies and equipment. By fundamentally altering the use of alternative energy, the organisation and management of ports, inland multimodal hubs and terminals and the related freight transport, digitalisation will propel ports & inland multimodal hubs towards a more sustainable, efficient and performant sector. Digitalisation & increased automation have the potential to reduce administrative burdens, lower operational barriers and so improve efficiency, productivity, interoperability of processes and competitiveness [11] in the multimodal freight transport nodes.

The PLOTO project follows the vision of Horizon Europe framework programme from 2021 to 2027, which aims at 'a sustainable, fair and prosperous future for people and planet' and reserves 35% of the budgetary target to tackle climate change in order to contribute towards Europe becoming a climate-resilient society by 2050 [12]. It targets Inland WaterWays (IWWs), focusing on their resilience and the development of tools to model, assess, forecast and mitigate the impact of natural hazards on port operability. Along the way, it aims to address multi-hazard risk understanding, smart prevention and preparedness, as well as faster, adapted and efficient response proposing a new integrated system to support operational and strategic adaptation and mitigation measures. It achieves its stated goals by better absorbing and efficiently recovering from Climate Change impacts, including extreme events, thus increasing the resilience of IWWs.

PLOTO Development Methodology

PLOTO is a pure technological project, but it is driven by the actual needs of the end users, mainly IWW operators, including inland ports, authorities and shipping companies. The pilot activities scheduled within the project lifetime have as the ultimate goal the achievement of a minimum technology readiness level of TRL7 concerning the technology components and the overall system developed in its context. This is even more strengthened through the adoption of an agile development that aims to provide the first prototypes early in the project. PLOTO will pursue to adopt a start-up mentality, targeting to get right the following tasks:

1. Identify the needed competences to succeed in its objectives.
2. Form a team that has the ability to perform efficiently and deliver high-quality results.

PLOTO will follow a process by iterating a series of activities, performing preliminary module and system assessments and validation campaigns well before the pilot demonstrations:

- The proposed approach will be realised in two cycles, allowing new data from the evaluations to be incorporated into the output development process, revising it whenever necessary. The first cycle will close with the successful delivery of the initial integrated PLOTO system. The second cycle will conclude through the final integrated system that will be based on the technical and operational assessments to be carried out in the first round of demonstration activities.

- PLOTO will adopt a fast-failure approach in terms of getting through the steps in the concept maturing and system evolution process. This will provide a means for rigorous assessment of options and the selection of the most suitable one based on balanced trade-offs that will not hamper the overall project progress. The project team's peripheral vision will be used in order to keep a live roster of opportunities, threats and challenges in the area of interest, allowing an effective alignment with the current conditions throughout PLOTO's lifecycle.
- The approach will allow the PLOTO consortium to respond to external or internal opportunities during the project's lifetime and will add to the project's agility to accommodate innovative solutions that will match emerging trends and needs at the actual time of implementation. Thus, the real potential of the final outputs ensures the uptake in the mid-term horizon after the end of the project.

In support of its agile/iterative approach, PLOTO is organised to perform smaller-scale intermediate validations, in which end users will be introduced to the developed and evolved solutions, so as to incorporate their feedback in a timely and resource-efficient manner, aligning their needs with project outcomes to the highest possible extent. This will also be a training opportunity for the end users, so that they can understand PLOTO's value proposition in improving IWWs' resilience through the integration of the proposed hazard-awareness solutions. Taking into consideration the high innovation and business potential of PLOTO, the iterative agile development will be executed with discipline, having as a clear aim the constant improvement of the components' and system's TRL. TRL7 is considered a critical threshold as a transition from experimental to real-life conditions is performed. From the demonstration, a feasibility study, business planning and economic viability will take place, which in combination with the ongoing developments and the pilot-scale study will produce a near-to-market product. This process will generate a gradually increasing TRL for the whole system and for each component individually.

PLOTO Technological and Scientific Basis

The technological backbone of PLOTO includes climate, atmospheric forcing and multi-hazard modelling, multi-hazard vulnerability modules and assessment toolkit for IWW assets, improved computer vision techniques and machine-learning techniques, remote sensing, including quick assessment damage maps, fore-now/casting weather predictions methods & tools, PLOTO middleware and data fusion, IWW assessment tool and IWW digital twin, enhanced visualisation common operational picture, incident management system and decision support system.

At the core of the PLOTO platform lies the digital twin (Fig. 11.1). A modular design is adopted to connect hazards, exposed assets and interconnected infrastructure networks to form a digital twin of the IWW that interacts with all PLOTO modules to efficiently transfer and process sensor data (Fig. 11.1). It is built upon pre-compiled data on-site hazard and asset exposure. Hazard information covers weather, climate, hydrological and seismic perils, while the exposed assets cover the entire IWW and hinterland infrastructure of each port, comprising both individual assets (piers, cranes, warehouses, etc.) as well as connected networks (power/transportation/etc.), and interconnectivity among said networks (i.e. power network influencing rail transportation).

Figure 11.2 presents the flow of information within the PLOTO platform modules. It comprises (1) the cloud-based digital twin incorporating all hazard and asset data, fed into (2) the multi-hazard vulnerability assessment tools projecting hazard scenarios to deliver (3) asset response characterisation. The results from individual assets and scenarios are combined

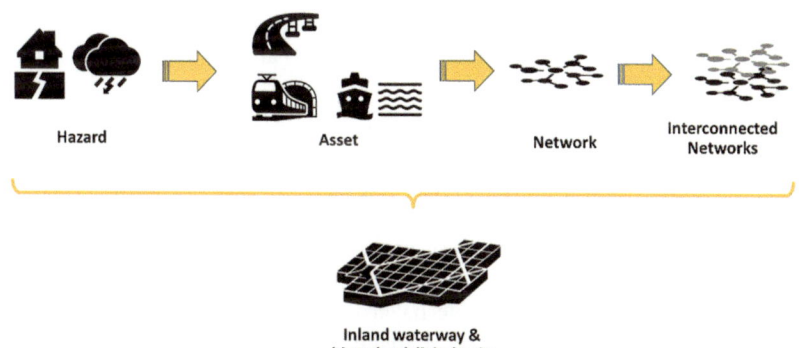

Hazard

Asset

Network

Interconnected Networks

Inland waterway & hinterland digital twin

Fig. 11.1   The proposed digital twin formulation that forms the core of PLOTO

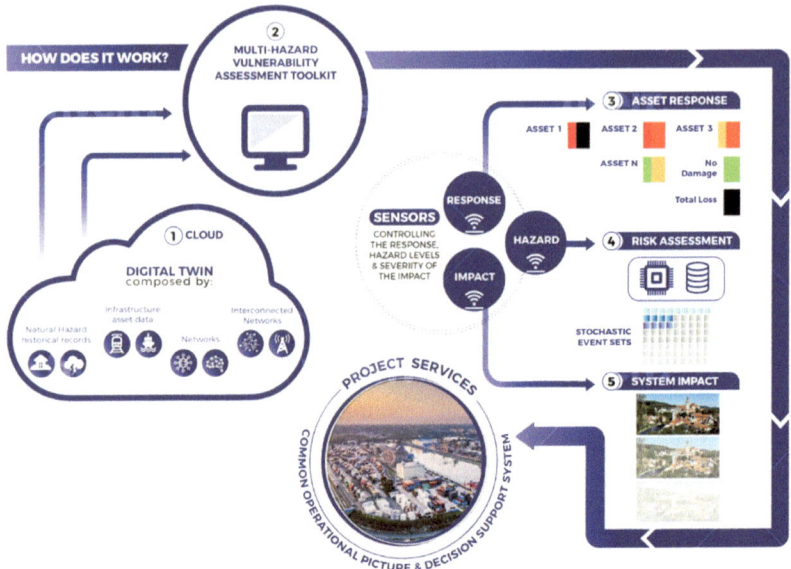

Fig. 11.2 The flow of information within the PLOTO platform modules, comprising (1) cloud-based data, (2) multi-hazard vulnerability assessment, (3) asset response characterisation, (4) risk assessment and (5) system impact evaluation, all fed by hazard, response and impact sensors, to offer decision support, incident management and a common operational picture

within (4) the system risk assessment, ultimately leading to (5) the system impact evaluation. Throughout this process, additional data by hazard, response and impact sensors is incorporated, serving to remove scenarios that are incompatible with current observations and to improve accuracy. All in all, this offers an integrated decision support system tied to incident management under a common operational picture.

USE CASES AND SCENARIOS

PLOTO will perform extensive tests in three different demo sites, Belgium, Hungary and Romania. The demonstration shall prove the suitability of the PLOTO platform for multiple-hazard assessment and optimised operational and strategic decisions for management and maintenance of IWW, considering hazards relevant for other sections of the same corridor, or for other critical parts of IWWs. The demonstration will focus on the following main objectives: (1) to improve multiple-hazard assessment and strategic management for protection of hotspots of the IWW ports and sections, (2) to improve strategic and operational decision-making, (3) to test the various PLOTO outcomes and the overall integrated decision support tool with actuation technologies in real-scale critical parts of the IWW.

Use Case A: Danube Area, Including the Waterways and Inland Ports

The Danube River sector located between Iron Gate II (RKM 863) and Călăraşi (RKM 375) (Fig. 11.3), belongs to the lower basin of the Danube and is characterised by a very dynamic riverbed with large flow rates (between 1600 and 15,000 m³/s). The geography of this sector is diverse. It includes mountains, large plains, sand dunes, forested or marshy wetlands. Similarly, climate and precipitation vary significantly, and they continuously form the basin's landscapes. The morphological processes in the Lower Danube can be classified as very dynamic. In the upper part of the section (km 863–km 730), the intensity of the erosion and accumulation is less than in the middle (km 730–km 500) or in the lower part of the section (km 500–km 375). Hundreds of kilometres at the left and the right bank of the Danube River are eroded.

Use Case B: Budapest Freeport and Railway Hub

The port is located in the Central region of Hungary, within the boundaries of the capital city Budapest (Fig. 11.4). The port has been established in the first third of the twentieth century, when that area was still considered a suburb. Then, a strong industrialisation and population growth

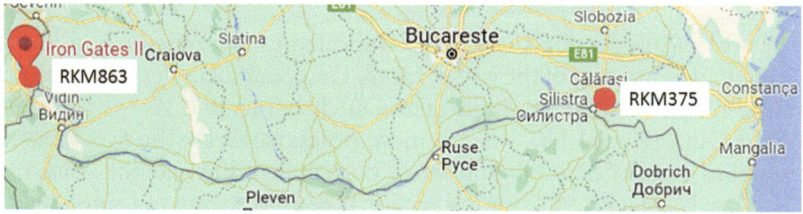

Fig. 11.3 Map of the Danube River sector between Iron Gate II (RKM 863) and Călăraşi (RKM375), use case A. (Figure processed from Google maps)

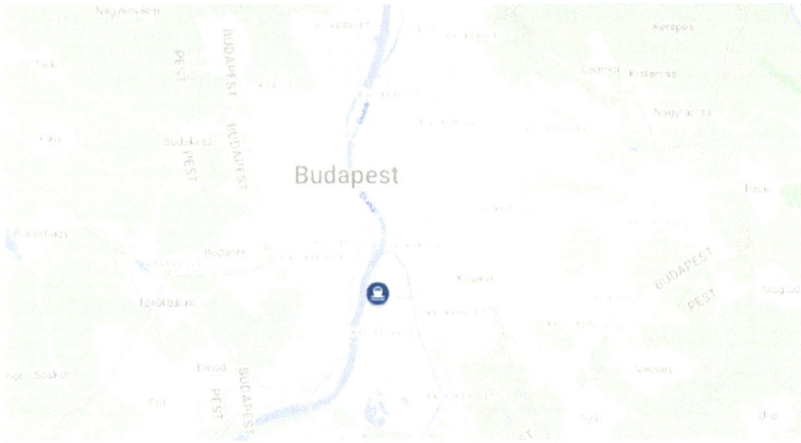

Fig. 11.4 Location of the Freeport of Budapest

occurred in the 1950s. Consequently, the city surrounded the port; thus, the port's development possibilities are limited recently, and its approach is difficult. The port is accessible via one railway connection and from two main road directions. Establishment of the M0 motorway (ring road around Budapest) in the 1990s significantly improved the road connection quality, but the traffic must pass through the 21st district of Budapest (called Csepel) causing harmful environmental effects there and congestion reaching the motorway. As the waterway network in Hungary is not very dense, the port serves not only the capital city with approximately 1.7 million inhabitants, but the entire central-Hungarian region with approximately 3.5 million inhabitants.

Use Case C: Region of Wallonie in Belgium

The area of interest for Use Case C is located in the Walloon region in Belgium downstream of the city of Liège. Here, our focus lies on a section of the river Meuse up to the Belgian-Dutch border and the part of the Albert Canal (Fig. 11.5b), which is more-or-less parallel to the river Meuse as well as the associated floodplain of both, where the relevant assets of Use Case C are located. Over this section (i.e. downstream of Liège), river Meuse itself is not navigable and shipping takes place on the Albert Canal. Just before reaching the Belgian-Dutch border, the Albert canal turns

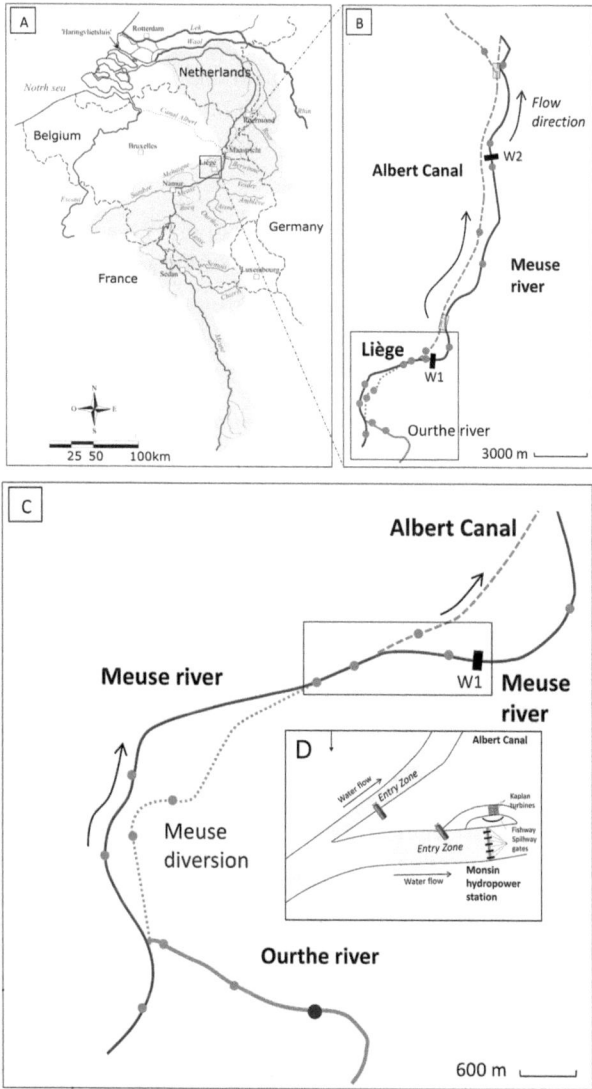

Fig. 11.5 Division between river Meuse and the Albert canal in Liège. (**a**) International Meuse basin covering parts of France, Belgium, Luxembourg, Germany and The Netherlands. (**b**) Studied Meuse river and Albert Canal stretches. (**c**) Meuse and Ourthe river stretches upstream of the entrance of the Albert canal. (**d**) Schematic representation of the division between river Meuse and the Albert canal downstream of Liege, with Monsin weir and hydropower plant. (Adapted from Renardy et al. [13])

towards the north-west to reach the sea harbour of Antwerp (Fig. 11.5a). Details of the configuration of the IWW stretches in and around Liège are visible in Fig. 11.5c and d. The floodplain and all its assets are especially interesting to analyse in the context of dike breaching scenarios, as such an event would have devastating consequences on infrastructure, buildings and people in the region.

Acknowledgements This work is a part of the PLOTO project. This project has received funding from the Horizon Europe innovation actions under grant agreement no. 101069941.

REFERENCES

1. European Commission. (2019). *Communication from the Commission to the European Parliament, the European Council, the Council, the European Economic and Social Committee and the Committee of the Regions* (COM(2019) 640 final).
2. https://climate.ec.europa.eu/eu-action/climate-strategies-targets/2030-climate-energy-framework_en
3. OECD – International Transport Forum. (2019). *ITF transport outlook 2019*, p. 38.
4. European Commission. (2020). *EU transport in figures: Statistical pocketbook 2020*, p. 25.
5. EC. (2007). *The EU's freight transport agenda: Boosting the efficiency, integration and sustainability of freight transport in Europe*, p. 2.
6. EC. (2020). *EU transport in figures – Statistical pocketbook 2020*, p. 21.
7. World Bank. (2018). *Connecting to compete 2018: Trade logistics in the global economy: The logistics performance index and its indicators*, p. 12.
8. OECD – International Transport Forum. (2019). *ITF transport outlook 2019*, p. 39.
9. European Commission. (2018). *In-depth analysis in support of Commission Communication COM(2018) 773 – 'A Clean Planet for all: A European long-term strategic vision for a prosperous, modern, competitive and climate neutral economy'*, p. 82.
10. Ugarte, G. M., Golden, J. S., & Dooley, K. (2016). Lean versus green: The impact of lean logistics on greenhouse gas emissions in consumer goods supply chains. *Journal of Purchasing and Supply Management, 22*(2), 98–109.
11. European Commission. (2018). *State of play and barriers to the use of electronic transport documents for freight transport options for EU level policy interventions: Final report – Study*, pp. 19–20.
12. https://climate.ec.europa.eu/eu-action/european-green-deal_en

13. Renardy, S., et al. (2021). Trying to choose the less bad route: Individual migratory behaviour of Atlantic salmon smolts (Salmo salar L.) approaching a bifurcation between a hydropower station and a navigation canal. *Ecological Engineering*, *169*, 106304. https://doi.org/10.1016/j.ecoleng.2021.106304

Open Access This chapter is licensed under the terms of the Creative Commons Attribution 4.0 International License (http://creativecommons.org/licenses/by/4.0/), which permits use, sharing, adaptation, distribution and reproduction in any medium or format, as long as you give appropriate credit to the original author(s) and the source, provide a link to the Creative Commons license and indicate if changes were made.

The images or other third party material in this chapter are included in the chapter's Creative Commons license, unless indicated otherwise in a credit line to the material. If material is not included in the chapter's Creative Commons license and your intended use is not permitted by statutory regulation or exceeds the permitted use, you will need to obtain permission directly from the copyright holder.

CHAPTER 12

Security Challenges in Critical Infrastructures in Transport: The PRECINCT Athens Use Case

Ioannis Lymaxis and Eftichia Georgiou

INTRODUCTION

EU Critical Infrastructures (CIs) are becoming more and more vulnerable to physical and cyberattacks as well as natural disasters. The focus of research and newly developed solutions is on the protection of individual CIs, however as most of the CI interrelationships have grown more intricate and rely on interconnected networks and devices, the failures in a critical sector may result in cross-sector—or even cross-border—cascading effects. The lack of proper awareness makes it difficult for operators to anticipate risks, protect the CI's critical services, and enable rapid recovery in the event of disruptions.

I. Lymaxis (✉)
Inlecom Innovation, Athens, Greece
e-mail: john.limaxis@inlecomsystems.com

E. Georgiou
Center for Security Studies, Athens, Greece
e-mail: e.georgiou@kemea-research.gr

© The Author(s) 2025
I. Gkotsis et al. (eds.), *Paradigms on Technology Development for Security Practitioners*, Security Informatics and Law Enforcement, https://doi.org/10.1007/978-3-031-62083-6_12

Through the application of PRECINCT project methodological framework and technological solutions developed, the work presented in this paper concentrates on improving the phases of crisis management that deal with the preparedness and response capabilities of interconnected CI's operating in the same geographical area. In this regard, the two out of three main tools, namely Digital Twin and its components, along with the Coordination Center for providing a common situational awareness picture to all relevant stakeholders involved during a crisis are described.

The findings and developments presented, are linked with the EU-funded project PRECINCT aiming to link private and public CI stakeholders in a geographic area to a common cyber-physical security management approach that will result in a protected area for people and infrastructure.

METHODOLOGY

PRECINCT [1] will develop a comprehensive Ecosystem Platform, connecting various stakeholders of interdependent CIs and Emergency Services, enabling them to collaboratively manage security challenges and enhancing their resilience against hybrid attacks. The overall goal is to provide new services and capabilities to CI operators, utilizing Artificial Intelligence (AI) and Machine Learning (ML) techniques, for early detecting and managing cyber-physical threats, thus strengthening CI defenses against vulnerabilities.

The following are the areas in which the project's research has been conducted along this line:

Understanding PRECINCT uses State of the Art (SOTA) modeling techniques to precisely determine the present and future risks in territory-based interdependent CIs under various multihazard conditions and configurations, to gain a deeper understanding of interdependent CIs. A key goal is to enable CI actors to anticipate sophisticated attacks, detect anomalies, and incentivize optimized command structures and coordinated responses between CIs and first responders, thereby minimizing cascading effects and allowing rapid recovery.

Improving The Digital Twins will help improve accuracy and automation in identification, remediation, and threat elimination. The application of Digital Twins to multihazard risk management yields a circular process of anticipating, preventing, protecting, and responding during events as well as recovering and learning after events.

Sustaining Modeling CI interdependencies to identify, forecast or simulate potential cascading effects has limitations in identifying vulnerabilities in complex and codependent CI threat contexts. The dynamic nature of the threat canvas reshapes based on new weekly exploits, the ingenuity of attackers in finding new and creative angles of attack, thus static and dynamic modeling approaches require considerable time and effort to maintain.

By the end of the project, through a series of validation scenarios demonstrated in four Living Labs (LL), the project will produce tools that are ready for use. The current paper will mainly focus on the demonstrations of the Athens LL, involving several operators, i.e., the Athens International Airport, the Attikes Diadromes S.A., and the Attiko Metro S.A., having as an end goal to increase the involved CIs' overall resilience against cyber-physical incidents affecting urban transport. Since the Athens LL participants represent the city's main transportation system (the rail and road network along the Athens Airport/Attiki Odos corridor as well as along the urban rail and road network), with a population of over 4 million and which is typically visited by more than 25 million people annually, the demonstration represents a difficult but essential case for resilience management of interconnected transportation systems as well as for demonstrating the efficiency and importance of PRECINCT tools.

PRECINCT LIVING LAB ATHENS: THREAT SCENARIO AND DIGITAL TWIN ARCHITECTURE

The main threat scenario created and tested by the end users of Athens LL revolves around a coordinated cyber-physical terrorist attack against the Athens International Airport, the cascading effects of these attacks in the surrounding vicinity, and to the other interconnected Critical Infrastructure in the area.

In more detail, the threat scenario starts when the terrorist party collaborates with an airport third party vendor and manages to install malware and infiltrate airport critical systems such as the Access Control (AC) and the Flight Information Display (FID). Following this, the attackers can move undetected within the airport, and strategically place baggage containing improvised Explosive Devices (IEDs) in critical areas as well as in the adjoining metro station. In the last step, the attackers in order to maximize their damages and causalities try to steer people suited in the airport premises close to the explosive locations through fake announcements and information displayed to them through airport's information display screens. Upon the confirmation of the attack, the airport's crisis management procedures are activated and all the relevant parties are notified including the critical infrastructures operating in the same geographical area (i.e., Attiko Metro and Attikes Diadromes).

In response to this situation, Attikes Diadromes displays through its road Video Messages Screens instructions to alert road users to avoid proceeding to the Airport as well as deploys and coordinates with the road patrols, through their Traffic Management Center, to reroute traffic away from road lanes, and exits close to the airport. Additionally, the Metro Operation Center advises train drivers to stop at the closest railway station and suspend operations on the suburban metro segment leading to and from the airport once the incident has been confirmed. Finally, the Airport's Police with the assistance of terminal and security staff proceed with the passengers' evacuation process and assist passengers to leave the airport premises safely.

Under this context and the identified Athens CI operators' needs, PRECINCT Digital Twin solution aimed at improving end users' capabilities of detecting cyber-physical threats in their installations/systems, supporting the efficient and in time exchange of information among CIs stakeholders, and finally the efficient decision-making. The developed Athens Digital Twin ecosystem offers an integrated solution, utilizing cutting-edge modeling and simulation technologies, for resilience management against cyber-physical threats and their cascading effects in the interconnected infrastructures as explained below.

To begin with, the deployed Digital Twin *Cyber Security and Detection* layer consists of the following three components: (i) Security and Privacy Monitoring, (ii) Root Cause Analysis (RCA), and (iii) Test and Simulation (TaS) components. The aforementioned component's goal is to monitor the status of the CIs' network traffic and Internet of Things (IoT) devices status for detecting anomalies or relevant cyberattacks against them by using Machine Learning (ML) techniques. Furthermore, the Athens Digital Twin consists of a *Knowledge Graph* acting as a database consuming and connecting the heterogeneous datasets and information provided by the participating CIs' sources as well as producing alerts upon receiving up-normal values as input from them.

The developed PRECINCT Athens DT *Complex Event Processing* component combines and fuses the different cyber-physical threats or alerts detected, by the other components, into a deeper level for helping the CIs' operators to have a more accurate understanding of the situation and the events occurring. The *Cascading Effect Simulation Engine* allows the user to model how a threat, or combined attacks may spread, have cascade consequences, and affects the Athens transport network/CIs region.

In addition, the Digital Twin incident reporting feature and the integration with the *Hellenic Coordination Center for Critical Infrastructure Protection (H3CIP)* [2] as demonstrated in the Athens LL sought to provide a common operational picture in near-real-time to all stakeholders connected to the 3HCIP platform; support the exchange of information among participating AMETRO, AIA, ATTIKES DIADROMES operators during an incident; and to facilitate coordination among the involved stakeholders during a crisis. Finally, the *Digital Twin Supervisory Resilience Control* component supports the end users' decision-making by providing a list of optimal mitigation actions, using Machine Learning (ML) and Artificial Intelligence (AI) algorithms, for minimizing the impact of the cascading effects and restoring the nodes of the network.

Concluding, as can be seen in Table 12.1, the developed DT comprises of various components clustered in three high-level modules: (i) a Cyber Detection module, where the network traffic and other CI data are analyzed to detect possible threats, (ii) a Preparedness and Alerting module, consolidating all information and providing the CI operators a single-user interface for increased situation awareness as well as simulation of threats

Table 12.1 PRECINCT LL digital twin components

Capabilities	Component name
Cyber detection	Security monitoring tool (SPM)
	Root cause analysis (RCA)
	Test and simulation (TaS)
Preparedness & alerting	Complex event processing (CEP)
	Cascading effects simulation engine (CESE)
	Knowledge graph (KG)
Response & coordination	Resilience supervisory control (RSC)
	Coordination center (H3CIP)

and their cascading effects, and (iii) a Response and Coordination module to support crisis management procedures with decision support and other tools.

PRECINCT ATHENS LL USER EXPERIENCE EVALUATION

The design of PRECINCT considered user satisfaction as one of the top priorities and one of the main goals of the Living Lab study was to thoroughly assess it. PRECINCT questionnaires were produced using standardized questionnaires for assessing user satisfaction, acceptability, and system usability during the PRECINCT project's LL research. The questionnaires have been designed by relying on existing standardized questionnaires such as the System Usability Scale (SUS) [3], the Software Usability Measurement Inventory (SUMI) [4], and the Computer System Usability Questionnaire (CSUQ) [5] that have been used in various studies [6]. The questionnaires were organized around the dimensions of usability as recommended by ISO/IEC TR 25060:2010 [7], ISO 9241-11:2018 [8], and the quality requirements as recommended by ISO/IEC 25010:2011 [9] namely effectiveness (accuracy and completeness with which users achieve specified goals); efficiency (resources expended in relation to the accuracy and completeness with which users achieve goals); satisfaction (extent to which the product or service meets the user's needs and expectations). Other metrics such as ease of learning were also considered.

The main objective of the PRECINCT User Experience evaluation was to collect end users' (transport operators, cyber-physical security professionals, etc.) feedback to identify needs for further optimizations of the PRECINCT framework. Participants could respond on a Likert scale ranging from 1 to 5, with 1 indicating "strongly disagree" and 5 indicating "strongly agree." Scores more than 3 are considered positive replies; nevertheless, each answer is evaluated independently based on the type of question. The findings reported in this section are based on responses from fourteen (14) survey participants in the Athens LL. The mean Likert-scale scores of the responses received in each question, as well as the Standard Deviation (SD) value, are used to draw conclusions. End users were given questionnaires to fill out after (posttest) utilizing the PRECINCT framework to assess their experience. The results are as follows:

Survey Results of the Entire PRECINCT Framework

Based on the survey, most participants found that PRECINCT is an acceptable solution to improve the capabilities of end users to manage cyber-physical threats more efficiently (4.20|SD = 0.79). Most of them agreed that PRECINCT could improve the operational resilience in the CIs' region (4.30|SD = 0.67), could increase the accuracy in cyber and/ or physical threats detection (4.30|SD = 0.82), could enhance the response and mitigation actions taken by the stakeholders during and after crisis (4.10|SD = 0.88) and that understanding the interdependencies and cascading effects among the various CIs engaged in a crisis could be improved by PRECINCT (4.30|SD = 0.67). This is very encouraging feedback because the end users attest to the PRECINCT framework's ability to achieve the project's primary goals. The worst-rated statement was the statement "It is easy to integrate the PRECINCT framework with the current systems in my organization" (3.40|SD = 0.70), which is still positive feedback. Given the difficulty of integration in present operating settings and the need for changes, it is anticipated that this be communicated.

Survey Results of Digital Twin (DT) Framework by End Users

On top of the entire PRECINCT framework evaluation documented in the previous section, the Digital Twin (DT) has been evaluated separately. The metrics considered were related to the perception of errors, comprehension of objectives, the level of completed objectives, the ease of learning, and basic subjective user satisfaction metrics.

Based on the survey, the majority of the participating end users understood the objectives of the DT (4.40|SD = 0.70), they agreed that the DT correctly identified (4.26|SD = 0.63) and correlated the cyber and or physical threats (4.20|SD = 0.63), and the statements of the threat scenarios were easy to understand (4.30|SD = 0.67). They also agreed that the results of the demonstration were appropriate to the threat scenario(s) presented (4.20|SD = 0.74) and that the DT is overall a significant improvement compared to their current methodology (4.40|SD = 0.70). Based on the answers to the basic subjective user satisfaction questions, the DT managed to meet user expectations. The majority of participants were satisfied with how simple it was to utilize the DT (4.00|SD = 0.82) and how simple it was to acquire significant insights from the data shown through the DT user interface (4.20|SD = 0.63).

Conclusions

Based on the feedback received from Athens Living Lab operators (e.g., operations, networks, security, crisis managers, et al.) and associated risk analysis, the key challenges and responsibilities they face are the early detection of threats and accessing the damages; ensuring business continuity and public safety in the event of an incident; mobilization of crisis management teams and emergency plans; coordination with first responders and implementation of recovery actions. Therefore, protecting critical infrastructure against complex and hybrid cyber-physical threats requires a comprehensive framework for modeling these attacks and their potential impact, as well as providing new tools and enhanced capabilities to the end users for managing them and coordinating.

The PRECINCT Framework aims to establish a unified and holistic approach to managing cyber-physical attacks in order to address the multifaceted challenges outlined above. This was achieved in the LL through the DT solution, which enables the various CIs and first responders to have a common and holistic situational awareness picture of the area. This is accomplished by integrating sophisticated algorithms, real-time data analysis, and simulation-based capabilities, as well as facilitating communication among them using a robust and secure platform to exchange messages and alerts.

According to the overall user experience evaluation findings, the PRECINCT framework is approved by the system's intended end users and managed to meet their expectations. Regarding the DT, the survey demonstrated that the DT performed all the tasks that it was designed for, supporting the operators in their further investigation and response actions in the context of the Athens LL threat scenarios. All the incidents identified by the DT were received with almost no delay and provided the operators with all the required details they needed to proceed with their standard operating procedures. The operators agreed that PRECINCT is an acceptable solution to improve the capabilities of end users to manage cyber-physical threats more efficiently and that the framework could enhance the response and mitigation actions taken by the stakeholders during and after a crisis. The validation results depict the usefulness of such a framework, improving the collaboration among the several stakeholders involved during a crisis compared to the current situation. At the same time, the PRECINCT interface was proven to be user-friendly and intuitive, as well as providing the right information to the users during their operations.

In conclusion, the PRECINCT framework has successfully brought together private and public stakeholders operating within the same geographic region within LL3 to effectively respond to and mitigate threats, thus enhancing the overall security of the citizens in the area.

Acknowledgments This project has received funding from the European Union's Horizon 2020 research and innovation program under grant agreement No. 101021668. This article reflects only the authors' views, and the Research Executive Agency and the European Commission are not responsible for any use that may be made of the information it contains.

REFERENCES

1. PRECINCT. [Online] https://www.precinct.info/en/about/
2. CIPROTECTION. [Online] [Cited: July 26, 2023] http://www.ciprotection.gr/index.php/en/
3. SUS. [Online] [Cited: July 26, 2023] https://www.usability.gov/how-to-and-tools/methods/system-usability-scale.html
4. SUMI. [Online] [Cited: July 26, 2023] https://sumi.uxp.ie/en/index.php
5. Lewis, J. R. (1992). Psychometric evaluation of the post-study system usability questionnaire: The PSSUQ. *Proceedings of the Human Factors Society Annual Meeting, 36*(16), 1259–1260.
6. Tullis, T., & Albert, B. (2013). Chapter 6: Self-reported metrics. In B. Albert & T. Tullis (Eds.), *Measuring the user experience* (2nd ed.). Morgan Kaufmann.
7. ISO. ISO 25060. *Systems and software engineering – Systems and software product Quality Requirements and Evaluation (SQuaRE).* [Online] https://www.iso.org/standard/35786.html
8. ISO 9241-11:2018. *Ergonomics of human-system interaction – Part 11: Usability.* [Online] https://www.iso.org/standard/63500.html
9. ISO/IEC 25010:2011. [Online] [Cited: July 26, 2023.] https://www.iso.org/obp/ui/#iso:std:iso-iec:25010:ed-1:v1:en

Open Access This chapter is licensed under the terms of the Creative Commons Attribution 4.0 International License (http://creativecommons.org/licenses/by/4.0/), which permits use, sharing, adaptation, distribution and reproduction in any medium or format, as long as you give appropriate credit to the original author(s) and the source, provide a link to the Creative Commons license and indicate if changes were made.

The images or other third party material in this chapter are included in the chapter's Creative Commons license, unless indicated otherwise in a credit line to the material. If material is not included in the chapter's Creative Commons license and your intended use is not permitted by statutory regulation or exceeds the permitted use, you will need to obtain permission directly from the copyright holder.

CHAPTER 13

PRAETORIAN: From Protection to Resilience of Critical Infrastructures

Eva Muñoz-Navarro, Juan José Hernández-Montesinos,
Antonio Marqués-Moreno, Lazaros Papadopoulos,
Antonios Karteris, and Konstantinos Demestichas

REGULATION AND THE PRAETORIAN SOLUTION: AN OVERVIEW

The new CER Directive [1] constitutes a considerable change as compared to the ECI Directive 2008/114/EC [2] since critical entities will have to meet specific obligations aimed at enhancing their resilience. Moreover, a wider sectoral scope will allow Member States and critical

E. Muñoz-Navarro (✉) • J. J. Hernández-Montesinos • A. Marqués-Moreno
ETRA Investigación y Desarrollo, Valencia, Spain
e-mail: emunoz.etraid@grupoetra.com

L. Papadopoulos • A. Karteris
School of Electrical and Computer Engineering, National Technical University of Athens, Athens, Greece

K. Demestichas
Department of Agricultural Economics and Rural Development, Agricultural University of Athens, Athens, Greece

© The Author(s) 2025   155
I. Gkotsis et al. (eds.), *Paradigms on Technology Development for Security Practitioners*, Security Informatics and Law Enforcement, https://doi.org/10.1007/978-3-031-62083-6_13

entities to better address interdependencies and potential cascading effects of an incident. European critical entities are more interconnected and interdependent, which makes them stronger and more efficient but also more vulnerable in case of an incident.

As requested by the CER Directive, critical entities will need to carry out risk assessments on their own, take technical and organizational measures to enhance their resilience and notify incidents. New tools will soon be demanded by Critical Infrastructure (CI) operators and innovative technologies will have to be used allowing the adoption of these measures. The CER Directive is complemented by the NIS2 Directive [3], thus becoming an updated and comprehensive legal framework to strengthen both the physical and cyber-resilience of critical infrastructure.

The goal of the H2020 PRAETORIAN project (https://praetorian-h2020.eu/) is to enable the security stakeholders of the CIs in Europe to manage the life cycle of security threats, from forecast, assessment and prevention to detection, response, and mitigation, in a collaborative manner with the security teams from related CIs, being the CIs in the same sector or not. PRAETORIAN proposes a toolset that:

(a) Makes use of data obtained from relevant legacy security systems of the CIs.
(b) Introduces novel sensors and innovative data analysis.
(c) Builds a model of the ecosystem of CIs.
(d) Improves the channels and quality of communication among stakeholders.
(e) Combines the emergency plans of those CIs.

The combination of these functionalities will support the decision-making process of CI operators to prevent major damages to the installations, neighboring population, and the environment while allowing a fast recovery after incidents.

THE PRAETORIAN FRAMEWORK

The PRAETORIAN toolset consists of four innovative products, which intend to provide the security managers with the capacity to protect the CIs from physical, cyber, and combined (physical and cyber) attacks. The Cyber Situation Awareness (CSA) system can recognize patterns within the network and generate corresponding events. The Physical Situation

Awareness (PSA) system can be integrated with existing sensors and legacy systems in the CI to collect meaningful data and combine them with information received from newly developed modules that implement drone detection and video analytics. Both the CSA and the PSA generate an alarm when cyber/physical threats are detected. The Hybrid Situation Awareness (HSA) system uses a digital twin of the related CIs to correlate the received alarms and estimate the cascading effects on own and related CIs. This information is processed in the Coordinated Response (CR) system which suggests an effective response to the threat, allowing notifications and information sharing through multiple channels.

Figure 13.1 highlights the flow of information between the aforementioned components. The HSA receives *events* and *alerts* generated by the PSA and CSA. The CR receives alerts from all components and generates relevant security *incidents* and proper notifications to operators and first responders, while it recommends mitigation actions.

The framework for developing this solution is largely based on the idea of interoperability of systems and components (which also allows focusing on scalability and replicability), therefore the PRAETORIAN back end relies on the *Interoperability Platform* (IOP) that interconnects all the elements, allowing: (i) the exchange of information between all the systems and modules, (ii) the storage of information, and (iii) avoiding the

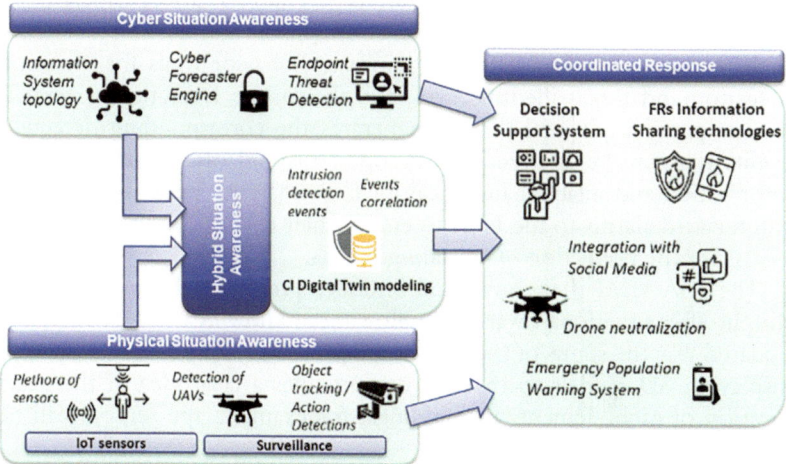

Fig. 13.1 PRAETORIAN platform

duplication of data between modules, the replication of changes, and possible inconsistencies. In this way, data is provided for the entire platform that can be processed and retrieved to offer useful and usable information for all its users, and therefore serving as a data-sharing infrastructure for all PRAETORIAN components.

The IOP offers a variety of connectivity methods, including RESTful Application Programming Interface (API), Datagram Delivery Protocol (DDP), and Advanced Message Queuing Protocol (AMQP). About the front end, the main PRAETORIAN Human Machine Interface (HMI) is the CR. However, each of the other systems (i.e., PSA, CSA, and HSA) provides a user-friendly HMIs, tailored to the needs of CI operators.

The following subsections describe the PRAETORIAN components, focusing on the features and the added value that they provide compared to the typical legacy systems. A detailed description of the modules and subcomponents of each system can be found in [4].

Creating Situation Awareness

Three PRAETORIAN systems have been developed to create Situation Awareness: the Physical Situation Awareness (PSA) System, the Cyber Situation Awareness (CSA) System, and the Hybrid Situation Awareness (HSA).

The PSA system focuses on the monitoring and control of critical areas. It receives and processes information from sensors and detection systems, such as the Video Analytics and Anomaly Behavior tools for intrusion detection and threat identification, and the Drone Detection system. It detects possible physical threats, generates the corresponding detection events that can be analyzed by the operator, and offers both real-time information and incident historical for later analysis. Moreover, the PSA will forward alarms to the HSA to enable their correlation with the cyber events for the prediction of cascading effects.

The PSA HMI is the graphical interface providing the user with the sought-after situational awareness, allowing to know in a real, precise, and updated way the status of the asset to be protected. The tool is capable of displaying 3D models so that the operator has a clear idea of the exact location of each element. In buildings, for example, the sensors will be located represented in height.

Figure 13.2 shows an example of the PSA map view. The icons represent the location of sensors, cameras, and agents. The operator can, at any

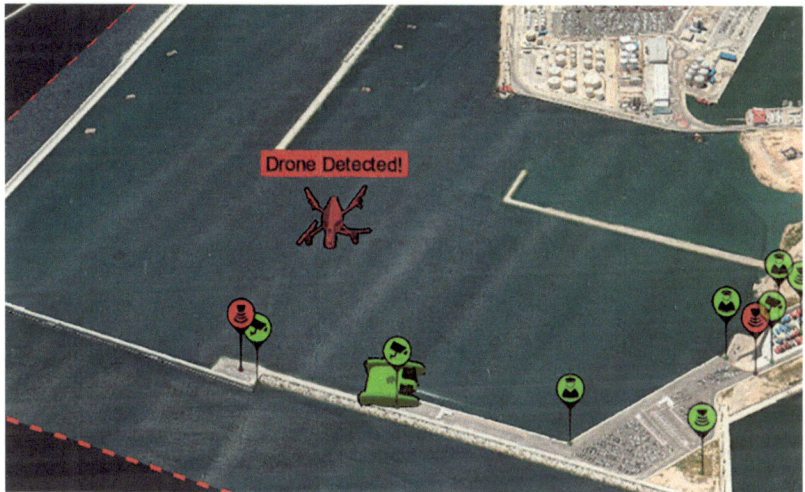

Fig. 13.2 The PSA map view

moment, retrieve information on measurements in real time, or get access to the streaming or recordings from cameras. Furthermore, unmanned vehicles are also represented in the map using 3D models, colored in green if they are assets in the CI, or in red if they are not recognized as part of the system.

The CSA system aims to improve the cyber situation awareness of the CI operator and to forward cyber events to the HSA to enable their correlation with the physical events for the prediction of cascading effects. The CSA has been implemented based on the development of Cybersecurity Digital Twins (CDT), which mimic some Critical Infrastructure's Operational Technology (OT) and Information Technology (IT) systems. The CDTs are designed so as not to interfere with the CI's real operational systems during the verification, validation, and demonstration activities. Cyber Assessment Tools (CAT) are used to simulate additional legitimate traffic, launch attacks, and collect cybersecurity logs. Finally, a Cyber Forecaster Engine (CFE) can forecast the end goal of the attacker.

The CSA HMI provides various visualization options to represent assets, alarms, detections, and attack goals, as predicted by the CFE. Figure 13.3 shows a timeline representation of attacks, showing the relation between the detections and the corresponding alerts. The

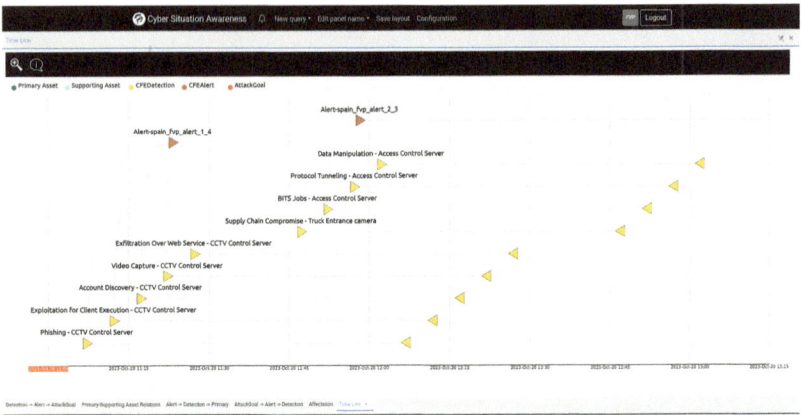

Fig. 13.3   CSA HMI: visualization of timeline

timeline visualization integrates the time dimension in the overall cyber perspective and allows for aggregating, in the same view, primary assets, supporting assets, detections, and alerts. In this kind of visualization, the operator can identify which detections triggered which alerts and how both detections and alerts spawned in time and also lasted during their life cycle. Primary assets are usually information or business processes while supporting assets can be hardware, software, and human resources [5].

Finally, the Hybrid Situation Awareness (HSA) system provides CI operators with accurate forecasts of potential cyber and physical consequences at the facilities, given any kind of physical and cyber alert detected by the PSA and the CSA respectively. Figure 13.4 reflects that physical and cyber domains cannot be understood and treated independently since the attacks on any of these dimensions may have also significant consequences in the other one. Moreover, the HSA will be able to calculate and show not only the cascading effects of attacks on a particular CI, but also on interrelated CIs.

The HSA back end includes two systems, the *Generic Digital Twin* (GDT) that models the relevant CI assets with their inter/intrarelationships and the *Threat Propagation Engine* (TPE) that simulates the consequences of incoming alerts including cascading effects.

The GDT consists of a graph-based representation in which the nodes represent the critical entities of the CIs, and the edges represent the dependencies among these assets. Each asset has a state which may change

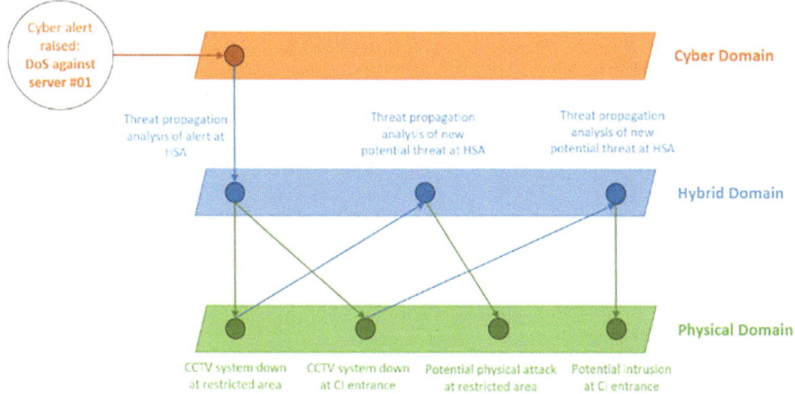

Fig. 13.4 Physical and cyber domains interdependency. (Source: PRAETORIAN D5.3 "HSA system Development and Functional Validation Report")

due to a cyber or physical event, as detected by the PSA or the CSA, respectively (e.g., fire, cyberattack etc.). When the state of a node changes, a notification is sent to each adjacent node, which may themselves react to the incoming notification and change their state, and, in turn, inform their own adjacent nodes. The GDT models can be created within a CI, to calculate cascading effects between assets of the particular CI, as well as between different CIs, for calculating the cascading effects between them.

Based on the GDT, the Threat Propagation Engine (TPE) describes the direct and indirect consequences of alerts generated by the PSA and the CSA, over time. In particular, for each alert forwarded to the HSA, the TPE is triggered and a set of interdependent threat propagation simulations is run. The simulation results are used to estimate the potential consequences of the threat on the overall network of interconnected CIs. The output of the TPE is a prediction of the propagation of the cascading effects, which is displayed on the HSA HMI.

The front end has two HMIs, designed to be reactive and show, in an autonomous manner, if new elements have been received (without user intervention) from the back end. One shows the cascading effects (HSA HMI) and the other one is the model of the CIs and the status of their assets (Synoptic Live Diagram, SLD HMI).

The HSA HMI shows on a map the predicted cascading effects, as calculated by the TPE simulations (Fig. 13.5). A graph-based representation

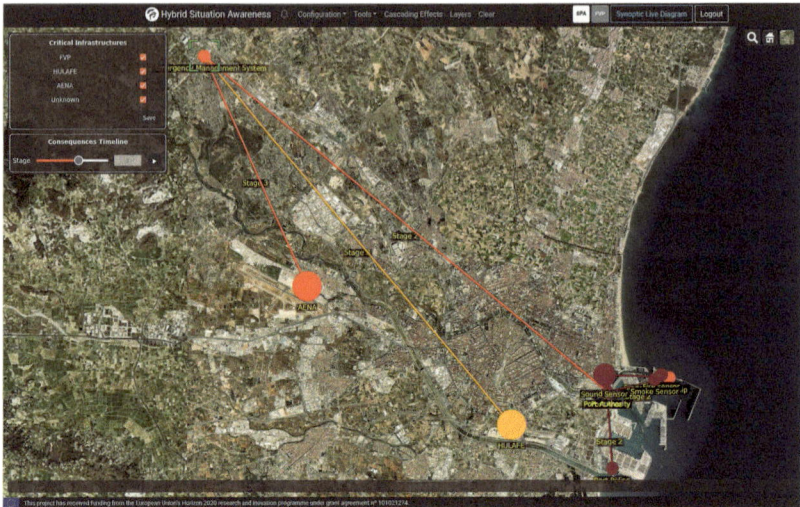

Fig. 13.5   PRAETORIAN HSA HMI: cascading effects visualization

is used, in which each node corresponds to a cyber or physical asset of a CI and each link corresponds to an interdependency. Different colors in each node indicate the corresponding impact of the threat (i.e., the degree by which the asset was affected). On the other hand, the SLD HMI allows designing 3D diagrams and linking its elements with the data in the back end, as well as monitoring the status of the data in near-real-time (Fig. 13.6).

Providing Coordinated Response

The main CR module is the Decision Support System (DSS). It acts as a hub, as it collects all alerts and events generated by the PSA, CSA, and HSA. Through a set of predefined *rules*, the DSS generates *events* (i.e., information potentially useful to operators) and security *incidents* (i.e., information that may require immediate action by operators), as can be seen in Fig. 13.7.

The rules determine under which condition an event generated in the PSA, CSA or HSA will trigger the generation of a security incident, or when the DSS will trigger another module (e.g., in case of a mitigation action).

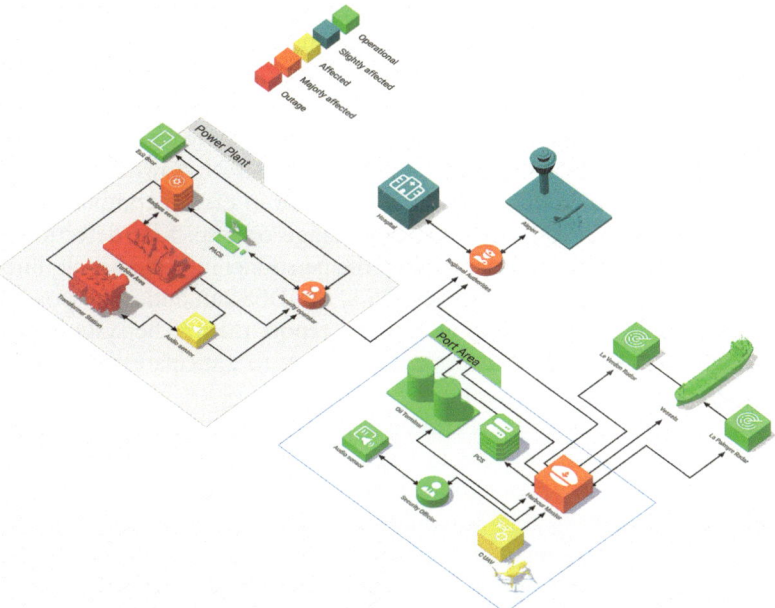

Fig. 13.6 Cascading effects displayed by the Synoptic Live Diagram

Fig. 13.7 The PRAETORIAN DSS interface

The DSS allows the CI operator to get real-time information about an incident, and decide about a possible mitigation action. If the proper rule is configured in the system, the mitigation action might be automatically launched, thus saving time and ensuring a rapid response. This is the case, for example, for drone detection and the corresponding countermeasures: Both systems have been integrated in PRAETORIAN. Any incident originating in both cyber, and physical domains, or in case of combination of both of them, due to some anomalous range or suspicious detection, is captured by the DSS in order to keep the personnel informed at all times. On the other hand, thanks to the HSA, the DSS can anticipate possible affected assets and cascading effects and report them. The system aims, through configurations and prior knowledge, to anticipate any incident and report it as soon as possible.

Once an incident has been created in the system, other operators can be notified in a variety of ways, including email, SMS or through a chat application, all of which are configurable through the *notifiers* page of the DSS.

PRAETORIAN is able not only to detect possible threats from the physical and cyber domains and generate the corresponding incidents, but also to analyze "weak signals" from different sources. This is the case for the *Social Media Security Threat Detection* (SMSTD) module [6], which utilizes text crawling techniques in order to monitor the entirety of the global Twitter stream and discern tweets that are potentially critical to the security of the CI, including tweets that mention data leaks, new vulnerabilities, and cyberattacks.

Moreover, another module provides a real-time feed of Twitter posts by the public during a crisis. The tool relies on machine learning techniques and identifies relevant informative-only tweets which can enhance the operator's situational awareness during crisis.

Communicating and Sharing Information

The PRAETORIAN system provides several mechanisms for the effective communication between operators and first responders, through the *Information Sharing & Communication with FRs* (ISC-FR) module [6]. It gathers information available on the PRAETORIAN platform and discerns the parts that are relevant for each type of first responder, for a particular incident type.

PRAETORIAN also supports the creation of EU-Alerts [7] to notify the population near the CI in case of an incident. Through the *Emergency*

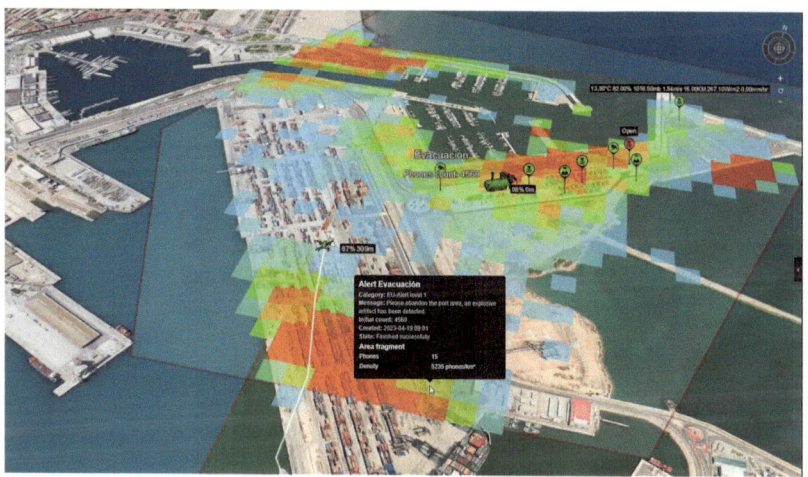

Fig. 13.8 EPWS tool and generation of EU-Alert

Population Warning System (EPWS) tool, the CI operators can select an area around the incident. After selecting a message from existing templates, the operator can potentially edit the message and send it to the cell phones of the population in the area. As shown in Fig. 13.8, a colored grid on the map indicates the number of cell phones in the area, which can be used to provide a rough estimation of the number of people and their distribution in the area of the incident.

Finally, PRAETORIAN offers connectivity with Twitter through the *Integration with Social Media* (IWSM) modules [6]. It offers a number of tweet templates that allow CI operators to generate messages for the public and share them on the social media platform with the press of a button.

PRAETORIAN IN THE CONTEXT OF RESILIENCE OF CIs

PRAETORIAN focuses on the interoperability of CIs' legacy systems together with new novel systems and sensors, aiming at improving the capability of CI security managers to prepare and apply in practice the Resilience Plan as requested by the CER Directive. This means taking technical, security, and organizational measures. Moreover, PRAETORIAN also allows the integration of additional information sources, such as signals from social media, agencies or any other open sources. Social media is indeed a valuable source of information during emergency situations, since it can be used to further improve the situation awareness of First Responders and rescue teams so they can act more effectively.

PRAETORIAN is also aligned to the Council Recommendation on a Union-wide coordinated approach to strengthen the resilience of critical infrastructure [8], which aims at maximizing and accelerating the work to protect critical infrastructure in three priority areas: preparedness, response, and international cooperation. Three out of the four PRAETORIAN systems allow to enhance the situational awareness, by means of threat detection and creation of alarms on the one hand, but also allowing correlation of multiple signals in order to understand the threat propagation ("preparedness"). The remaining system focuses on the "response," based on a Decision Support System providing automated reaction to the detected threat and promoting that the CIs take a unified response that ensures less harm and extent of damage and maintains business continuity. The last aspect to consider is the "international cooperation," and in that sense, PRAETORIAN has explored quite interesting mechanisms that can be used by CI operators to share information about incidents. This is a relevant area, as well as difficult to implement; while CI managers have to decide about the nature and quantity of the information to be shared, due to the sensitiveness of it, the new regulations promote cooperation among different actors (public sector, public-private entities). This can only be achieved when this process is envisaged as part of the strategy for enhancing the resilience of critical entities, which will have to be defined by Member States.

CONCLUSIONS

The PRAETORIAN framework is a significant contribution to addressing the challenges of CI protection from combined cyber and physical attacks. It provides an advanced toolset which can be customized to the requirements of each particular type of CI. It focuses a lot on the prediction of the cascading effects of attacks and on the impact of these effects on interdepended CIs. Finally, it provides user-friendly interfaces for effective use by the CI operators. The toolset is designed and developed to focus on resilience of critical infrastructures, which is defined by the CER Directive as "a critical entity's ability to prevent, protect, against, respond to, resist, mitigate, absorb, accommodate, and recover from an incident."

Acknowledgments PRAETORIAN has received funding from the European Union's Horizon 2020 research and innovation program under grant agreement No. 101021274.

REFERENCES

1. Directive (EU) 2022/2557 of the European Parliament and of the Council of 14 December 2022 on the resilience of critical entities and repealing Council Directive 2008/114/EC (CER Directive).
2. Council Directive 2008/114/EC of 8 December 2008 on the identification and designation of European critical infrastructures and the assessment of the need to improve their protection.
3. Directive (EU) 2022/2555 of the European Parliament and of the Council of 14 December 2022 on measures for a high common level of cybersecurity across the Union, amending Regulation (EU) No 910/2014 and Directive (EU) 2018/1972, and repealing Directive (EU) 2016/1148 (NIS 2 Directive).
4. Papadopoulos, L., Karteris, A., Soudris, D., Muñoz-Navarro, E., Hernandez-Montesinos, J. J., Paul, S., Museux, N., Koenig, S., Egger, M., Schauer, S., Gómez, J. H., & Hadjina, T. (2023). PRAETORIAN: A framework for the protection of critical infrastructures from advanced combined cyber and physical threats. In *Proceedings of the 18th international conference on availability, reliability and security (ARES '23)*. Association for Computing Machinery.
5. ISO/IEC 27005:2022. *Information security, cybersecurity and privacy protection – Guidance on managing information security risks.*
6. Karteris, A., Tzanos, G., Papadopoulos, L., Demestichas, K., Soudris, D., Philibert, J. P., & Gómez, C. L. (2022). A methodology for enhancing emergency situational awareness through social media. In *Proceedings of the 17th international conference on availability, reliability and security (ARES '22)* (Article No.: 130) (pp. 1–7). Association for Computing Machinery.
7. ETSI TS 102 900 V1.3.1 – Emergency Communications (EMTEL); European Public Warning System (EU-ALERT) using the Cell Broadcast Service.
8. Council Recommendation on a Union-wide coordinated approach to strengthen the resilience of critical infrastructure. Document 15623/22 of the Council of the European Union.

Open Access   This chapter is licensed under the terms of the Creative Commons Attribution 4.0 International License (http://creativecommons.org/licenses/by/4.0/), which permits use, sharing, adaptation, distribution and reproduction in any medium or format, as long as you give appropriate credit to the original author(s) and the source, provide a link to the Creative Commons license and indicate if changes were made.

The images or other third party material in this chapter are included in the chapter's Creative Commons license, unless indicated otherwise in a credit line to the material. If material is not included in the chapter's Creative Commons license and your intended use is not permitted by statutory regulation or exceeds the permitted use, you will need to obtain permission directly from the copyright holder.

CHAPTER 14

Improved Resilience of Critical Infrastructures Against Large-Scale Transnational and Systemic Risks

Gabriele Giunta, Carmen Stira, Jolanda Modic,
Denis Čaleta, Theodoros Semertzidis, Theodore Zahariadis,
and Charalampos Skianis

BACKGROUND

Reliable operation of Critical Infrastructures (CIs) is a pre-requisite for the integrity and resilience of vital elements in our society that help to ensure the security, well-being and economic prosperity of Europe, its citizens and businesses. Nowadays, CIs have become very complex, operating in a rapidly evolving societal, technological and business environment. Moreover, since CIs are becoming more interconnected and

G. Giunta (✉) • C. Stira
Engineering Ingegneria Informatica Spa, Palermo, Italy
e-mail: Gabriele.Giunta@eng.it

J. Modic • D. Čaleta
Institute for Corporative Security Studies, Ljubljana, Slovenia

© The Author(s) 2025
169
I. Gkotsis et al. (eds.), *Paradigms on Technology Development for Security Practitioners*, Security Informatics and Law Enforcement, https://doi.org/10.1007/978-3-031-62083-6_14

reliant upon one another, disruptions in one CI can have severe and long-lasting cascading effects in other CIs that are essential for the continuity of critical societal and economic activities, even in multiple sectors and countries. This increases the attack surface as well as the scale and significance of the impacts of attacks. Growing digitalisation and interconnectedness of CIs, based on novel technologies and technologically complicated applications, may result in the emergence of new risks, generating new vulnerabilities, including those carried through people and employees, either intentionally through insider threats or through human errors and social engineering.

In this emerging safety-security landscape, the protection and resilience of CIs are of paramount significance. Critical Infrastructures in Europe are increasingly becoming the targets of new categories of hybrid threats and attacks powered by technological innovations. However, limited research has been conducted on large-scale, transnational and cross-domain coordinated attacks. More importantly, large-scale vulnerability assessment and systemic risks analysis of CIs, considering the risks derived by major man-made or natural hazards and complex Cyber-Physical-Human (CPH) threats as well as consequences of the entire system collapse, have never been addressed before. A fundamental challenge to governing systemic risks is to understand the system as a complex network of individual and institutional actors with different and often conflicting interests, values and worldviews.

The ATLANTIS project evaluates and addresses systemic risks against major natural hazards and complex attacks that could potentially disrupt vital functions of the European society. The mission of ATLANTIS is to improve the resilience of the interconnected CIs in Europe exposed to ever-evolving, existing and emerging, large-scale, combined, CPH threats and hazards. By providing future-proof, sustainable security solutions, ATLANTIS supports public and private actors in guaranteeing the continuity of vital operations while minimising cascading effects in the infrastructure itself, the environment, other CIs and the involved population.

T. Semertzidis
Centre for Research and Technology Hellas, Thessaloniki, Greece

T. Zahariadis • C. Skianis
Synelixis Solutions SA, Athens, Greece

The mission of ATLANTIS will be achieved by pursuing the following four (4) Strategic Goals (SG): (1) improving *knowledge* on large-scale vulnerability assessment and long-term systemic risk management in CIs; (2) improving the *systemic resilience* of CIs in Europe through novel, adaptive, flexible and customisable security solutions based on AI; (3) facilitating effective *cooperation* among CI operators and authorities while preserving CI autonomy and sovereignty; and (4) delivering *AI-based solutions* (TRL7) for increased awareness, capability and cooperation in managing systemic threats.

OBJECTIVES AND EXPECTED OUTCOMES

The fundamental objective of ATLANTIS is to ensure the continuation of operations while limiting cascading impacts on the infrastructure, the environment and other CIs, as well as on the population concerned. The project intends to allow public and commercial players to adopt sustainable security solutions to solve current and future issues. ATLANTIS will deploy a composite resilience/security indicator-based approach that enables CI operators to make educated choices on CPH risks, systemic threats, natural hazards, catastrophes (including those caused by extreme weather) and their cascading effects. By defining the CI assets/systems and by identifying/classifying their susceptibilities to systemic, critical and hybrid risks, the approach and indicators will produce a paradigm shift in systemic risk management.

ATLANTIS aspires to transition from the operator level to a fully integrated European protection strategy, broadening the vision beyond the boundaries of individual CI (or even individual assets) and placing a greater emphasis on societal and cross-operator concerns at the EU level. ATLANTIS will necessitate the openness and accountability of the risk management procedure, as well as its efficacy and long-term viability. Understanding the system as a complex network of human and institutional players with varied and sometimes opposing interests, beliefs and worldviews is essential to regulating systemic risk in a networked situation.

The goal of ATLANTIS is to address the growing complexity of cross-CI, cross-sector and cross-border CI interdependence. It intends to facilitate the modelling of high-impact strikes with potentially devastating repercussions and the investigation of their cascading ramifications, which is impossible in real-world contexts. In addition, ATLANTIS enables operators and stakeholders to assess the consequences of potential and ongoing attacks either within the CI under consideration or on other, interconnected CIs in Europe by identifying criteria to assess the CPH risk, forecasting emerging and future systemic threats, and analysing cascading/interconnection effects. It also increases the risk management capability by fostering a better risk and resilience assessment culture and effective risk-related knowledge exchange among all stakeholders, seeing risk assessment from an intra-sector, cross-sector and cross-country perspective.

ATLANTIS will be validated and demonstrated in three large-scale cross-border and cross-sector pilots (LSPs), with a focus on improving the security of the information exchange at different levels of operation: inside individual CIs, across CIs in a national security environment and across borders between CI operators. The three (3) large-scale cross-border and cross-sector pilots are the following:

- LSP#1: *Cross-Border/Cross-Domain LSP in Transport, Energy, and Telecoms* (Fig. 14.1). Validation in (i) multimodal cross-country transport encompassing sea transport with two international seaports, rail transport with two national railway operators, and road transport with a national highway operator, (ii) energy (oil) and (iii)

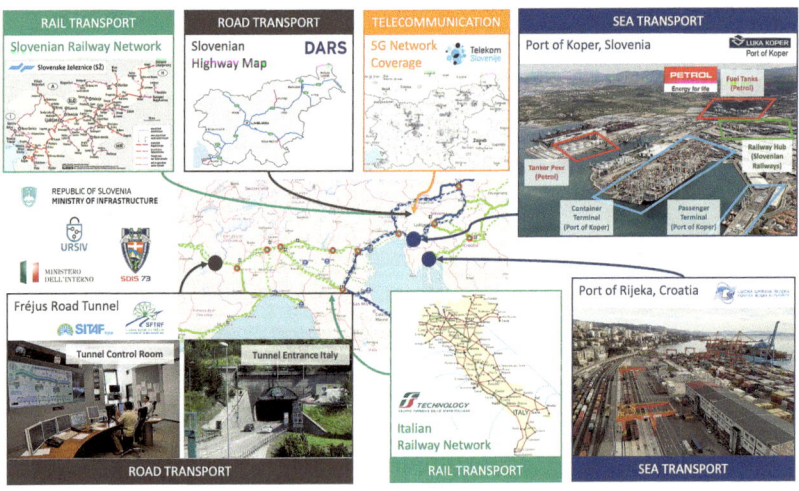

Fig. 14.1 Cross-Border/Cross Domain Large-Scale Pilot in Transport, Energy, and Telecoms

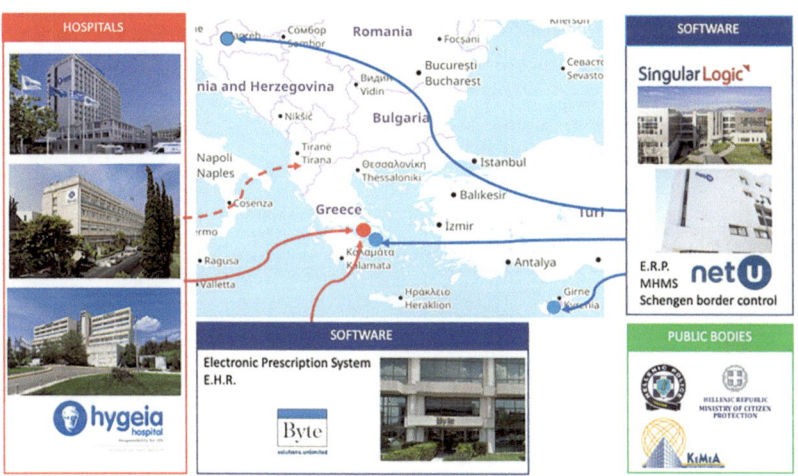

Fig. 14.2 Cross Domain Large-Scale Pilot in Health, Logistics/Supply Chain and Border Control

telecoms in four neighbouring countries: Slovenia, Croatia, Italy and France.

- LSP#2: *Cross-Domain LSP in Health, Logistics/Supply Chain and Border Control* (Fig. 14.2). Validation in (i) the health sector covering physical protection of hospitals and cybersecurity of Electronic Health Records with a group of three hospitals in Greece, (ii) logistics/supply chain covering logistics and Enterprise Resource Planning (ERP) platforms with one of the largest ERP tool providers in Greece and Cyprus and (iii) border control with a focus on the Schengen II Information System for border control of Cyprus, Greece and Croatia.
- LSP#3: *Cross-Country LSP in FinTech/Financial* (Fig. 14.3). Validation in the financial sector covering cybersecurity incidents and systemic threats with an independent investment house, a bank and technology providers specialised in developing technology, infrastructure and business solutions for the financial sector.

The results of ATLANTIS will provide a strong basis for achieving impacts, which will be realised through the various exploitation,

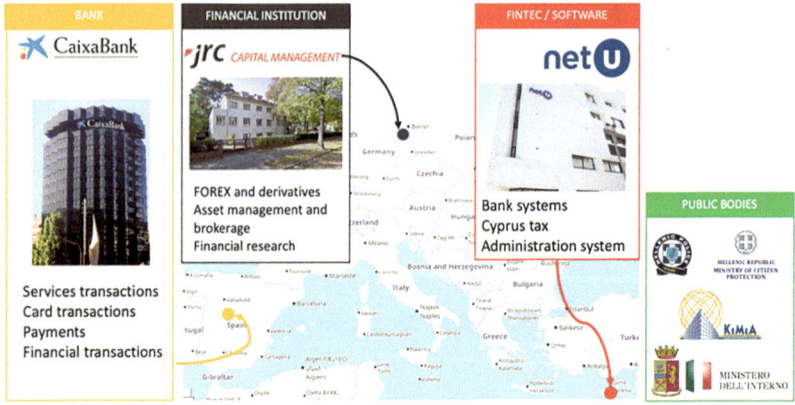

Fig. 14.3 Cross Domain Large-Scale Pilot in FinTech/Financial

dissemination and communication measures. The key expected outcomes of ATLANTIS are as follows:

- EO1: Improved *large-scale vulnerability assessment* of key infrastructures covering six (6) critical sectors in seven (7) EU Member States.
- EO2: Improved *cooperation* to counter Hybrid Threats and subsequent large-scale disruptions of infrastructures in Europe, allowing for operational testing in real scenarios or their realistic simulations with specific regard to the cross-border dimension (intra-EU as well as non-EU).
- EO3: Improved concepts and tools for comprehensive *long-term systemic risk assessments* to European Critical Infrastructure with regard to climate change, technological trends and foreign direct investment (FDI).
- EO4: Improved *risk, vulnerability and complexity related assessments* for cyber-physical CIs aiming to increase security, resilience preventiveness and mitigating against and cascading effects.
- EO5: *Terrestrial back-up/alternative PNT solutions* to ensure continuous operation of CIs in case of the disruption of GNSS services.
- EO6: Enabling the *decentralisation of large infrastructure* to mitigate vulnerability in case of large-scale disruptions.
- EO7: Enhanced *preparedness and response* by the definition of operational procedures of both private and public infrastructure operators

as well as public authorities considering citizens' involvement (needs and vulnerabilities) in case of large-scale infrastructure disruptions also with a view of assessing the combined physical and cyber resilience.

CONCEPT AND APPROACH

The key challenges of security, resilience and privacy need to be encapsulated in a user-driven three-dimensional approach in order to achieve the strategic objectives that lead to holistic and systemic security. The three traditional security elements of technology, processes and humans implement a *Technology-Humans-Process symbiotic relationship*, supplemented by a fourth "node" of collaborative security strategy to create a *3D ATLANTIS security model*. In this "pyramid," the technology is specifically assigned to develop and implement tools focused on the protection of CI, which requires advancing the technology itself, but also improving the collaboration between vendors and users to achieve optimal security. Processes represent explicit, formal means by which CI operation is performed. It requires support from technology but also formal governance and policies that should involve all levels within an organisation. Humans need to actively participate in the security process, not only as technology operators but also as human sensors enabling the technology in a *coexistent Intelligence Amplification approach*. At the pyramid's top is the collaborative security, which focuses on the need to architect technological solutions that create competitive advantages, enable knowledge sharing and collaborated risk mitigation, imposed governance structures and policies that go beyond a single CI, domain or country to collaborative security and a culture that elevates security to the first priority, pervasive throughout the CI organisation. ATLANTIS utilises and extends technology, processes and humans though it primarily focuses on the collaborative security to offer cross-CI systemic security.

ATLANTIS forms a three-layered architecture:

Layer 1 is the *CI-Specific Incident Detection System (IDS)* or civilian *Command, Control, Communications, Computers and Intelligence (C4I)* systems. It is responsible for CI-specific information gathering and CI observation, including physical (i.e. sensors, video surveillance, drones), cyber (i.e. complex data from PLC, SCADA, IDC and network connectivity systems such as routers and firewalls) and *Humans in Vicinity (HiVIC)* as human sensors.

Layer 2 is responsible for local incidents processing, systemic risk patterns extraction, situation awareness, threat prediction/early-detection and automatic countermeasures' enforcement. *Situation Awareness & Comprehension Framework (ACF)* enables ATLANTIS to analyse the environment, detect abnormal events or patterns and predict or foresee long term and systemic risks and understand their size, magnitude and severity. The core of the framework is a hybrid *CI Digital Twin (DT)* component that co-models the physical and cyber CI and enables *Systemic and Continuous Risk Analysis* based on machine learning (ML). It is important to underline that for confidential or classified datasets, model training may be quite complicated as ML training may implicitly leak information about the training data. Thus, ATLANTIS adopts a novel confidentiality preserving Federated ML (FML) [1, 2] model training framework, while zero-knowledge proof [3] is used to validate that the ML training has been realised on a proper dataset, without revealing, disclosing, moving or copying any data of the original training dataset, and thus reducing the risk of information leakage, along with any legacy constraints. Moreover, ACF employs *Intelligence Amplification* (IA) [4] to actively engage the humans (i.e. operators, stakeholders, HiVIC) and the processes (i.e. operation, control, maintenance), in *Technology-Humans-Process symbiotic relationship* implementing "*Situation Awareness.*" Results are forwarded to the ACF *Decision Support System* (DSS) which based on a multi-criteria decision support tool and ATLANTIS FML framework, aims to detect/predict short-term incidents (that occurred or are about to happen) or foresee/forecast long-term (natural or manmade) systemic risks, implementing "*Situation Comprehension.*" In addition, *Risk Reduction & Incidents Mitigation Framework (RRIM)* enables incident mitigation and countermeasures enforcement via several technology innovations. Based on ACF output, the *Risk Reduction DSS* will propose several mitigation actions and countermeasures considering specific Service Level Agreements (SLAs), priorities and the Physical Security as a Service (PSaaS) business model in order to minimise downtime, reduced production and cascading effects. Mitigation strategies include *information provision to the CI operators* for preventive maintenance and detection of problems before they happen, location of problems (e.g., cracks), intruders' presence in critical areas using Copernicus data for security application, *information provision to the citizens in vicinity* including early and trusted warnings using secure smartphones applications.

Layer 3 is a federated *Cross-CI collaborative Knowledge Sharing, Risk Assessment, State Awareness and Incidents Mitigation (CCI-SAAM)* platform between collaborative cross-border and cross-domain CIs. ATLANTIS collaborative risk mitigation system targets not only a single CI but also aims to be able to trigger simultaneous risk assessment and collaborative management of cyber-physical threats over multiple inter-connected CIs. Starting from H2020 PHOENIX [5] *Incidents Information Sharing Platform (I2SP)*, which collects and shares data from various energy utilities, and platforms such as MISP [6] and MeliCERTes [7], ATLANTIS creates a CCI-SAAM platform that covers the compete for CI ecosystem and coordinates risk assessment, state awareness, collaborative incidents mitigation and countermeasures enforcement over multiple cross-connected CI, such as Energy 5G telecoms and ground satellite segment.

ATLANTIS offers interconnection with external systems (e.g., other IDS and CI Security Systems, weather forecast, Earth Observation systems) along with external Information Sharing and Analysis Centers (ISACs) (e.g., MeliCERTes).

Within ATLANTIS, we strongly support that a *case-by-case techno-economic analysis is needed* for each site to match the costs with the severity/cascading effects of a potential incident or accident. Thus, the ATLANTIS toolbox is quite modular and different components may be applied per CI site. To meet the resiliency, survivability, high availability, and minimal delay requirement for systemic security, we consider moving the ATLANTIS framework to the *micro-services mesh model*. Moreover, for increased survivability and fast recovery in case of mass attacks or large-scale incidents, we are studying reallocation of critical control functions to the *5G Edge Cloud* thus utilising the 5G inherently enabled sophisticated security (via e.g. intensive processing and scaling by default) and resiliency features (e.g. network slicing, offloading to Mobile Edge Cloud).

TOOLS AND IMPLEMENTATION

ATLANTIS relies on several preventive, protective and remediation technologies. Specifically, it defines the ATLANTIS *"Cross-CI Risk Assessment & Incidents Mitigation Strategies"* to consistently identify the security systemic threats related to natural hazards, physical and cyberattacks. Moreover, it implements "*Preventive Technologies to reduce systemic risk by design*" realising Earth Observation (EO) and physical protection, resiliency and self-healing, resilient Positioning, Navigation and Timing (PNT) services and geolocation as GNSS [8] fall-back and Information & metadata Traceability using inter-DLT [9] technology. In addition, it implements "*Protective Technologies to reduce systemic risks by innovation*" including a Unified Monitoring System (UMS), tools to fight disinformation, the Situation Awareness & Comprehension Framework (ACF), Systemic Risks Foresight and Incidents Detection DSS and the Risk Reduction & Incident Mitigation Framework (RRIM), along with solutions to for "Humans in Vicinity Sensing and Engagement" (HiVIC). Finally, it implements "*Cooperative prevention, anticipation and mitigation of systemic risks*" including strategies & tools for cooperative remediation, mitigation and response along with the cross-domain, cross-CI, cross-border knowledge sharing, risk assessment, threat analysis and countermeasures mitigation (CCI-SAAM) platform. A DevSecOps continuous integration approach, setup is realised to perform real-life Cross-CI LSP validation & penetration testing.

ATLANTIS follows an agile and incremental approach of iteration cycles. In the *first phase*, the aim is to build the baseline solution and provide the initial ATLANTIS proof of concept, by integrating and validating all key components and tools. The *second phase* of implementation will increase the TRLs and introduce additional functionalities, where stronger integration with 5G networks will be realised and validated in the LSPs. Finally, the *third phase* will mature the solutions and focus on validation and optimisation in realistic conditions, while the impact creation and stakeholder engagement activities will be strengthened.

To achieve these goals, a Continuous Integration/ Continuous Deployment/Continuous Piloting (CI/CD/CP) will be followed, based on SCRUM [10] methodology, allowing concurrent research, design, development, integration, deployment, testing, validation and qualification throughout the whole project, gradually providing an increasingly refined set of features, ultimately delivering the measurable KPIs as defined in the ATLANTIS objectives (as per section "Objectives and Expected Outcomes").

IMPACT AND TARGETED END USERS

The importance of CIs' resilience against large-scale transnational and systemic risks cannot be overstated. It is not merely an option for CI operators, but a strategic imperative for governments, policymakers and other stakeholders. By investing in resilient CIs, countries can safeguard their national security, protect their economies, ensure the well-being for their society and address environmental challenges.

Increasing International and Cross-Sectorial Collaboration

Our *collaborative, cross-organisational/-sectorial/-border approach to vulnerability assessment and anticipation of systemic risks to the CIs in Europe* allows for a comprehensive long-term risk assessment concerning various challenges. The assessment includes the analysis of the impact of climate change and increasingly frequent and severe natural disasters, as well as the relentless pace of technological advancements. The CI operators often struggle to keep up with the evolving technologies, while malicious attackers continually refine their tactics and capabilities, making it increasingly challenging to safeguard the infrastructure effectively. Additionally, this risk assessment considers other crucial factors, such as the EU's dependence on critical supplies from non-EU countries, as well as human factors like ageing population, lack of skills and even potential acts of sabotage.

By jointly and holistically identifying and understanding these complex and interrelated issues, we can (i) better equip the CI operators, technology developers and researchers working in this field to adequately protect and safeguard our vital infrastructure from evolving threats and (ii) support governments and policymakers to make informed decisions when developing new regulatory mechanisms.

Facilitating Coordinated and Effective Protection and Response

By having a unified understanding of (i) interdependencies, (ii) pertaining systemic risks, hybrid threats, and natural hazards and (iii) the possible large-scale, pan-European cascading effects of local disruptions, organisations across Europe representing different sectors can *cooperate in countering these threats more efficiently and effectively*. On a strategic as well as an operational level. Namely, apart from facilitating more effective communication and information sharing among the CI stakeholders, ATLANTIS is also fostering and driving business innovation within the CI sector by *developing and offering cutting-edge solutions for the protection, response and recovery from incidents and attacks.* These solutions are the result of collaborative efforts and support from diverse stakeholders across Europe. By tapping into the knowledge and best practices shared by representatives from different economies, geographies, cultures, backgrounds, interests and worldviews, ATLANTIS is creating a powerful synergy that not only actively supports the CI operators and other stakeholders (e.g., Civil Protection Agencies, Law Enforcement Agencies, CERTs) in enforcing better security mechanisms and approaches but also drives advancements and improvements in the CI domain in terms of research and development.

Improving Security, Well-Being, Skills and Opportunities for EU Citizens

In its innovation, ATLANTIS is taking a *user-driven approach* to encapsulate the needs and expectations of society, business owners, regulators and policymakers, as well as the skills of the CI employees. With this, ATLANTIS is ensuring that the delivered solutions will be easily used by the CI operators and offer wider benefits to the entire society in terms of safety, security, well-being and quality of life. Moreover, with cutting-edge technologies, ATLANTIS is also creating new fields of investment and generating new employment opportunities.

Embracing the ATLANTIS approach forms a resilient foundation upon which *societies and economies can thrive,* even in the face of unprecedented challenges.

CONCLUSIONS

The envisioned ATLANTIS system is intended to be used by critical infrastructures in order to increase their *awareness, capability* and *cooperation* in managing systemic threats to their physical and digital security considered as the pillars of the new legal framework for critical infrastructures, as known as the Directive on Resilience of Essential Entities (CER Directive) [11] and the revised Network and Information Systems (NIS2) Directive [12].

Consequently, the mission for ATLANTIS is to create a platform and tools (the ATLANTIS system) capable of improving the resilience and the security of interconnected critical infrastructures and ensuring the continuity of their operations, while minimising potential cascading effects. The ATLANTIS system is also intended to enable public and private actors to adopt sustainable solutions which allow them to improve (1) their knowledge on vulnerability assessment and long-term systemic risks; (2) their systemic resilience through customisable security measures ("by design") and tools ("by innovation"); and, (3) their effective cooperation with other critical infrastructures as well as with government security stakeholders.

Acknowledgements This project has received funding from the European Union's Horizon Europe research and innovation programme under grant agreement No. 101073909. This article reflects only the authors' views and the Research Executive Agency and the European Commission are not responsible for any use that may be made of the information it contains.

References

1. Seeliger, A., Pfaff, M., & Krcmar, H. (2019). Semantic web technologies for explainable machine learning models: A literature review. In *CEUR Workshop Proceedings*.
2. Zhang, Q. S., & Zhu, S. C. (2019). Visual interpretability for deep learning: A survey. *Frontiers of Information Technology and Electronic Engineering, 19*, 27–39.
3. Goldreich, O., & Oren, Y. (1994). Definitions and properties of zero-knowledge proof systems. *Journal of Cryptology, 7*, 1–32.
4. Dobrkovic, A., Liu, L., Iacob, M.-E., & van Hillegersberg, J. (2016). Intelligence amplification framework for enhancing scheduling processes. In *BERAMIA 2016* (pp. 89–100). Springer.
5. https://phoenix-h2020.eu/
6. https://www.misp-project.org/
7. https://github.com/melicertes/csp
8. Zhu, N., Marais, J., Bétaille, D., & Berbineau, M. (2018). GNSS position integrity in urban environments: A review of literature. *IEEE Transactions on Intelligent Transportation Systems, 19*(9), 2762–2778.
9. Chowdhury, M. J. M., Ferdous, M. S., Biswas, K., Chowdhury, N., Kayes, A. S. M., Alazab, M., & Watters, P. (2019). A comparative analysis of distributed ledger technology platforms. *IEEE Access, 7*, 167930–167943.

10. Schwaber, K. (1997). Scrum development process. In *Business object design and implementation: OOPSLA'95 workshop proceedings 16 October 1995, Austin, Texas*. Springer.
11. European Commission. (2022). Directive (EU) 2022/2557 of the European Parliament and of the Council of 14.12.2022 on the resilience of critical entities and repealing Council Directive 2008/114/EC (CER Directive). *Official Journal of the European Union, L 333*, 164.
12. European Commission. (2022). Directive (EU) 2022/2555 of the European Parliament and of the Council of 14 December 2022 on measures for a high common level of cybersecurity across the Union, amending Regulation (EU) No 910/2014 and Directive (EU) 2018/1972, and repealing Directive (EU) 2016/1148 (NIS 2 Directive). *Official Journal of the European Union, L 333*, 80.

Open Access   This chapter is licensed under the terms of the Creative Commons Attribution 4.0 International License (http://creativecommons.org/licenses/by/4.0/), which permits use, sharing, adaptation, distribution and reproduction in any medium or format, as long as you give appropriate credit to the original author(s) and the source, provide a link to the Creative Commons license and indicate if changes were made.

The images or other third party material in this chapter are included in the chapter's Creative Commons license, unless indicated otherwise in a credit line to the material. If material is not included in the chapter's Creative Commons license and your intended use is not permitted by statutory regulation or exceeds the permitted use, you will need to obtain permission directly from the copyright holder.

CHAPTER 15

Supporting Maintenance Tasks and Upgrading Roadworks Through an Integrated Automated System

Ilias Gkotsis, Aggelos Aggelis, Leonidas Perlepes,
Antonis Kostaridis, Theodora Karali, Dimitrios Bilionis,
Stefanos Camarinopoulos, Yannis Handanos,
and Solon Molcho

INTRODUCTION

Given the increasing interconnection among infrastructures, networks, and operators delivering essential services across the internal market, it is necessary to fundamentally reinforce the resilience and operation of the

I. Gkotsis (✉) • A. Aggelis • L. Perlepes • A. Kostaridis
Satways Ltd, Neo Irakleio, Greece
e-mail: i.gkotsis@satways.net

T. Karali • D. Bilionis • S. Camarinopoulos
RISA Sicherheitsanalysen GmbH, Berlin, Germany

Y. Handanos • S. Molcho
Olympia Odos Operation SA, Megara, Greece

© The Author(s) 2025   183
I. Gkotsis et al. (eds.), *Paradigms on Technology Development*
for Security Practitioners, Security Informatics and Law
Enforcement, https://doi.org/10.1007/978-3-031-62083-6_15

critical entities that manage and operate them [1]. One of the main infrastructure domains that provide essential services is that of transport, which can result in significant and lasting degradation of ecosystems and habitats. Considering that Europe has the highest transport infrastructure density in the world, there is an urgent need to address this rapidly increasing challenge. Road infrastructure tops the list of all transport infrastructures and public assets. Roads are crucial for economic development and growth, providing access to education, health, and employment. The maintenance, repair, and upgrade of roads is one of the most important parts of their high-level service provision and business continuity.

At a time of zero tolerance (zero accidents, zero operating restrictions, etc.), it is increasingly necessary to control risks and to improve the knowledge of the condition of structures to organize preventive and/or predictive maintenance that minimizes risks at an acceptable cost. In specific, instrumentation and risk analysis approaches allow to better understand the behavior of structures, to know their condition, and thus to provide reliable input data for robust risk analysis.

To meet the above needs, the HERON project will develop an integrated automated system to perform the maintenance and upgrading roadworks, such as sealing cracks, patching potholes, asphalt rejuvenation, autonomous replacement of CUD elements, and painting markings, but also supporting the pre/post-intervention phase including visual inspections and dispensing and removing traffic cones in an automated and controlled manner. In turn, this will reduce accidents, lower maintenance costs, and increase road network capacity and efficiency.

More specifically, for coordinating maintenance works, the project will design an autonomous ground robotic vehicle supported by autonomous drones. Sensors and scanners for 3D mapping will be used in addition to artificial intelligence toolkits to help coordinate road maintenance and upgrade workflows. All the above components combined with several other technologies will be integrated into a decision support system (DSS), providing decision-makers, operators, and field crew with all the information required to organize their operational procedures and execute successful road inspection and decision-making activities.

The needs and expectations of the road infrastructure operators are presented in the following sections, based on which the overall architecture of the HERON solution has been designed. More details are provided for those components responsible for the decision-making process and the operational picture during the missions.

END USER NEEDS AND EXPECTATIONS—THE GREEK PILOT CASE

The main end users of HERON are ACCIONA[1] and OLYMPIA ODOS[2] the latter of which constitutes a Motorway Concession Project of particular strategic importance on the national and regional level for the development of the Peloponnese and Western Greece, as it connects Athens with North Peloponnese, Western Greece, and the Port of Patras (Fig. 15.1).

OLYMPIA ODOS provides operation and maintenance services, such as toll collection, traffic management & safety and routine maintenance. Their main needs include [2]:

- to prevent its early wear and restore any damage, wear, or malfunction may be presented in an effective and efficient way;
- to operate at a high level of service and to keep a smooth and continuous traffic flow under normal operation conditions, maintain these flow conditions, and minimize delays;
- to zero the incidents with implication of its personnel;
- to save natural resources, prevent pollution and reduce its negative environmental impact, and protect third parties' assets, in the areas of the company's operation.

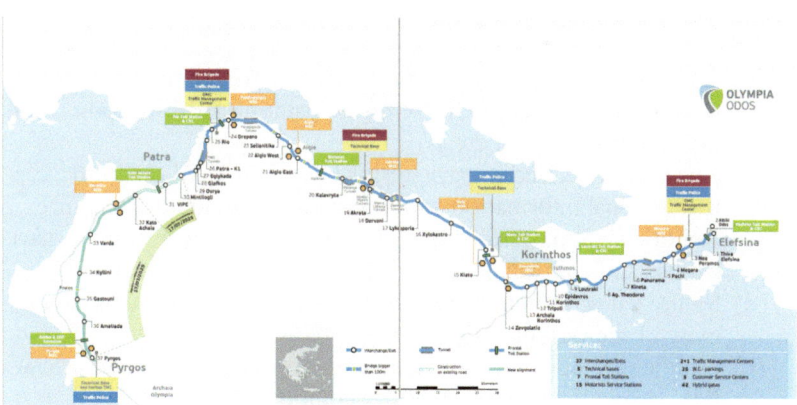

Fig. 15.1 Map of Olympia Odos Motorway

[1] https://www.acciona.com/solutions/
[2] https://www.olympiaodos.gr/

Table 15.1 HERON specifications

Functional specifications	Non-functional specifications
Speed of the ground autonomous vehicle; Autonomy of the drones; Images and video quality; Communications systems; Area of intervention; Operational capacity (i.e., minimum patching materials storage capacity, autonomy, compaction capacity, cutting and milling capacity, pumping/praying/conveying capacity).	Personnel safety; Data integrity; Performance monitoring; Combined inspection and maintenance works for different assets—less traffic disruption as less working time is needed

Based on the end user needs, the following key specifications have been identified (Table 15.1).

Based on the above needs and specifications, the following architecture and technical components have been identified, defined, developed, and will be evaluated through three piloting activities.

HERON SOLUTION

To coordinate maintenance works, HERON project will design an integrated automated system which consists of the following components as further depicted in Fig. 15.2 [3].

(a) UGV actuators perform various actions related to maintenance missions (cone placement/removal, painting road markings, pothole fill, crack repair, etc.)

(b) Sensors, embedded as payloads of UGV and UAV platforms, capturing visual information with regard to the road infrastructure

(c) Robotic system, performing all actions related to planning, navigating, and controlling the robotic system and its actuators

(d) Secure data communication, enabling secure and seamless connectivity of the various HERON subsystems

(e) Middleware & data fusion, enabling information flow and interaction among the various HERON components

(f) Sensing Interface & AI, including data acquisition from the various UGV/UAV inspection sensors, data analysis, and information extraction, to perform object detection and semantic segmentation of the road infrastructure

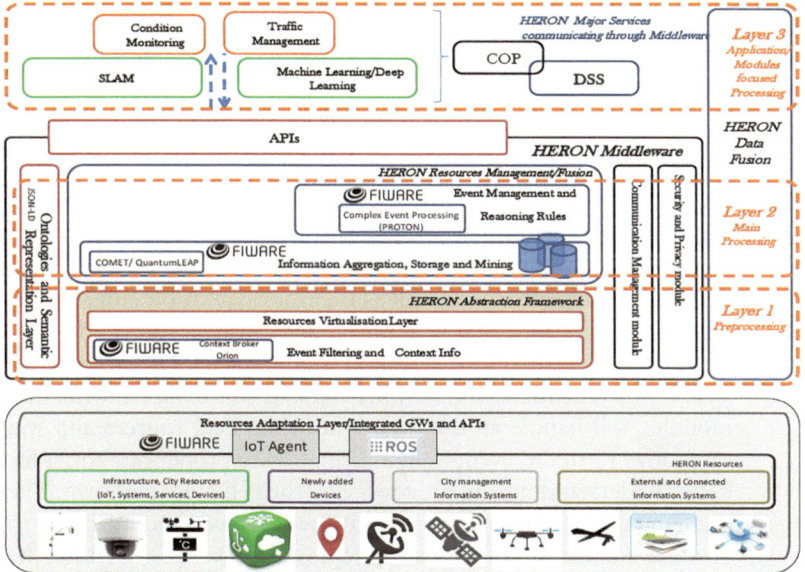

Fig. 15.2 HERON high-level architecture

(g) Incidence Management (IMS) & Decision Support System (DSS), enhanced with Augmented Reality (AR), providing decision-makers, operators, and field crew with all the information required to organize their operational procedures and execute successful road inspection and decision-making activities.

The following sections focus on those components that aim to enhance their situational awareness and execute successful road maintenance missions and decision-making activities.

HERON Middleware

HERON middleware will incorporate information from various HERON components and sensors and will interact with the application layer (Layer 3) in an appropriate format while ensuring high availability and scalability. It consists of two layers:

I. *Layer 1—Preprocessing*: At this layer, the necessary data preprocessing actions take place. Middleware will be accepting information from various sources that will be consolidated in common smart data models keeping a unified data scheme. HERON will utilize available FIWARE data models using JSON Schemata for the so-called key-value representation of context data.

II. *Layer 2—Main Processing*: At this layer, the various data coming from Layer 1 are stored, processed, and served to Layer 3, which constitutes the application level. Layer 2 also includes the virtualization of data into objects as well as their normalization and storage. Moreover, the processing is tightly coupled with resource management to produce further events and handle the data fusion process. Through Event Management, filtering and contextual information modules will handle all the events from the data sources and will categorize retrieved events and information to proper categories for better understanding and processing by the HERON platform. The events and data will be also transformed, stored, and analyzed to produce additional events and data aggregations that will enrich the system. The fused information will be available through a corresponding developed API to the DSS system and the other modules that require additional data. Finally, raw data will be available to all high-level modules and applications for further and more application-driven processing.

Following Fig. 15.3, the Middleware will stand at the core of the HERON system and its ultimate role is to ensure data integrity by accepting and storing sensor data from trusted sources and data security by allowing access only to authorized requests. It includes a tailored policy-based management framework along with suitable enforcement mechanisms dealing with data encryption, access control, privacy, and anonymity. Furthermore, this block will include intrusion detection and prevention mechanisms, such as tools dealing with protocol analysis, detection of anomalous behavior, security events, intrusion detection, vulnerability assessment, and honeypots. It also includes knowledge repositories and distributed threat registries.

The appropriate interfaces/protocols (e.g., sftp, kafka broker, HTTP/HTTPS, MQTT/AMQP) for communication with the different data sources and user services will be created. The module will handle

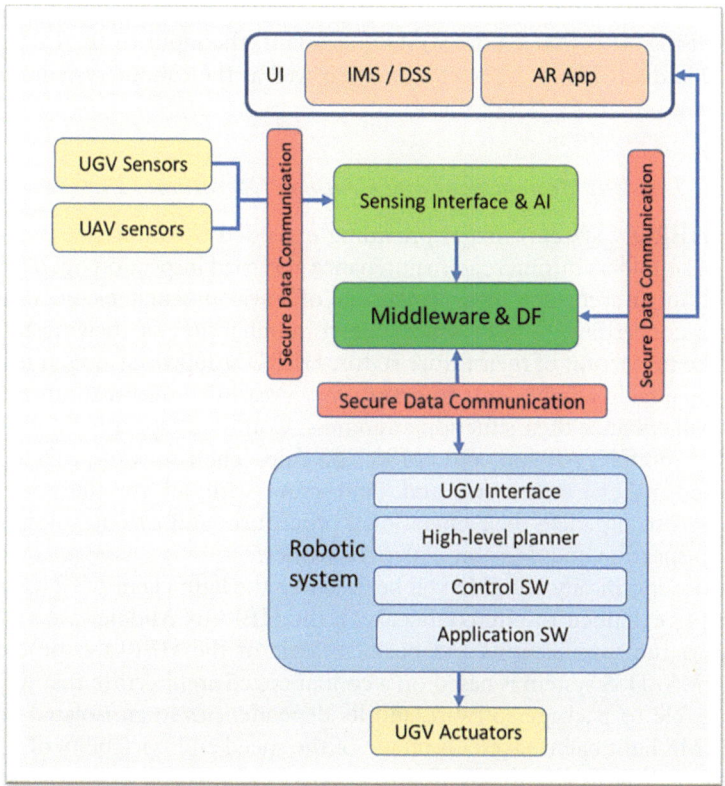

Fig. 15.3 Overall middleware and data fusion architecture of HERON [3]

seamlessly aspects such as time synchronization, scheduling, selection of communication paths, fault tolerance, and traffic shaping.

The main input for the middleware will be the information coming from the Sensing Interface and AI component where the information from the sensors from UGV and UAV is acquired and analyzed, along with data provided directly by the road operator such as inventory data, traffic data, and meteorological data. Especially in OLYMPIA ODOS pilot, various types of data provided by the RI operator, such as Traffic Data, Meteorological Data, data coming from Dynamic Message Signs system and maintenance work data will be transferred to the middleware via an appropriate developed sftp protocol. All this data will be further processed creating the output to IMS/DSS.

Finally, the output produced by the HERON middleware will constitute, using APIs that follow a JSON standard, the input to the IMS and
the DSS discussed in the next section, as well as the robotic system interface (e.g., the UAV/UGV mission).

Incident Management and Common Operational Picture

The HERON system aims at providing extended situation awareness to
key stakeholders during road maintenance and road inspection operations.
Situation awareness is the perception of environmental elements and
events concerning time or space, the comprehension of their meaning,
and the projection of their future status. HERON approach aims at utilizing three main components to grant users a real-time information stream
that will enhance their situation awareness: Common Operational Picture
(COP) module, AR app, and IMS&DSS App. The aim is to provide the
decision-makers, operators, and field crew with all the information
required to organize their operational procedures and execute successful
road inspection and decision-making activities.

More specifically, the IMS will be based on the light client of ENGAGE
IMS [4] extended to interconnect with the HERON Middleware and to
support the specifications and business logic of the HERON use cases.
The IMS/DSS system is based on a containerized architecture that makes
it possible to package software and its dependencies in an isolated unit.
The IMS light client takes advantage of the speed and portability of web-
based applications to display all the necessary information to the user with
all information processing handled on the server-side of the system. The
connections are handled with the reverse proxy server as the most efficient
way to link to a web server from a remote location. This approach protects
the HERON DSS/IMS server from direct connections from an outside
source, performs load balancing, keeps track of requests made, and provides a level of anonymity for the server for cybersecurity purposes. The
storage of the data will be handled with a database server that is capable to
receive heterogeneous data from multiple sources, processing it to obtain
normalized and aggregated data and storing it in a distributed resilient file
system to be ready for consumption by IMS/DSS system. With these
modules, the IMS will generate and share a COP among RI personnel and
relevant road authorities permitting the collaborative response of all
involved relevant local and regional partners when needed. For maintaining effective communication, facilitating the process, and ensuring unity

of effort, the IMS will utilize protocols for multi-level and multi-actors' interaction.

The COP will act as the central and virtual representation of the HERON Robotic platform controller, providing the Robot operators and decision-makers of the RI companies with all the information required to successfully organize their tasks. The various COP elements will be decomposed into information layers and categories to allow for a flexible system that permits the "need-to-know" principle as different users and roles are envisaged to interact with the HERON tools.

The AR system will provide real-time visual information on the surrounding environment of the robot operators. The AR app software will visualize the number of existing defects (automated detection of pavement defects and classification of severity) and will use overlays of 3D models to display possible hidden structural elements which can affect the maintenance process or additional damages. Display of functional elements will be available through appropriate commands as well.

The above components (see Fig. 15.4) will be interconnected with HERON Middleware to support the specifications and business logic of

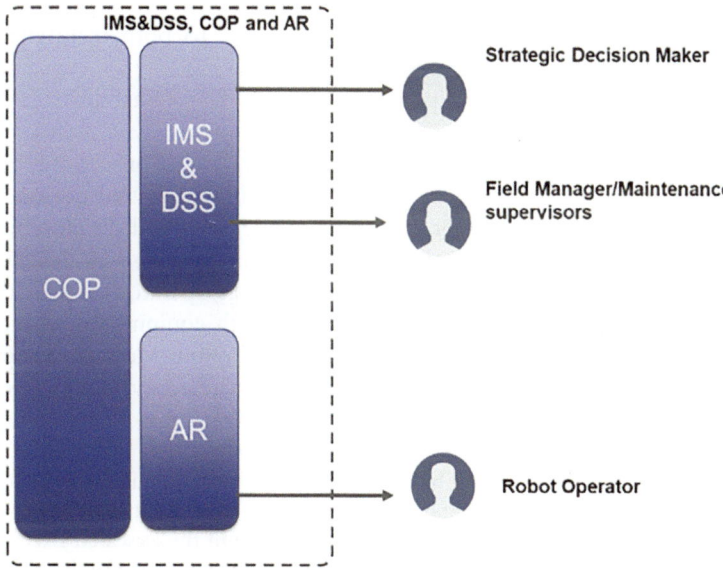

Fig. 15.4 IMS, COP, AR user, and roles

the use cases. Furthermore, the middleware will process data related to UAVs and UGVs (robots) and sensor telemetry data, as well as data from other digital assets in the maintenance field. This data will be processed, analyzed, and used, combined with 3D data of the scene on the field, to provide the necessary information, on the maintenance and inspection, through user interfaces with 3D visualization and GIS capabilities. In addition, it will provide the high-level mission and tasks (through Middleware) to the robot operators for further actions to be accomplished.

CONCLUSIONS

In this paper, the authors present the concept and added value of an integrated and autonomous system for the inspection and maintenance of road infrastructures. In specific, the proposed system called HERON is expected to:

- improve the cost of maintenance activities, by reducing mainly the required human resources;
- reduce the time period of road/lane closures and the relevant road users' annoyance;
- minimize personnel's exposure to risks due to both maintenance activities and adjacent traffic;
- minimize environmental pollution and ensure sustainability.

Respective KPIs will be used as the basis for evaluation procedures both at the Spanish and at the Greek pilot case which are planned during the next (and last) phase of the project. Meanwhile, technical developments are continuously implemented, fulfilling the user needs, as described above, but also integrating the legacy system of the operators to provide a holistic solution through a unique interface. Such equipment and infrastructure include fleet management systems, Variable Message Systems (VMS), Traffic Management Center, CCTV, meteorological sensors, Tunnel monitoring systems, etc.

Acknowledgments This project has received funding from the European Union's Horizon 2020 research and innovation program under grant agreement no. 955356. This article reflects only the authors' views, and the Research Executive Agency and the European Commission are not responsible for any use that may be made of the information it contains.

REFERENCES

1. European Commission. (2020). *Proposal for a Directive of the European Parliament and of the Council of 16 December 2020 on the resilience of critical entities.* Available: https://eur-lex.europa.eu/legal-content/EN/TXT/HTM L/?uri=CELEX:52020PC0829
2. HERON. (2023, July). *D2.1 | End-user needs and KPIs report* [Online]. Available: https://www.heron-h2020.eu/wp-content/uploads/2023/03/ D2.1_EndUserNeedsKPIs_PU.pdf
3. HERON. (2023, July). *D2.2 | Architecture specification* [Online]. Available: https://www.heron-h2020.eu/wp-content/uploads/2023/03/D2.2_ ArchitectureSpecification_CO_Redacted.pdf
4. Satways Ltd. (2023, July). *ENGAGE IMS/CAD* [Online]. Available: https:// satways.net/products-sw/engage-ims-cad/

Open Access This chapter is licensed under the terms of the Creative Commons Attribution 4.0 International License (http://creativecommons.org/licenses/ by/4.0/), which permits use, sharing, adaptation, distribution and reproduction in any medium or format, as long as you give appropriate credit to the original author(s) and the source, provide a link to the Creative Commons license and indicate if changes were made.

The images or other third party material in this chapter are included in the chapter's Creative Commons license, unless indicated otherwise in a credit line to the material. If material is not included in the chapter's Creative Commons license and your intended use is not permitted by statutory regulation or exceeds the permitted use, you will need to obtain permission directly from the copyright holder.

PART III

Better Protect the EU and Its Citizens Against Crime and Terrorism

PERIVALLON: Improved Intelligence Picture and Operational Capacities to Combat Organised Environmental Crime

Eduardo Villamor Medina, Theodora Tsikrika,
Piero Fraternali, Luigi Caldararu, Vasiliki Efstathiou,
Sandra Balbierz, Efstathios Skarlatos, Eva Korenjak,
Anastasios Karakostas, Luca Di Nuovo, Federico Benolli,
Dario Bellingeri, Dries Borloo, Renato Sciunnach,
Jimmy Berggren, Ovidiu Manolache, Ioannis Petropoulos,
Radu Bors, Nir Haimov, Andrew Staniforth,
and Jordan Thompson

E. Villamor Medina (✉)
ETRA Investigación y Desarrollo, SA, Valencia, Spain
e-mail: evillamor.etraid@grupoetra.com

© The Author(s) 2025   197
I. Gkotsis et al. (eds.), *Paradigms on Technology Development*
for Security Practitioners, Security Informatics and Law
Enforcement, https://doi.org/10.1007/978-3-031-62083-6_16

INTRODUCTION

Environmental crime and, more specifically, organised environmental crime are identified as one of the key crime threats faced by the EU, being undeniably on the rise. As part of the EMPACT (2022–2025) priorities [1] and having a 5–7% yearly growth in number of offences [2], environmental crime has turned into one of the leading crimes on the European and global stage. Intentional dumping of polluting substances, illegal disposal of (hazardous) waste, (cross-border) illegal trafficking of waste and illegal trade of hydrofluorocarbons (HFCs) are examples of organised environmental crime. Such forms of crime can be challenging to detect and difficult to investigate by conventional means, highlighting the need for more sophisticated solutions enabling remote identification and

T. Tsikrika
Information Technologies Institute, CERTH, Thessaloniki, Greece

P. Fraternali
Dipartimento di Elettronica Informazione e Bioingegneria, Politecnico di Milano, Milan, Italy

L. Caldararu
Set Mobile Srl, Bucharest, Romania

V. Efstathiou
MarineTraffic, Athens, Greece

S. Balbierz
University of Applied Sciences for Public Service in Bavaria, Department Police, Fürstenfeldbruck, Germany

E. Skarlatos
Center for Security Studies (KEMEA), Hellenic Ministry of Citizen Protection, Athens, Greece

E. Korenjak
Department of Innovation and Digitalisation in Law, University of Vienna, Vienna, Austria

A. Karakostas
DRAXIS ENVIRONMENTAL SA (DRAXIS), Thessaloniki, Greece

L. Di Nuovo
DYLOG HITECH SRL, Torino, Italy

F. Benolli
Fondazione SAFE, Soave, Italy

evidence collection, as well as multimodal analysis and correlation of the information obtained. Moreover, significant disparities among Member States regarding the legal and judicial administration of different forms of environmental crime and their sanctioning, along with the lack of data and comparable EU statistics lead to an incomplete intelligence picture of organised environmental crime activities.

PERIVALLON aims to address these challenges by delivering an environmental crime observatory aiming to provide an improved and comprehensive intelligence picture of organised environmental crime and by developing an environmental crime detection and investigation platform at the forefront of technological innovation, while improving capacity building and international cooperation of security practitioners through enhanced investigation processes. Through this, the capacities of Police Authorities, Border Guards, and National and Regional Authorities will be improved by the means of extensive training, hands-on experience, joint exercises and testing of key technologies in relevant environments, boosting the uptake of

D. Bellingeri
ARPA Lombardia, Milan, Italy

D. Borloo
R&D Department De Watergroep, Brussels, Belgium

R. Sciunnach
Ministerio della Difesa, Arma dei Carabinieri, Rome, Italy

J. Berggren
Swedish Police Authority – National Forensic Centre, Linkoping, Sweden

O. Manolache
General Inspectorate of Border Police, Bucharest, Romania

I. Petropoulos
Hellenic Police HeadQuarters, Athens, Greece

R. Bors
General Inspectorate of Police, Chisinau, Moldova

N. Haimov
Tamar Group LTD., Caesarea, Israel

A. Staniforth
Saher (Europe) OÜ, Pudisoo, Kuusalu Parish, Estonia

J. Thompson
CENTRIC, Sheffield Hallam University, Sheffield, UK

the PERIVALLON technological stack. To this end, the application of PERIVALLON capabilities will be validated in four transnational operational demonstrations, including one EU Agency, as well as authorities from Italy, Greece, Belgium, Sweden, Romania and Moldova.

Second section describes the "Environmental Crime Observatory" and the approach carried out to obtain an improved intelligence picture of organised environmental crime. Section "PERIVALLON Technologies" describes the PERIVALLON platform's high-level architecture and its main components. Section "Conclusions" presents the main conclusions and next steps ahead.

ENVIRONMENTAL CRIME OBSERVATORY

The aim of the environmental crime observatory is to provide an improved and comprehensive intelligence picture of organised environmental crime activities across Europe, the modus operandi of such criminal organisations and networks, both online and offline, as well as comparable EU statistics regarding such types of crime. The goal will be to reveal the different push, pull and facilitating factors in order to provide a more sophisticated picture of the drivers and motivations behind environmental crime. An overview of applicable legislative and judiciary structures, the type of enforcement action, and its effectiveness on the local, national and European level will also be delivered.

Therefore, a holistic approach has been developed which includes desk research, along with questionnaires to relevant stakeholders (e.g., police authorities, border guards, environmental regional and national authorities/agencies, think tanks, NGOs, etc.), as well as to the society at large, with data obtained through the analysis and correlation of available data sources, such as EUROSTAT databases and relevant online sources. To get a comprehensive understanding of the criminological phenomenon, five research clusters have been created:

(i) *Cluster 1: Academic research.* Analysis of scientific papers and handbook articles, such as articles on Green Criminology and Transnational Organised Crime published between 2015 and 2022.

(ii) *Cluster 2: Reports.* Analysis of reports, such as the EUROPOL report on 'Environmental Crime in the Age of Climate Change' (2022) [3], the IPEC report on 'Environmental Crime in Europe' (2015) [4] and the EUROPOL report 'Serious and Organised Crime Threat Assessment' (SOCTA) (2021) [5].

(iii) *Cluster 3: Police Authorities, Border Guards and Regional and National Authorities.* Survey with the stakeholders in the cluster to learn more about the impact and the challenges of environmental crime (e.g., organisational, technological, knowledge and skills) on the law enforcement level.

(iv) *Cluster 4: Legislation.* Analysis of the existing legislative and judiciary structures and bodies on environmental crime on the European and national level.

(v) *Cluster 5: Former EU-funded projects.* Analysis of the results of former EU-funded projects, such as EFFACE [6] and AMBITUS [7].

The outcome of this research will support policy recommendations and inform the development of the PERIVALLON platform that will enhance investigation processes and methodologies. The results will be used to implement decision-support processes and to facilitate a continuous monitoring and review of the current landscape of environmental crime activities in Europe.

Some of the key challenges on addressing environmental crime described in academic research, as well as in reports by relevant authorities include the following characteristics: (a) victimless, (b) grades of invisibility, (c) 'low risk–high profit'. These three characteristics reflect the 'nature' of environmental crime. Does 'victimless' mean that water, air, earth, and soil, as well as climate, flora and fauna are not victims in a classical sense of human victims who are able to report an offence? Does 'invisibility' mean that damages caused by environmental crime acts are 'often part of an accumulative process' that is not 'immediately visible' [4]? In addition to that, environmental crime activities can be very profitable and less risky for offenders—shaped by low detection rates and low sanctions. These characteristics of environmental crime create challenges for the investigation processes and call for further developments of technologies detecting incidents at an early stage and decision support tools to identify different levels of risk, harm and types of causes.

One idea stemming from academic research is to use the 'conceptual framework for environmental horizon scanning' [8] as a tool to assess and analyse risks and harms. Although environmental crime is complex in nature and its consequences on human and non-human health are difficult to measure, PERIVALLON also aims to focus on the impact on the social level. These frameworks will thus be considered to perform the analysis that will inform the development of the PERIVALLON technologies, discussed next.

PERIVALLON Technologies

PERIVALLON starts with approximately 28 components, most around TRL 5 (technology validated in relevant environment) and aims to deliver them in TRL 7 (system prototype demonstration in operational environment). These components will be integrated to build a unique platform providing a single-entry point for the end users: the PERIVALLON platform.

PERIVALLON Platform

The PERIVALLON platform integrates a collection of components to a single-entry point delivered to end users that exploits the latest advancements in artificial intelligence (AI) in the fields of geospatial intelligence, remote sensing, online monitoring and multimodal analytics for combatting organised environmental crime. As described in Fig. 16.1, the PERIVALLON platform builds upon the concept of multidimensional integration of heterogeneous multimodal sensor data.

The capabilities of the PERIVALLON platform include automatic detection of waste disposal and pollutants on land and water based on

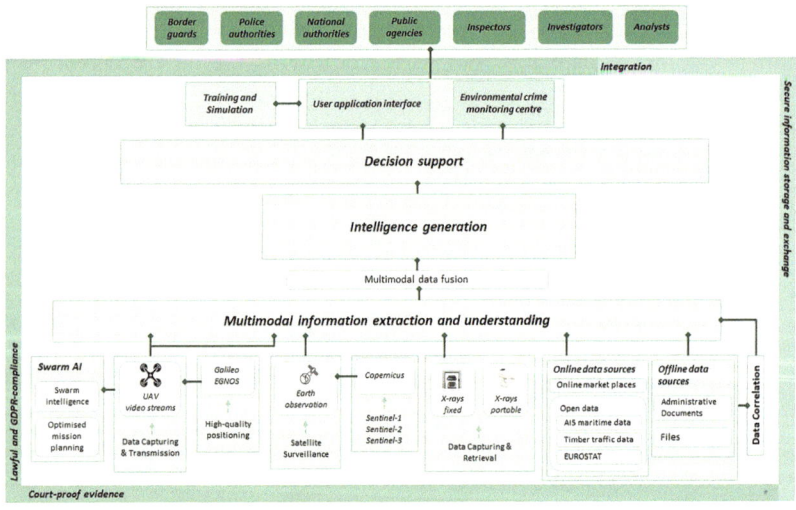

Fig. 16.1 PERIVALLON platform architecture

satellite imagery, optimal inspection and characterisation of sites of interest based on imagery captured by (swarms of) Unmanned Aerial Vehicles (UAVs), optimised X-ray scanning of concealed objects, multimedia-multilingual online content monitoring and analysis, maritime routes prediction, pattern recognition, real-time risk assessment, predictive analytics, audit trail and secure evidence collection and exchange, and holistic situational awareness.

Multidimensional integration of multimodal sensor data, ranging from satellite images, video streams from cameras mounted on UAVs, to information gathered from publicly available online sources and related administrative documents, is at the core of the PERIVALLON platform. Through the analysis and correlation of such multimodal information, the platform will provide explainable decision support to all relevant security practitioners towards detecting, investigating and preventing environmental crimes. Moreover, international cooperation and secure evidence collection will be established through improved data sharing and blockchain technologies.

Additionally, the PERIVALLON platform will provide a secure and user-friendly interface that will allow relevant stakeholders, such as law enforcement agencies, environmental organisations, government bodies and researchers, to seamlessly exchange information, insights and best practices. Furthermore, interactive dashboards and visualisations will present complex information in a clear and intuitive manner. This will empower users to gain valuable insights, make informed decisions and effectively communicate findings to stakeholders.

AI-Based Geospatial Intelligence, Remote Sensing and Scanning

Geospatial Intelligence is the discipline that exploits Earth Observation to enhance territory monitoring, e.g., to detect garbage dumped violating waste management laws. PERIVALLON exploits Geospatial Intelligence by designing, implementing and validating a pipeline for territory monitoring that exploits both remote sensing images, such as the ones collected by the Copernicus satellite constellations, and also images acquired at a short range by means of UAVs. The objective of the use of Geospatial Intelligence in PERIVALLON is to aid environmental agencies, Police Authorities and Border Guards in scanning the territory to detect clues of such criminal activities, such as illegal waste dumping in land and water.

Artificial intelligence and computer vision techniques are used in the design of supervised image processing models for detecting waste items in land and pollutants in water, as shown in Fig. 16.2. This requires the construction of Deep Learning components trained with images annotated by experts with a binary label (waste/no waste) and other information (e.g., the type of material and storage container); a relevant image dataset has

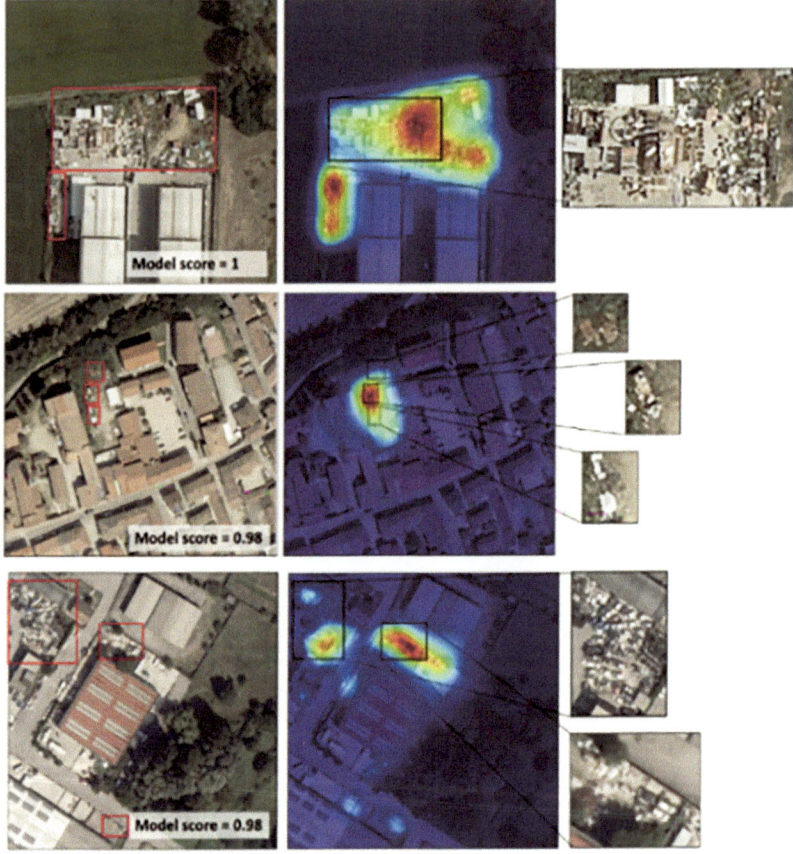

Fig. 16.2 Examples of waste detection in remote sensing images by a Deep Learning model: images annotated by the detector with the confidence value (model score) about the presence of waste (left); images overlaid with Class Activation Maps that highlight the region where the detector has found waste (centre); zoom on the regions of interest containing waste dumps (right)

been made public [9]. The detection approach consists of building binary classifiers for discriminating suspicious sites and multi-label classifiers able to recognise the type of materials. Such a scalable and semi-automatic approach enables the fast detection and prioritisation of sites where the investigation should focus, thus saving time and optimising operations.

Next, short-range images are acquired in UAV missions, with the aim of mapping the terrain in 3D, classifying the type of the visible materials more precisely, also exploiting the European waste codes as categories [10, 11], and quantifying volumes and growth rates. UAV mission management exploits advanced flight control techniques enabling the coordination of drone swarms for better site coverage and evidence acquisition, addressing such challenges as obstacles and no-fly zones.

Finally, the use of computer vision and AI also tackles the analysis of X-ray images, such as those acquired in marine ports and customs, in search of illegally transported materials, such as, for example, containers of ozone-depleting gases. This task required the development of yet another family of image-processing models, coping with the specific characteristics of X-ray imagery.

Online Monitoring for Environmental Crime Detection

PERIVALLON's online monitoring capabilities for detecting environmental crimes encompass several crucial components that work in synergy, as illustrated in Fig. 16.3. These components are designed to acquire

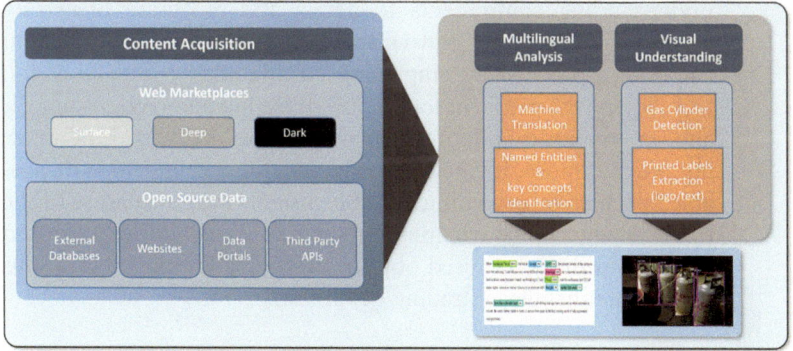

Fig. 16.3 The diagram illustrates the seamless integration of content acquisition, multilingual analysis, and visual understanding components

pertinent content (left), conduct multilingual analysis on the collected information (right), and employ advanced visual understanding (right) techniques to detect potential indicators of environmental crime. Here, we provide a detailed description of each component:

(i) *Content Acquisition*: The PERIVALLON platform employs robust content acquisition components to collect relevant information from diverse data sources. These sources include Surface/Deep/Dark Web marketplaces, external databases, data portals, websites and third-party APIs. Specifically, it focuses on detecting activities such as the production and sale of forged documents, illegal trade of ozone-depleting substances and HFCs, and dissemination of advertisements related to local illegal dumping areas. Continuous monitoring of identified pages, rapid relevance classification of posts, metadata extraction and periodic updates from waste crime data sources ensure the platform proactiveness.

(ii) *Multilingual Analysis*: Leveraging state-of-the-art machine translation techniques, multilingual analysis components enable semantic understanding of collected multilingual textual data. They automatically identify Named Entities and key concepts, disambiguate and resolve co-reference issues associated with concepts and entities, and enhance semantic understanding through online lexical resources. By conducting a comprehensive multilingual analysis, the platform identifies potential indications of environmental crime across languages.

(iii) *Visual Understanding*: PERIVALLON leverages advanced visual understanding components to detect and recognise objects of interest within images and videos collected through the content acquisition process. By employing cutting-edge computer vision techniques, the platform focuses on detecting and recognising gas and/or oxygen cylinders that may be associated with HFC trading. Furthermore, it categorises the identified cylinders as disposable or refillable, providing valuable insights into the nature of the trade. AI algorithms are utilised to extract textual information and logos from the recognised objects, offering additional indications of the potential illegal trading of HFCs.

Maritime Traffic Monitoring for Vessel Route Detection

The *Automatic Identification System* (AIS) is extensively used in the maritime world for the exchange of navigational information between AIS-equipped terminals. PERIVALLON partner MarineTraffic owns an extensive global network of AIS terrestrial receivers that capture vessels within coastal ranges at any given time, complemented by satellite AIS data for areas beyond coastal range.

In the scope of PERIVALLON, vessel mobility data reflected in the AIS-transmitted signal will be used in order to develop data-driven models for representing commonly sailed sea routes, as depicted in Fig. 16.4. The underlying algorithms will leverage vast collections of historical AIS data in order to derive maritime traffic statistics at a fine-grained spatial granularity. The underlying modelling methodology will focus on accurately extracting 'origin to destination' connections and their spatial characteristics at a global scale automatically, without the reliance on any additional information sources (e.g., nautical maps) or a priori knowledge. To this end, the implemented algorithms will succeed in overcoming big data challenges that pertain to huge volumes of uncertain data and transform them into representative models of vessel traffic patterns.

Subsequently, these patterns will be used in order to furnish the PERIVALLON platform with route prediction capabilities and will also serve as the basis of normality upon where deviations may indicate

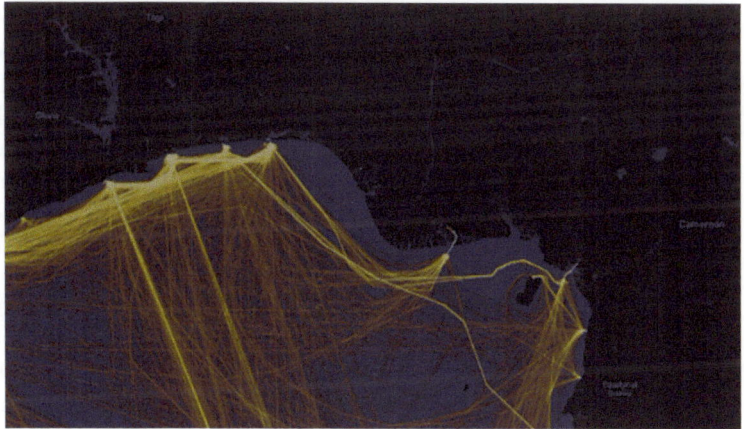

Fig. 16.4   Container Vessel Routes in the Gulf of Guinea

abnormal sailing behaviour. The resulting routes will reflect typical voyages per different vessel categories and will focus on voyages connecting European ports with sub-Saharan African areas of interest, where organised operations of electronic waste smuggling are commonly carried out. Additional gap-filling mechanisms for vessel route reconstruction via vessel detection in satellite imagery will be employed in order to compensate for information loss in cases where vessels cannot be detected via AIS, for instance in cases of intentional AIS switch-off performed by the vessel crew during illicit operations.

Intelligent Decision Support & Secure Information Management

With the ultimate goal of supporting earlier, informed and optimised decisions of security practitioners, the PERIVALLON platform includes Intelligent Decision Support and Secure Information Exchange capabilities that work in an integrated fashion. The related components are developed to (i) perform real-time risk assessment of criminal activities, (ii) identify patterns and trends in multimodal data, (iii) make predictions to anticipate short- and long-term risks, (iv) monitor and analyse environmental crime activities through user-friendly interfaces and (v) store and exchange multiple evidence types with full auditing and chain of custody features.

In particular, the related components are the following:

(i) *Multimodal Fusion and Risk Assessment*: PERIVALLON platform performs the ingestion and transformation of the diverse geospatial, remote sensing, scanning and online data via an ETL (Extraction, Transformation and Loading) process. Based on the insights extracted by the geospatial intelligence (Section "AI-Based Geospatial Intelligence, Remote Sensing and Scanning") and online monitoring (Sections "Online Monitoring for Environmental Crime Detection" and "Maritime Traffic Monitoring for Vessel Route Detection") components, as well as available historical data, a risk assessment module supports the practitioners in the identification, evaluation and prioritisation of criminal activities. Dynamic risk assessment processes, triggered by the assessment of impact and probability of occurrence, formulate optimised mitigation strategies.

Fig. 16.5 Monitoring and analysis of different waste disposal sites on land in the Environmental Crime Monitoring Centre

(ii) *AI-Based Pattern Recognition and Trend Detection*: Leveraging the diverse multimodal data containing spatio-temporal information, computationally efficient AI algorithms are used to reveal hidden correlations, detect irregularities and identify data trends. The component provides red-flag indicators about environmental crime activities and is capable to adapt to the ever-evolving modus operandi of criminal behaviour.

(iii) *Predictive Analytics*: PERIVALLON also develops a proactive approach to crime prevention by exploiting multiple sources of information, such as previous cases, EUROSTAT data, and socio-economic factors, combined with information extracted by tools developed within PERIVALLON to effectively forecast future crime events.

(iv) *Environmental Crime Monitoring Centre*: A user-friendly dashboard will allow practitioners participating in the investigation to better exploit the available information through the geolocation and representation of results of waste detection and visualisation of real-time monitoring, as illustrated in Fig. 16.5. After the analysis, the practitioner will be able generate a report with selected pieces of evidence.

(v) *Secure Information Management, Audit Trail and Evidence Exchange*: The diverse data are stored in a secure database allowing full auditing and chain of custody, alerting the user in case of unusual access patterns. Furthermore, a blockchain-based system provides secure and reliable data exchange between the authorities involved in assessing the evidence of the crime.

Conclusions

This paper introduced the main expected results from the PERIVALLON project, with a key focus on the environmental crime observatory and the main technological components of the integrated PERIVALLON platform, thus presenting an approach to improve the current intelligence picture and initial findings about key factors describing organised environmental crime activities across Europe. Furthermore, the main technological components were described taking into account the provided improvements on the operational capacities of security practitioners. As the project evolves, the work described in this paper will be implemented, demonstrated and evaluated by the practitioners, aiming to provide long-term benefits on their daily practice.

Acknowledgements This project has received funding from the European Union's Horizon Europe research and innovation programme under grant agreement No 101073952. Views and opinions expressed are, however, those of the author(s) only and do not necessarily reflect those of the European Union or the European Research Executive Agency (REA). Neither the European Union nor the granting authority can be held responsible for them.

References

1. Colantoni, L., & Bianchi, M. (2020). *Fighting environmental crime in Europe. Preliminary report.* IAI.
2. EUROPOL. (2022). *EU Policy Cycle – EMPACT* [Online]. Available: https://www.europol.europa.eu/crime-areas-and-statistics/empact.   Accessed April 2023.
3. EUROPOL. (2022). *Environmental crime in the age of climate change* [Online]. Available: https://www.europol.europa.eu/publications-events/publications/environmental-crime-in-age-of-climate-change-2022-threat-assessment

4. EnviCrimeNet and Europol. (2015). *Report on environmental crime in Europe* [Online]. Available: https://www.europol.europa.eu/publications-events/ publications/report-environmental-crime-in-europe
5. EUROPOL. (2021). *Serious and organised crime threat assessment* [Online]. Available: https://www.europol.europa.eu/publications-events/main-reports/socta-report
6. Faure, M., Philipsen, N., del Castillo, T. F., et al. (2016). *European Union Action to Fight Environmental Crime – EFFACE: Conclusions and recommendations.*
7. Colantoni, L., & Bianchi, M. (2020). *Fighting environmental crime in Europe – Preliminary report. AMBITUS project report.* IAI.
8. Heckenberg, D., & White, R. (2020). *Innovative approaches to researching environmental crime.* Routledge.
9. Torres, R. N., & Fraternali, P. (2023). AerialWaste dataset for landfill discovery in aerial and satellite images. *Scientific Data, 10*, 63.
10. 2014/955/EU: Commission Decision of 18 December 2014 amending Decision 2000/532/EC on the list of waste pursuant to Directive 2008/98/EC of the European Parliament and of the Council Text with EEA relevance.
11. Commission of the European Communities – EUROSTAT. (2010). *Guidance on classification of waste according to EWC-Stat categories. Supplement to the Manual for the Implementation of the Regulation (EC). No 2150/2002 on Waste Statistics.*

Open Access   This chapter is licensed under the terms of the Creative Commons Attribution 4.0 International License (http://creativecommons.org/licenses/by/4.0/), which permits use, sharing, adaptation, distribution and reproduction in any medium or format, as long as you give appropriate credit to the original author(s) and the source, provide a link to the Creative Commons license and indicate if changes were made.

The images or other third party material in this chapter are included in the chapter's Creative Commons license, unless indicated otherwise in a credit line to the material. If material is not included in the chapter's Creative Commons license and your intended use is not permitted by statutory regulation or exceeds the permitted use, you will need to obtain permission directly from the copyright holder.

CHAPTER 17

A Semantic Engine for Fighting Cultural Goods Crime

Emmanouil Daskalakis, Theodoros Alexakis,
Nikolaos Peppes, Konstantinos Demestichas,
and Evgenia Adamopoulou

INTRODUCTION

According to Europol [1], cultural goods' crime revolves around three main phenomena:

- Theft, when original cultural goods are robbed from their owners or caretakers;

E. Daskalakis (✉) • T. Alexakis • N. Peppes • E. Adamopoulou
Institute of Communication and Computer Systems, School of Electrical and
Computer Engineering, National Technical University of Athens, Athens, Greece
e-mail: edaskalakis@cn.ntua.gr; talexakis@cn.ntua.gr; npeppes@cn.ntua.gr;
eadam@cn.ntua.gr

K. Demestichas
Department of Agricultural Economics and Rural Development, Agricultural
University of Athens, Athens, Greece
e-mail: cdemest@aua.gr

© The Author(s) 2025   213
I. Gkotsis et al. (eds.), *Paradigms on Technology Development*
for Security Practitioners, Security Informatics and Law
Enforcement, https://doi.org/10.1007/978-3-031-62083-6_17

- Looting, which refers to the removal of ancient relics from archeological sites and old buildings;
- Forgery, which involves the illegal imitation of cultural goods.

Looting and trafficking of cultural property can seriously undermine cultural heritage on a global scale. The revenues from such illicit activities are extremely high and are estimated to be billions of dollars annually. These phenomena are very usual in countries facing crises and conflicts, while the money from such activities is also used, in many cases, to support terrorist activities [2]. On the other hand, the demand for illicit antiquities often comes from economically and politically secure states which fund looters and sellers from less politically/economically secure states. The social political and economic structure of the antiquities market which formulates the way different actors are involved in illicit trades is quite complex [3]. Robbery and trafficking of cultural property are often characterized by traits met in organized crime [4]. Illicit online sales on the dark web using cryptocurrencies, the utilization of false documentation, and person-to-person trade are some examples of how the aforementioned activities take place [5].

Thus, the prevention of looting and illicit trafficking of cultural objects is of paramount importance for the protection and preservation of the global cultural heritage. Fighting such phenomena requires cooperation among different actors as well as a clear understanding of how illicit antiquities and trafficking networks work [6]. Any loss of time is in favor of the looters/traffickers and can seriously undermine the global cultural heritage. The highly promising Artificial Intelligence (AI) and advanced Machine Learning (ML) techniques and tools can increase the research capabilities of Law Enforcement Agencies (LEAs) and drastically reduce the time required for combatting trafficking cases [7].

The semantic engine presented herein will apply rule-based reasoning and will reveal hidden relations relevant to the activities of looters/traffickers. Such relations can be identified in the source and destination points of cultural artifacts, in the traits and activities of illicit traders, in the distribution channels of cultural property, in illicit online listings referring to artifacts, etc. Furthermore, the tool under consideration will create unified graphs which can help LEAs, archaeologists or other practitioners and end users to timely detect suspicious activities related to the illicit trading of antiquities and cultural property. Thus, the envisioned toolset is expected to play an active role in the fight of LEAs against trafficking of cultural heritage.

The remainder of this paper is organized as follows: Section "Related Works" features related research works. Section "Semantic Engine" provides a detailed description of the semantic engine together with a brief overview of the ANCHISE project. Finally, section "Conclusion-Future Work" draws conclusions and discusses future work related to the Semantic Engine.

RELATED WORKS

The usefulness of image analysis for tracking looting marks as well as for detecting illegal excavations is presented by Agapiou et al. in [8]. In a use case implemented in a village of Cyprus, the authors utilized Google Earth images, WorldView-2 images and images from the Cyprus' Department of Land and Surveys. Extensive editing and transformation of images, then, took place for improving the extracted results (e.g., linear histogram enhancements, extraction of vegetation indices, spectral transformations, brightness and contrast corrections). Based on the results of their study, the authors highlighted the capabilities of image analysis in facilitating the detection of looting and illegal excavations. In [9], Tapete and Cigna highlighted how the use of Synthetic Aperture Radar (SAR) images can play a vital role in detecting, monitoring, and assessing the condition of cultural heritage sites due to the impact of certain activities such as looting, mining or agriculture. More specifically the authors explored the use of COSMO-Skymed data whose characteristics (e.g., high spatial resolution, frequent site revisit) make it particularly useful for archaeologists. Different use cases from archeological sites in Syria, Italy, Iraq, and Italy were showcased. The authors noted that Digital Elevation Models (DEM) making use of SAR data can be particularly important for surveying archeological features. Additionally, the spatial and temporal interpretation of SAR backscatter can also play an important role toward this direction. As mentioned before, optical remote sensing plays an important role in the detection of looting. However, it can be ineffective in areas with very dense vegetation. Danese et al. [10] showcased how LIDAR (LIght Detection And Ranging) data can be of particular usefulness in such cases and how it can importantly benefit looting detection cases. Towards this direction, they utilized spatial visualization methodologies as well as the geomorphon method for landform extraction. A relevant use case was implemented in Lazio, Italy, where the authors' methodology yielded very satisfactory results for looting prediction, reaching a 95% detection rate in one of the four test areas.

Solutions for cultural heritage protection encompass large amounts of data which need to be collected and associated. In this light, the generation and adoption of a common representational model or data retrieval, processing, and storage system can be of benefit. Oreste Signore [11] mapped the available technologies and ontologies that can lift the restrictions imposed by the heterogeneity and scarcity of cultural heritage data. According to his work the CIDOC Common Representational Model (CRM), which is based on an ontology, can aid academics, experts as well as LEAs in storing, retrieving, and processing data effectively. Ontologies consist an effective way to deal with heterogeneous data sources. They can capture properties and metadata of a certain artifact or archeological site, as well as complex relationships through predefined classes. Based on the ontology scheme and by examining the need for better annotation and correlation processes, Bobasheva et al. [12] proposed a solution which engaged and combined semantic reasoning and machine learning. Their solution mainly aimed to improve the annotation process and enrich the metadata of artifacts contained in museums in picture format. Even though their solution referred mainly to museum curators, it has many similarities with the solution presented later in this paper. More specifically, they introduced a combination of semantic reasoning techniques alongside machine and deep learning approaches. Their results indicated that this combination could lead to enhanced metadata, automated annotation procedures, and better search results based on visual relevance and search criteria. For their experiments, they used the Joconde knowledge base which was accessed by SPARQL queries [13].

SEMANTIC ENGINE

ANCHISE is a Horizon Europe project, aiming to tackle the following problems: (i) looting and trafficking of cultural goods affecting different countries which can have a devastating impact on the global cultural heritage; (ii) trafficking of cultural heritage adopting new, harder-to-trace, digital methods (e.g., social networks, dark web); and (iii) the fact that data related to cultural heritage protection is dispersed, under various formats and mostly lacking a clear ontological description that would allow large-scale interoperability. In addition to the problems that ANCHISE project is called to address, there are some pertinent end user (experts, academics, LEA officers, etc.) requirements that are envisioned to be addressed. More specifically, end users can benefit from the integration of cross-domain knowledge and expertise in order to lift traditional barriers and exploit available technologies alongside the combination of state-of-the-art technologies in order to create a common framework for fighting looting and trafficking of cultural goods. This leads to another crucial demand of the end users that of proactiveness in looting and trafficking procedures in order to decrease the time required for tackling such criminal acts.

The ANCHISE project aims at offering European societies efficient methods, knowledge, and a toolkit to enhance the protection of cultural heritage against looting and illicit trafficking, by facilitating four main aspects: (1) understand; (2) prevent; (3) act; and (4) repair. More specifically, ANCHISE consists of:

- a hub of social science, politics and economics (for in-depth results likely to lead to structural evolutions in heritage protection),
- a large-scale evaluation of technologies and needs,
- a toolkit of innovative solutions,
- pilot experimentation areas (museums, border control, archeological sites), and,
- a unique and wide network of practitioners.

The aim and the concept of ANCHISE underline the importance of the acquisition, correlation and processing of heterogeneous data in an efficient way. In this light, semantic reasoning techniques and tools will be a major part of the ANCHISE toolkit. Thus, in the next paragraphs, two different components of the semantic technologies, which are currently being developed in the ANCHISE project, are presented, namely:

- ART-CH: An Advanced Reasoning Tool for Fighting Trafficking of Cultural Heritage.
- CTD-TRAC: A Complex Threat Detection Tool for Detecting Illicit Trafficking of Cultural Artefacts.

ART-CH

ART-CH which is presented herein will apply rule-based reasoning with a view to reveal hidden relations relevant to the activities of looters/traffickers. Such relations can be found in the source and destination points of cultural artifacts, in the traits and activities of illicit traders, in the distribution channels of cultural property, in illicit online listings referring to artifacts, etc. Thus, the envisioned tool is expected to play an active role in the fight of LEAs against the trafficking of cultural heritage.

ART-CH, based on the SPARQL Query language, will apply rule-based reasoning in the existing data, will be able to infer logical conclusions from stated facts, and evaluate if these conclusions are complete/consistent. In addition to this, ART-CH will use predefined rules to detect potential relations among different events or entities which may be present in a considered database. ART-CH can drastically increase the investigation and anticipation capabilities of LEAs regarding the illicit trading of cultural property, underpinning activities for identifying the traffickers' modus operandi, the source and destination places of looted or stolen artifacts, the distribution channels of traffickers, illicit marketplaces, flows of cultural property, etc.

The integration of SPARQL queries will enable the user to explore more results in the ontology. Both the predefined rules as well as the SPARQL queries will be further assessed by practitioners, in order to insure that they are suitable for the analysis of a given data input format. Soon after that, the tool will infer useful results for the end user. It should be noted that both logical as well as probabilistic rules can be integrated in ART-CH.

Figure 17.1 presents a high-level architecture of ART-CH.

As a first step, the expert defines the SWRL [14] logical and probabilistic rules. The SWRL rules definition is a crucial part for this methodology

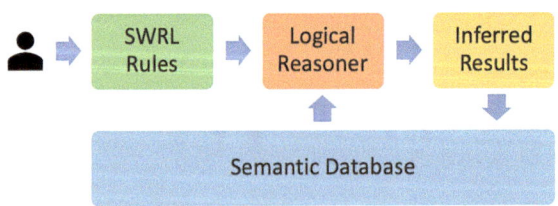

Fig. 17.1 High-level architecture of the ART-CH tool

as in this step the knowledge of experts is practically integrated into the ART-CH tool. In this way, the proposed solution is directly connected with the end users' needs. Then, the Logical Reasoner integrates these rules and applies them on the data stored in the semantic database, which is an ontology. The Logical Reasoner is a component that integrates advanced semantic methods as well as SPARQL queries that reflect the rules proposed by the experts in the previous step. So, through automated processes taking place in the backend, the Logical Reasoner concludes new knowledge deriving from the predefined rules and the data stored in the semantic database. This newly inferred knowledge then is stored back to the semantic database enhancing in this way the previously stored data which at first sight seemed irrelevant and unconnected. This procedure then provides the capability to the end users to gain better insights, discover hidden patterns and correlate different pieces of information in a timely manner and without consuming valuable personnel and time resources.

CTD-TRAC

CTD-TRAC is a threat detection tool, which will generate diverse unified graphs related to illicit cultural heritage trading activities. Through this kind of visualization, end users will be able to easily identify correlations between different entities/activities and detect suspicious illegal trading activities as well as suspects for these activities. In addition to this, the tool will generate alerts and broadcast them via Kafka topics. Through these alerts, LEAs can timely detect traffickers/looters and stop them before they laterally move.

More specifically, the types of inputs which can be used by the tool include, but are not limited to, databases related to the illicit trading of antiquities, illicit listings of cultural property on sites or the social media, coordinates of places where illicit activities frequently take place (e.g., near the borders of countries), data from suspicious financial transactions, travel data of suspects of illegal cultural heritage trading, etc.

The tool can also be connected to other tools (e.g., make use of the inferred results based on the declared rules of a semantic reasoner or make use of the data deriving from a data fusion tool tailored to the needs of cultural trafficking detection). For creating the unified graphs,

CTD-TRAC will integrate weights calculated based on the existing connections between entities/activities and the number of existing alerts.

CTD-TRAC will offer an innovative way of identifying diverse types of suspicious/illegal trading activities which undermine the global cultural heritage. The incidents will be detected in real-time or in near-real-time. The above can potentially increase the investigation capabilities of LEAs under the execution of advanced Machine Learning (ML) algorithms, responsible for identifying structured relationships among the different types of entities/activities involved.

Figure 17.2 contains an indicative screenshot from the User Interface (UI) of CTD-TRAC. In this figure, the connections among different entities are visible, together with some integrated controls which will be available to the user in order to customize the visualization results. It should be noted that the UI may be modified according to the feedback received by end users.

Alpha: Depicts Simulation activity

Center: Shifts the view, so the graph is centered at this location.

x .5

y .3

☑ **Charge:** Attracts (+) or repels (-) nodes to/from each other.

strength -30

distanceMin 1

distanceMax 2000

☑ **Collide:** Prevents nodes from overlapping

strength .6

radius 4

iterations 1

Anastasios Galanis

Crime Commited: FALSO

Anastasios Smyrlis

Crime Commited: True

Fig. 17.2   CTD-TRAC user interface

CONCLUSION-FUTURE WORK

This paper presents the main two components (i.e., ART-CH and CTD-TRAC) of a semantic engine which is currently being developed in the context of the ANCHISE project for fighting the trafficking of cultural property. ART-CH will apply rule-based reasoning and reveal previously unknown relations between the source and destination points of stolen artifacts, among diverse distribution channels, among different activities of traffickers, etc. The tool will be based on the SPARQL Query Language and future developments will encompass the inclusion of more semantic rules tailored to the end users' needs. In addition to this, the system will also generate alerts and will publish them via Kafka topics. CTD-TRAC will create unified graphs illustrating potential illicit trading cases and connections among different entities/activities. It will also generate alerts for suspicious trafficking activities of cultural property. These alerts will be provided in real-time and/or in near-real-time to LEAs in order to stop illicit trading before it is completed. Both components are currently under development and future steps include the integration of more input sources, the generation of more types of real-time alerts, and the completion of an API of the tool.

REFERENCES

1. A Total of 52 Arrests in Operation across 28 Countries Targeting Trafficking in Cultural Goods. Available online: https://www.europol.europa.eu/media-press/newsroom/news/total-of-52-arrests-in-operation-across-28-countries-targeting-trafficking-in-cultural-goods. Accessed 9 May 2023.
2. United Nations International Cooperation against Trafficking in Cultural Property. Available online: https://www.unodc.org/documents/treaties/trafficking_cultural_property/CCPCJ_Resolution_27_5.pdf. Accessed 28 June 2023.
3. Mackenzie, S., Brodie, N., Yates, D., & Tsirogiannis, C. (2019). *Trafficking culture: New directions in researching the global market in illicit antiquities.*. ISBN:978-1-315-53221-9.
4. Dietzler, J. (2013). On 'Organized crime' in the illicit antiquities trade: Moving beyond the definitional debate. *Trends in Organized Crime, 16*, 329–342. https://doi.org/10.1007/s12117-012-9182-0
5. Paul, K. A. (2018). Ancient artifacts vs. digital artifacts: New tools for unmasking the sale of illicit antiquities on the dark web. *Arts, 7*. https://doi.org/10.3390/arts7020012

6. Zeynep, B. (2023). *Fighting the illicit trafficking of cultural property: A toolkit for European Judiciary and Law Enforcement* (Rev ed.). UNESCO. ISBN:978-92-3-100289-2.

7. Abate, D., Paolanti, M., Pierdicca, R., Lampropoulos, A., Toumbas, K., Agapiou, A., Vergis, S., Malinverni, E., Petrides, K., Felicetti, A., et al. (2022). Significance. Stop illicit heritage trafficking with artificial intelligence. *The International Archives of the Photogrammetry, Remote Sensing and Spatial Information Sciences, XLIII-B2–2022*, 729–736. https://doi.org/10.5194/isprs-archives-XLIII-B2-2022-729-2022

8. Agapiou, A., Lysandrou, V., Hadjimitsis, D.G. (2017). Optical Remote Sensing Potentials for Looting Detection. Geosciences 7:. https://doi.org/10.3390/geosciences7040098

Tapete, D., Cigna, F. (2019). COSMO-SkyMed SAR for Detection and Monitoring of Archaeological and Cultural Heritage Sites. Remote Sensing 11:. https://doi.org/10.3390/rs11111326

Danese, M., Gioia, D., Vitale, V., et al. (2022). Pattern Recognition Approach and LiDAR for the Analysis and Mapping of Archaeological Looting: Application to an Etruscan Site. Remote Sensing 14:. https://doi.org/10.3390/rs14071587

9. Signore, O. (2007). *The semantic web and cultural heritage: Ontologies and technologies help in accessing museum information.*

10. Bobasheva, A., Gandon, F., & Precioso, F. (2022). Learning and reasoning for cultural metadata quality: Coupling symbolic AI and machine learning over a semantic web knowledge graph to support museum curators in improving the quality of cultural metadata and information retrieval. *Journal on Computing and Cultural Heritage, 15*. https://doi.org/10.1145/3485844

11. SPARQL 1.1 Overview. https://www.w3.org/TR/sparql11-overview/. Accessed 25 Apr 2023

12. SWRL: A Semantic Web Rule Language Combining OWL and RuleML. Available online: https://www.w3.org/Submission/SWRL/. Accessed 26 Apr 2023.

Open Access This chapter is licensed under the terms of the Creative Commons Attribution 4.0 International License (http://creativecommons.org/licenses/by/4.0/), which permits use, sharing, adaptation, distribution and reproduction in any medium or format, as long as you give appropriate credit to the original author(s) and the source, provide a link to the Creative Commons license and indicate if changes were made.

The images or other third party material in this chapter are included in the chapter's Creative Commons license, unless indicated otherwise in a credit line to the material. If material is not included in the chapter's Creative Commons license and your intended use is not permitted by statutory regulation or exceeds the permitted use, you will need to obtain permission directly from the copyright holder.

CHAPTER 18

Assessing AI Technologies for LEA Use: The ALIGNER Methodology

Donatella Casaburo, Mathilde Jarlsbo, Daniel Lückerath, and Norea Normelli

INTRODUCTION: THE ALIGNER METHODOLOGY

The world is changing at an unprecedented rate, and artificial intelligence (AI) is at the forefront of this change. While providing numerous benefits, many have raised concerns over the impact AI has or will have on matters

D. Casaburo (✉) • N. Normelli
KU Leuven Centre for IT & IP Law – imec, Leuven, Belgium
e-mail: donatella.casaburo@kuleuven.be

M. Jarlsbo
Swedish Defence Research Agency – Totalförsvarets forskningsinstitut, Stockholm, Sweden

D. Lückerath
Fraunhofer Institute for Intelligent Analysis and Information Systems IAIS, Sankt Augustin, Germany

© The Author(s) 2025   225
I. Gkotsis et al. (eds.), *Paradigms on Technology Development for Security Practitioners*, Security Informatics and Law Enforcement, https://doi.org/10.1007/978-3-031-62083-6_18

such as security. The EU-funded ALIGNER[1] project aims to unite European actors who have concerns about AI, law enforcement, and policing to jointly identify and discuss how to enhance Europe's security whereby AI strengthens law enforcement agencies (LEAs) while providing benefits to the public.

To achieve this goal, ALIGNER has established expert groups from policing and law enforcement, civil society, policymaking, research, and industry, who work together in regular workshops to identify opportunities and risks arising from law enforcement use of emerging AI technologies, which capability enhancement needs are associated with the increased use of AI (both by LEAs and criminals), and which ethical, legal, and technical/operational impacts the deployment of AI technologies by LEAs might have.

To provide a basis for the work of ALIGNER's expert groups, a sound methodological approach is necessary for identification of promising (emerging) AI technologies for LEAs and their potential implications in the technological, ethical, legal, and organisational dimensions at LEAs. To the knowledge of the authors, such a comprehensive, integrated assessment approach for AI technologies specific for law enforcement is currently missing. ALIGNER fills this gap by pursuing a collaborative technology watch process, followed by three assessments, as shown in Fig. 18.1. In particular, the assessment phase begins with the *AI Technology Impact Assessment* (section "AI Technology Impact Assessment"); if this leads to significant scores, the *Risk Assessment* follows (section "Risk Assessment"); if this leads to significant scores, the assessment phase ends with the *Fundamental Rights Impact Assessment* (section "Ethical and Legal Assessment"). These assessment approaches are described in detail in the following sections.

AI Technology Impact Assessment

The first step in assessing law enforcement AI technologies is conducting the AI Technology Impact Assessment. By doing this evaluation, LEAs can identify which AI technologies they are most interested in using and prioritise the ones to consider for research or investment.

The ALIGNER AI Technology Impact Assessment and its associated criteria have been inspired by an existing method for assessing

[1] https://aligner-h2020.eu/

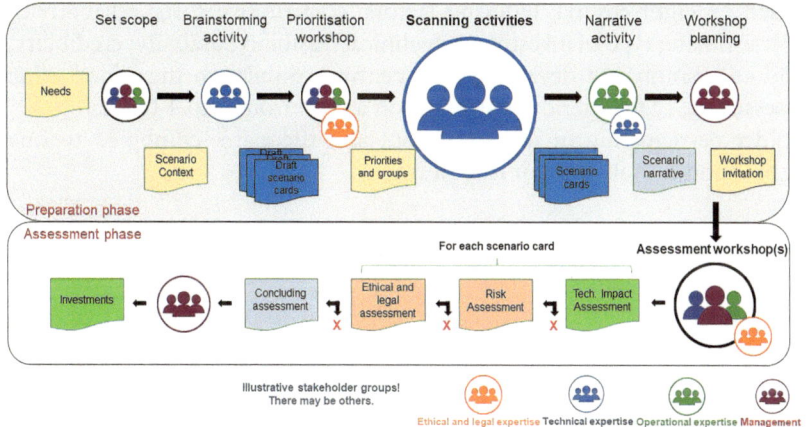

Fig. 18.1 The ALIGNER technology watch and assessment process. In grey: Technology watch process; in white: the assessment process

technologies [1], which already combined an assessment of added value and feasibility to recommend technology investments. However, the ALIGNER AI Technology Impact Assessment further expands this approach: as both added value and feasibility are assessed on discrete and non-linear ordinal scales rating 1–4, the investment recommendations will result from discrete combinations of these assessments. This allows more detailed recommendations, reflecting different levels of potential investments.

The AI Technology Impact Assessment consists of three steps: (i) assessing the added value; (ii) assessing the feasibility of an AI technology; and (iii) recommending AI technology investments [2].

The goal of the *added value* evaluation is to estimate to what degree an AI capability will improve the LEA capacity. It is assessed based on six criteria: need; comparative effectiveness; beneficial side effects; flexibility; future proofness; and future additional potential. Each criterion is assessed on a scale from 1 to 4 by each LEA stakeholder participating in the assessment. The stakeholders' ratings of each criterion are then considered to determine an overall score of added value. LEAs may decide how to combine each individual's rating to one overall and agreed score.

The goal of the *feasibility* assessment is to estimate how likely it is that the AI technology will work as intended and/or advertised. It is assessed based on eight criteria: blocking factors; comparative costs; reliability of cost estimate; type of investment; technical feasibility; usability; credibility; and acquisition and development incentives. Similar to the added value assessment, each criterion is assessed on a scale from 1 to 4 by each stakeholder participating in the assessment and thereafter combined to one agreed and overall score of feasibility.

Example, Workshop Session ALIGNER
The assessment conducted during a workshop in the project ALIGNER can serve as an example for how to determine an overall score. First, each stakeholder assesses each criterion individually by holding up 1 to 4 fingers. If the discrepancy of the highest and lowest score in the group is more than one step, those participants put forward arguments to motivate their stance. Then, the voting process is repeated. The average score is used for added value and feasibility. However, if significant differences still exist, the participants are encouraged to document each side's strongest argument.

The overall score of added value and feasibility are used to recommend AI technology investment decisions. Figure 18.2 shows how the rating for feasibility combined with the rating for added value gives an output L1–6.

As shown in Fig. 18.3, the L-number provides guidance in whether to prioritise the research or development of an AI technology or not. The higher the L-number, the more prioritised the research or investment should be.

Fig. 18.2 Reference table for recommended AI technology investment [2]

		Feasibility			
		1	2	3	4
Added value	1	L1	L1	L2	L2
	2	L1	L3	L4	L5
	3	L3	L4	L5	L6
	4	L4	L5	L6	L6

Investment recommendation	TRL 1-6	TRL 7-9
L1	No research investment for now.	No investment.
L2	No research investment –Put the AI technology on a watch list and revisit in a few years.	No investment – Put the AI technology on a watch list and revisit a few years.
L3	High-risk research – Be prepared to fund a few projects if there is a high potential gain. Put the rest on a watch list and revisit in a few years.	High-risk investment – Be prepared to fund a few projects if there is a high potential gain. Put the rest on a watch list and revisit in a few years.
L4	Sound research – Be prepared for a reasonable investment in R&D, especially if the costs can be shared with other partners.	Suggested investment – Be prepared to invest in R&D to either improve feasibility or find alternative applications.
L5	Priority research – Be prepared to invest in R&D.	Priority investment – Be prepared to prioritise relative to other needs.
L6	High priority research – Be prepared to invest big in R&D.	High priority investment – Be prepared to prioritise it ahead of other projects.

Fig. 18.3 Legend for recommended AI technology investment decision [2]

Risk Assessment

Although AI technologies provide many benefits for LEAs and society in general, there is an urge to evaluate the associated risks. Therefore, AI technologies that have achieved a significant score in the AI Technology Impact Assessment will then go through the Risk Assessment.

The ALIGNER Risk Assessment instrument is still under interdisciplinary development between the technology, ethics, and legal experts of the project, but its main idea is shortly described below.

The Risk Assessment will pursue two main objectives. First, it will identify the risks or unintentional unwanted impacts, posed by AI in the context of law enforcement. Additionally, it will support AI-based technologies improving the EU's resilience against emerging, 'classical' and 'new' AI-supported threats. Consequently, the capacities of LEAs at national and at EU level will be reinforced.

By using a combined methodological approach, applying research literature, AI policies, and AI regulations, the objective is to integrate the latest technical research and best practices with European law. To do so, the first step is conducting a review of the risks associated with development of AI technologies, while the second step consists of a review of the existing instruments for AI risk assessment and technical and organisational measures for risk mitigation. This will result in the ALIGNER Risk Assessment instrument that will be carried out during future project workshops.

Ethical and Legal Assessment

If both the AI Technology Impact Assessment and the Risk Assessment have obtained significant scores, the AI technology is subject to a final ethical and legal assessment. This is performed using the ALIGNER Fundamental Rights Impact Assessment (AFRIA) templates [3].

The AFRIA is a tool addressed to LEAs aiming to deploy AI technologies for law enforcement purposes within the European Union. The AFRIA is a reflective exercise, seeking to enhance the ethical and legal governance systems of LEAs. Hence, the AFRIA has two main functions. First, it helps LEAs identify and mitigate the impact of the deployment of a certain AI technology on ethical principles and those fundamental rights of individuals most likely to be impacted. Second, it is a suitable instrument for LEAs to explain and record their decision-making processes.

LEAs should perform a single AFRIA for a single AI technology, deployed for a single law enforcement purpose or connected law enforcement purposes.

An AFRIA should be performed by LEAs prior to the deployment of the AI technology, to inform the decision-making process on the *if, when, why,* and *how* of the deployment. However, performing an AFRIA should be considered as an iterative process: the AFRIA needs to be recorded, reviewed, and updated throughout the whole lifecycle of the AI technology, to reflect eventual changes in the functioning of the technology and/ or its circumstances of deployment.

To perform an AFRIA, LEAs should establish a diverse and multidisciplinary team. This should include members of the organisation with legal, operational, and technical expertise.

The AFRIA consists of two different, but connected, templates: the *Fundamental Rights Impact Assessment* (section "Fundamental Rights Impact Assessment Template") and the *AI System Governance* (section "AI System Governance Template").

Fundamental Rights Impact Assessment Template

The Fundamental Rights Impact Assessment template helps LEAs identify and assess the impact that the AI technology may have on the fundamental rights of individuals.

The template focuses on those fundamental rights most likely to be impacted by law enforcement AI [4], i.e., presumption of innocence and right to an effective remedy and to a fair trial; right to equality and non-discrimination; freedom of expression and information; and right to privacy and data protection. Accordingly, the template is divided into four sections and, in each one of them, a group of fundamental rights is used as benchmark for the following assessment.

As shown in Fig. 18.4, each section of the template is divided into three columns. The first column lists some '*challenges*', namely some possible characteristics embedded in the assessed AI technology that may have a negative impact on the considered fundamental right. In the second '*evaluation*' column, LEAs need to precise whether and, if so, to what degree the listed challenges relate to the assessed AI technology. In the last '*estimated impact level*' column, LEAs need to estimate the level of negative effect the deployment of the AI technology may have on the considered fundamental right. To do so, LEAs should use an impact matrix based on

1. Presumption of innocence and right to an effective remedy and to a fair trial		
Everyone charged with a criminal offence must be presumed innocent until proved guilty according to law. Everyone whose rights and freedoms are violated has the right to an effective remedy before a tribunal. Everyone is entitled to a fair and public hearing within a reasonable time by an independent and impartial tribunal previously established by law, including rights: ❖ to be informed promptly of the nature and cause of the accusation; ❖ to bring their arguments and evidence as well as scrutinise and counteract the evidence presented against them; and to obtain an adequately reasoned and accessible decision.		
Challenge	Evaluation	Estimated impact level
1.1 The AI system does not communicate that a decision/advice or outcome is the result of an algorithmic decision		
1.2 The AI system does not provide percentages or other indication on the degree of likelihood that the outcome is correct/incorrect, prejudicing the user that there is no possibility of error and therefore that the outcome is undoubtedly incriminating		
1.3 The AI system produces an outcome that forces a reversal of burden of proof upon the suspect, by presenting itself as an absolute truth, practically depriving the defence of any chance to counter it		
1.4 There is no explanation of reasons and criteria behind a certain output of the AI system that the user can understand		
1.5 There is no indication of the extent to which the AI system influences the overall decision-making process		

Fig. 18.4   Section of Fundamental Rights Impact Assessment template

two factors: the severity of the prejudice and the number of affected individuals.

AI System Governance Template

The AI System Governance template helps LEAs identify, explain, and record possible measures to mitigate the negative impact that the deployment of the AI technology would have on AI ethics principles and fundamental rights of individuals.

The template relies on the seven key requirements that a trustworthy AI technology should fulfil, as identified by the high-level expert group on artificial intelligence [5]. The requirements are human agency and oversight; technical robustness and safety; privacy and data governance; transparency; diversity, non-discrimination and fairness; societal and environmental well-being; and accountability. Accordingly, the template is divided into seven sections and, in each one of them, a key requirement for trustworthy AI is used as benchmark for grouping the minimum standards an AI technology should achieve.

As shown in Fig. 18.5, each section of the template is divided into seven (groups of) columns. In the first '*components*' column, the building blocks of the considered key requirements are listed. The second column reports some '*minimum standards*', namely characteristics an AI technology

		1. Human autonomy						
Component	Minimum standards to be achieved	Initial impact estimate		Additional mitigation measures implemented	Final assessment		Responsible department	Timeline
		Challenge no.	Impact level		Final estimated impact level	Further actions		
Human agency	❑ The task allocation between the AI system and the user allows meaningful interactions	[1.2]						
		[1.5]						
	❑ There are procedures to describe the level of human involvement and the moments for human interventions	[1.5]						
		[2.2]						
		[4.1]						
Human oversight	❑ The AI system does not affect human autonomy by interfering with the user decision-making process	[1.2]						
		[1.3]						
		[1.5]						
		[4.1]						

Fig. 18.5 Section of AI System Governance template

should embed or possible governance procedures the organisation should implement for the deployment of the technology to be considered trustworthy. The third group of columns, '*initial impact estimate*', connects the minimum standard with previously estimated challenges and impact levels. In the fourth '*additional mitigation measures implemented*' column, LEAs need to precise how the minimum standard is implemented and how it mitigates the initial impact estimate. In the fifth group of column, '*final assessment*', LEAs need to estimate the final level of negative effect the deployment of the AI technology may have on fundamental rights and list possible further actions to improve the implementation of the minimum standard. In the last two columns, LEAs need to specify the '*responsible department*' and the '*timeline*' for the implementation of the mitigation measures.

Conclusion

Both the AI Technology Impact Assessment method and the Fundamental Rights Impact Assessment have already been successfully validated with stakeholders from ALIGNER's expert groups, including practitioners from law enforcement, industry professionals, civil society representatives, and researchers. Similar validation exercises will be held for the Risk Assessment method, once fully operational.

The complete integrated assessment approach will be used to assess around 20 emerging AI technologies around different future scenarios that cover both the criminal misuse of AI as well as the use of AI by LEAs for the benefit of society. These scenarios include, amongst others: (i) identifying and countering disinformation and social manipulation; (ii) AI-supported cybercrime against individuals and organisations; (iii) the use of AI-supported vehicles, robots, and drones; and (iv) future organisation of LEAs with deeply integrated AI processes.FundingThe ALIGNER [Artificial Intelligence Roadmap for Policing and Law Enforcement] project is funded by the European Union's Horizon 2020 research and innovation programme, under grant agreement no. 101020574. However, the opinions expressed herewith are solely of the authors and do not necessarily reflect the point of view of any EU institution.

References

1. Peters, C. E., Grönwall, C., Bronkhorst, A., & Adlakha-Hutcheon, G. (2019). From foresight to impact for technologies at low technology readiness level. In *Proceedings of the 13th NATO Operations Research and Analysis (OR&A) conference: Challenges for NATO OR&A in a Changing Global Security Environment* (NATO STO-MP-SAS-OCS-ORA-2019). NATO Science and Technology Organization.
2. Westman, T., Svenmarck, P., & Chandramouli, K. (2022). ALIGNER D3.1 – Impact assessment of AI technologies for EU LEAs. *H2020 ALIGNER, GA no. 101020574.*
3. Casaburo, D., & Marsh, I. (2023). ALIGNER D4.2 – Methods and guidelines for ethical & law assessment. *H2020 ALIGNER, GA no. 101020574.*
4. Eren, E., Casaburo, D., & Vogiatzoglou, P. (2022). ALIGNER D4.1 – State-of-the-art reports on ethics & law aspects in law enforcement and artificial intelligence. *H2020 ALIGNER, GA no. 101020574.*
5. High-Level Expert Group on Artificial Intelligence. (2019). *Ethics guidelines for trustworthy AI.* https://ec.europa.eu/newsroom/dae/document.cfm?doc_id=60419

Open Access This chapter is licensed under the terms of the Creative Commons Attribution 4.0 International License (http://creativecommons.org/licenses/by/4.0/), which permits use, sharing, adaptation, distribution and reproduction in any medium or format, as long as you give appropriate credit to the original author(s) and the source, provide a link to the Creative Commons license and indicate if changes were made.

The images or other third party material in this chapter are included in the chapter's Creative Commons license, unless indicated otherwise in a credit line to the material. If material is not included in the chapter's Creative Commons license and your intended use is not permitted by statutory regulation or exceeds the permitted use, you will need to obtain permission directly from the copyright holder.

CHAPTER 19

Online Child Grooming Detection: Challenges and Future Directions

Nikolaos Mylonas, Nikolaos Stylianou,
Despoina Chatzakou, Theoni Spathi, Stefanos Alevizos,
Annika Drandaki, Alexandros Koufakis, George Kalpakis,
Theodora Tsikrika, and Stefanos Vrochidis

INTRODUCTION

The growing number of reported crimes suspected of involving online sexual abuse [1], along with the proliferation of online social networking communities which mainly contain young people between 15 and 24 years old [2], has made the growing phenomenon of child sexual abuse (CSA) and especially online grooming activities, even more prominent. In particular, the increase in online risky behaviours such as sexting, grooming,

N. Mylonas (✉) • N. Stylianou (✉) • D. Chatzakou • T. Spathi • A. Koufakis •
G. Kalpakis • T. Tsikrika • S. Vrochidis
Information Technologies Institute, Center for Research and Technology Hellas,
Athens, Greece
e-mail: myloniko@iti.gr; nstylia@iti.gr

S. Alevizos • A. Drandaki
The Smile of the Child, Athens, Greece

© The Author(s) 2025 237

I. Gkotsis et al. (eds.), *Paradigms on Technology Development for Security Practitioners*, Security Informatics and Law Enforcement, https://doi.org/10.1007/978-3-031-62083-6_19

and child prostitution has raised concerns among parents, educators, and mental health professionals [3]. Among these activities, grooming specifically poses a significant risk to the safety and well-being of children. Grooming is defined as the process of preparing a child, significant individuals, and the environment for the purpose of sexually abusing the child [4], with specific goals such as gaining access to the child, ensuring compliance, and maintaining secrecy to prevent disclosure. Overall, grooming not only reinforces the abusers' patterns but can also be used to justify or deny their actions.

In the context of child grooming, Information and Communication Technologies (ICTs) are commonly utilised to recruit and exploit young individuals for sexual purposes within relationships based on trust between minors and adults [5]. The grooming process often begins with the perpetrator engaging in inappropriate online sexual activities or sending explicit content. To create a safer online space for children, machine learning methods have been developed to enable the automatic detection of grooming activities in online platforms. In this chapter, we analyse the current methods created to deal with online grooming and explore their challenges. Based on our findings, we propose future directions, as part of CESAGRAM project's response to online child sexual exploitation and abuse, in improving online grooming detection and allowing for its use in many languages; thus far, the focus has been on English content only. CESAGRAM[1] is a two-year European-funded project (GA No. 101084974) which aims at tackling online child sexual exploitation and abuse through enhancing the understanding of the process of grooming, and more particularly, the way it is facilitated by technology, as well as its link to CSA and missing-children's cases, a sector currently under-researched. Research, training, and awareness raising, development of a set of artificial intelligence (AI) tools which will facilitate the detection and prevention of grooming content online, and advocacy are the main pillars of the project activities per se during its 2-year lifespan.

---

[1] Towards a Comprehensive European Strategy Against tech-facilitated Grooming And Missing; https://cesagramproject.eu/

BACKGROUND

Child grooming commonly starts in an online setting, instigated by adults by forwarding inappropriate content or employing sexual activities to children. These actions aim to desensitise the child and increase the likelihood of future sexual abuse [6]. Although grooming methods may vary, certain constants can be observed throughout the process. The perpetrator intentionally desensitises the child both physically and psychologically, making them more susceptible to engaging in sexual activities. Techniques such as active involvement, power dynamics, and control are employed to manipulate the child and reduce their inhibitions [7]. A comprehensive understanding of the nature and characteristics of grooming is crucial in addressing the risks associated with online activities and ensuring the protection of young children. By recognising the complex aspects of grooming and its manifestation in the digital realm, strategies to effectively prevent and respond to this form of abuse can be developed.

In order to not only prevent and combat child grooming, both offline and online, but also, to protect children's rights, several legislative efforts have been proposed and adopted on national, European, and international level. The United Nations Convention on the Rights of the Child (UNCRC),[2] the Universal Declaration of Human Rights,[3] along with the Charter of Fundamental Rights of the European Union and the European Convention on Human Rights (ECHR),[4] have been crucial treaties that ensure among others, the proper respect and protection of children's rights and well-being. Parallel to those, the Council of Europe Convention (Lanzarote Convention)[5] has adopted specific measures on the Protection of Children against Sexual Exploitation and Sexual Abuse, complemented with the Directive 2011/93/EU of the European Parliament,[6] while a new Regulation[7] has been proposed to empower the prevention, detection, reporting, and removal of child sexual abuse material and grooming online, and the further support of the victims. Furthermore, the European

---

[2] https://www.unicef.org/child-rights-convention/convention-text

[3] https://www.un.org/en/about-us/universal-declaration-of-human-rights

[4] https://www.echr.coe.int/documents/d/echr/convention_eng

[5] https://www.coe.int/en/web/children/lanzarote-convention

[6] https://eur-lex.europa.eu/legal-content/EN/TXT/?qid=1574272335934&uri=CELEX:32011L0093

[7] https://eur-lex.europa.eu/legal-content/EN/TXT/?uri=COM%3A2022%3A209%3AFIN&qid=1652451192472

Union has proposed a five-year strategy (2020–2025)[8] focusing on the need for better coordination among responsible stakeholders through multi-stakeholder cooperation, with the goal of having a strong legal framework in place and establishing a strengthened law enforcement response that facilitates Member States in addressing the new challenges stemming from emerging technological advancements.

LANDSCAPE OF AVAILABLE GROOMING DATA

Machine learning has been extensively leveraged to develop solutions that could enable effective detection of potential online child grooming activities. To allow for a robust creation of machine and deep learning models, the need for qualitatively and quantitatively labelled (annotated) datasets is more than mandatory. However, the availability of publicly available datasets containing grooming examples is rather limited possibly due to the sensitivity of the subject under study. Nevertheless, thus far there have been some initial attempts to create datasets that could be exploited by machine/deep learning models to tackle to the extent possible the problem at hand. One of the largest sources of predatory conversations comes from Perverted-Justice (PJ),[9] which contains chat logs of individuals convicted of grooming, conversing with decoy operators rather than actual victims. However, the majority of logs are over a decade old, with 2016 being the most recent, while the largest part of them is from earlier than 2010. Having data from so many years back can negatively affect the effective detection of potential grooming activities, as the models developed may struggle or even fail to capture recent changes in dialogue and predatory tactics, due to their potential outdated content.

ChatCoder2 [8], another source of data, consists of only predatory conversations extracted from Perverted-Justice. Overall, it contains 497 conversations (chats) and was mainly built for studying the semantic segmentation of grooming chats, characterising each segment as predatory or not. Moreover, the PAN12 [9] dataset consists of non-grooming conversations, obtained from the logs of IRC (Internet Relay Chat) channels and of the chatting site Omegle,[10] in conjunction with grooming ones. Non-grooming conversations also include cybersex between consenting adults

---

[8] https://eur-lex.europa.eu/legal-content/EN/TXT/?uri=CELEX%3A52020DC0607
[9] http://www.perverted-justice.com
[10] https://www.omegle.com

among other non-predatory ones, while similarly to the PJ conversations, grooming chats are from decoy operations, whereas the non-grooming chats are with real people. Conversations are split into segments, with the dataset containing 222 k segments with only 2.58% of them being grooming; this distribution aims to mimic the real distribution of grooming chats on the Web. Finally, since the dataset was introduced in 2012, the conversations comprising it are up to that date.

In addition, a dataset combining the aforementioned two is known as PANC [10] and consists of non-predatory segments from PAN12 and full-length predator chats from ChatCoder2 divided into segments. Finally, PJZC [11] contains data originating from PAN12, organised in JSON format and re-organised by the authors in a way to fit their task of early grooming detection. Specifically, the authors combined predatory segments belonging to the same conversation and labelled the entire conversation as predatory, rather than the individual segments, with the aim of detecting early signs of grooming attempts in entire conversations.

Based on the above, several limitations can be observed with the grooming-related datasets available. First, the data come from earlier years, hindering the process of effectively identifying more recent manifestations of grooming activities. Additionally, they all contain decoys and not real victims, which can also hinder the effectiveness of machine/deep learning models when trying to detect grooming in conversations consisting of real victims and perpetrators; it is particularly difficult to imitate the real behaviour of other persons as each person has a unique way of reacting to a situation and consequently expressing their feelings. Finally, another less apparent limitation is the lack of multilingual grooming datasets, as all existing ones contain only conversations written in English, thus restricting their use in grooming detection for other languages.

Machine and Deep Learning Methods for Grooming Detection

As mentioned, machine and deep learning have been leveraged thus far to enable the detection of online grooming activities. In particular, grooming detection is tackled as a text-based binary classification problem over a set of chat messages, typically split into segments, with the goal of identifying whether a segment contains grooming or not. Each text consists of a sequence of words that must be converted into machine-readable representations before being fed to a machine/deep learning model. One of the most commonly used methods to this end is Term Frequency-Inverse Document Frequency (TF-IDF) [12, 13]. However, more recently, to enable a better representation of textual data, pre-trained word embeddings obtained from Word2Vec or GloVe [14] have been exploited that allow also capturing the relationships between words in a sequence.

Focusing on the models themselves, most works apply traditional machine learning-based solutions, including Support Vector Machines (SVM) [13, 15], k-Nearest Neighbors (kNN) [15], and Logistic Regression (LR) [16]. Additionally, feature extraction and grooming characteristic detection inside the examined chats were shown to aid in the classification process. In detail, for instance, it was found that providing the model with a binary vector denoting the existence of seventeen distinct grooming characteristics extracted from the text at hand (e.g. asking questions to know the risk of conversation and asking if the child is alone or under adult or friend supervision), instead of the TF-IDF vector representation of the actual text, leads to increased detection performance [15].

Deep learning-based models have also been employed for grooming detection, including multi-layer perceptron (MLP) [12] and convolutional neural networks (CNN) [14]. The former followed an author-based approach, where all messages in a conversation originating from the same author were grouped together to deduce whether there is any grooming activity. In the latter, it was first explored using recurrent neural networks (RNN), concluding that their performance would be inadequate when dealing with large segments of conversation as is typical in the field of grooming detection. To this end, they instead proposed the use of a CNN-based model whose performance is not degraded with that issue. Finally, through experimentation, they additionally found that providing CNN with the input data directly so that the model can learn the embeddings itself, can help increase the performance. In particular, in such a case the model will be able to learn task-specific word representations, especially for words commonly used in grooming chats, which are not present in pre-trained embedding models (such as Word2Vec or GloVe) that are often used in classification tasks.

OPEN CHALLENGES

Despite the ever-increasing efforts to develop effective methods to deal with online grooming activities, the field still faces significant challenges that impede progress. First, as pointed out earlier, publicly available datasets are scarce and mostly come from a single source (namely, Perverted-Justice). However, even for the existing ones, their suitability is somewhat questionable, as they include data from even more than a decade ago, and therefore, there is an increased possibility that they cannot effectively capture today's way of expressing (e.g. higher prominence of transliterations in recent years and different slang terms) and manifesting grooming overall. Additionally, as mentioned, the currently available datasets do not contain conversations conducted by real victims but only decoy operators, which raises questions as to whether the models that will be developed will be able to be effective in real-life scenarios. Finally, the absence of multilingual grooming datasets makes it difficult to apply grooming detection to non-English conversations, giving rise to another open challenge. As a countermeasure, language models (LMs) can be used to translate the existing datasets into the desired language; however, this approach could introduce bias to the dataset, or fail to capture the unique idioms of each language, thereby hindering detection effectiveness overall.

Focusing on the models themselves, and in particular on the important step of text representation, thus far existing approaches mostly make use of simple solutions, such as TF-IDF or non-contextual embeddings like Word2Vec and GloVe. With such approaches, the structure of a text (sentence) is not taken into consideration, but instead the representations are extracted for each utterance regardless of the context being used. These representations, while effective in certain scenarios, may not be the most effective in grooming detection, where the context in which each word is used could be vital in determining whether or not a conversation potentially contains grooming attempts. Thus, to facilitate the detection process, a possible solution could be to train and use contextual embeddings, such as BERT [17], that consider the context of a word in a sentence in contrast to the non-contextualised ones, while also enabling the representation of slang words commonly used in chats, that may not be present in pre-trained embedding models such as GloVe [18]. However, training models to provide contextual embeddings for grooming detection is a challenging task, as the amount of grooming-related data is limited, and such models require a large amount of instances as well as resources to provide high-quality embeddings.

FUTURE DIRECTIONS

In the online world, individuals can maintain multiple identities across different platforms, or even within the same one, with the goal of either deceiving a wider range of individuals or better concealing and maintaining their online identity; e.g. even if an account is detected for infringing behaviour, their activity can seamlessly continue [19]. As mentioned, often perpetrators resort to a similar course of action with the aim of deceiving their victims [20], e.g. through victim isolation and trust development, making it difficult to identify accounts that are managed by the same person in a timely manner. However, each individual's personality is made up of a unique set of behaviours, experiences, and feelings, which is also reflected in the way of writing. To that end, the writing blueprint could be leveraged by automatic mechanisms known as identity resolution that allow for the uncovering of potential links among the unprecedented high number of online user accounts [21]. So far, identity resolution has been employed by law enforcement as a way to uncover previously unknown connections between actors that share common characteristics (e.g. similar address) [22], thus paving the way for its use in the fight against grooming activities as well. Stylometric attributes (e.g. vocabulary diversity or writing idiosyncrasies), as well as contextualised distributional semantic features (e.g. captured by BERT) can be leveraged in an attempt to identify multiple accounts likely to be operated by the same perpetrator [23]. Ultimately, unknown, well-hidden relationships can be revealed, thus allowing identification of further potential victims at early stages.

Similarly, adapting the way language is perceived by LMs through fine-tuning [24] to better reflect current trends in written language in online settings, such as the change in word meanings over time, slang terms, and transliteration, will be an invaluable asset in the development of more effective grooming detection systems. While such approaches require unlabelled and generic data, they do not circumvent the lack of training data for grooming detection. As such, in-depth experimental investigation is required in annotating new gold-standard data, creating synthetic data that simulate real behaviours to the extent possible [25], or considering transfer learning approaches such as few-shot and meta-learning [26, 27].

Acknowledgements This work was supported by the CESAGRAM project funded by the European Union (Internal Security Fund) under Grant Agreement No. 101084974. Views and opinions expressed are, however, those of the author(s) only and do not necessarily reflect those of the European Union. The European Union cannot be held responsible for them.

REFERENCES

1. Negreiro, M. (2022, December). [Online]. Available: https://www.europarl. europa.eu/RegData/etudes/BRIE/2022/738224/EPRS_BRI(2022)738224EN.pdf
2. Petrosyan, A. (2023). *Worldwide digital population 2023*. Retrieved from https://www.statista.com/statistics/617136/digital-population-worldwide/
3. Estefenon, S. G. B., & Eisenstein, E. (2015). La sexualidad en la Era Digital. *Adolescencia e Saude, 12*, 83–87.
4. Craven, S., Brown, S., & Gilchrist, E. (2006). Sexual grooming of children: Review of literature and theoretical considerations. *Journal of Sexual Aggression, 12*, 287–299.
5. Wachs, S., Wolf, K., & Pan, C.-C. (2012). Cybergrooming: Risk factors, coping strategies and associations with cyberbullying. *Psicothema, 24*, 628–633.
6. Quayle, E., Allegro, S., Hutton, L., Sheath, M., & Lööf, L. (2014). Rapid skill acquisition and online sexual grooming of children. *Computers in Human Behavior, 39*, 368–375.
7. Berson, I. R. (2003). Grooming Cybervictims. *Journal of School Violence, 2*, 5–18.
8. Chat Coder 2 dataset. https://www.chatcoder.com/data.html. Accessed 27 June.
9. Inches, G., & Crestani, F. (2012). *PAN12 deception detection: Sexual predator identification*. Zenodo.
10. Vogt, M., Leser, U., & Akbik, A. (2021). Early detection of sexual predators in chats. In *Proceedings of the 59th annual meeting of the Association for Computational Linguistics and the 11th international joint conference on natural language processing* (Long Papers) (Vol. 1) Online.
11. Milon-Flores, D. F., & Cordeiro, R. L. F. (2022). How to take advantage of behavioral features for the early detection of grooming in online conversations. *Knowledge-Based Systems, 240*, 108017.
12. Bours, P., & Kulsrud, H. (2019). Detection of cyber grooming in online conversation. In *2019 IEEE international Workshop on Information Forensics and Security (WIFS)*.
13. Sulaiman, N. R., & Siraj, M. M. (2019). Classification of online grooming on chat logs using two term weighting schemes. *International Journal of Innovative Computing, 9*.

14. Ebrahimi, M., Suen, C. Y., & Ormandjieva, O. (2016). Detecting predatory conversations in social media by deep Convolutional Neural Networks. *Digital Investigation, 18,* 33–49.
15. Gunawan, F. E., Ashianti, L., Candra, S., & Soewito, B. (2016). Detecting online child grooming conversation. In *2016 11th international conference on Knowledge, Information and Creativity Support Systems (KICSS).*
16. Pranoto, H., Gunawan, F. E., & Soewito, B. (2015). Logistic models for classifying online grooming conversation. *Procedia Computer Science, 59,* 357–365.
17. Devlin, J., Chang, M.-W., Lee, K., & Toutanova, K. (2018). BERT: Pretraining of deep bidirectional transformers for language understanding. *CoRR, abs/1810.04805.*
18. Borj, P. R., Raja, K., & Bours, P. (2023). Online grooming detection: A comprehensive survey of child exploitation in chat logs. *Knowledge-Based Systems, 259.*
19. Reuters Staff. (2019). *Twitter suspends 100k accounts for creating new ones after suspension.*
20. RAINN. (2018). *Grooming: Know the warning signs.*
21. Tsikerdekis, M., & Zeadally, S. (2014). Multiple account identity deception detection in social media using nonverbal behavior. *IEEE Transactions on Information Forensics and Security, 9,* 1311–1321.
22. Homeland Security. (2018). *The role of identity resolution in criminal investigations.*
23. Chatzakou, D., Soler-Company, J., Tsikrika, T., Wanner, L., Vrochidis, S., & Kompatsiaris, I. (2020). User identity linkage in social media using linguistic and social interaction features. In *Proceedings of the 12th ACM conference on Web Science.*
24. Gururangan, S., Marasović, A., Swayamdipta, S., Lo, K., Beltagy, I., Downey, D., & Smith, N. A. (2020). Don't stop pretraining: Adapt language models to domains and tasks. In *Proceedings of the 58th annual meeting of the Association for Computational Linguistics.* Online.
25. Stylianou, N., Chatzakou, D., Tsikrika, T., Vrochidis, S., & Kompatsiaris, I. (2023). Domain-aligned data augmentation for low-resource and imbalanced text classification. In *European conference on information retrieval.*
26. Parnami, A., & Lee, M. (2022). Learning from few examples: A summary of approaches to few-shot learning. *arXiv preprint arXiv, 2203.04291.*
27. Vanschoren, J. (2019). Meta-learning. *Automated machine learning: methods, systems, challenges,* 35–61.

Open Access This chapter is licensed under the terms of the Creative Commons Attribution 4.0 International License (http://creativecommons.org/licenses/by/4.0/), which permits use, sharing, adaptation, distribution and reproduction in any medium or format, as long as you give appropriate credit to the original author(s) and the source, provide a link to the Creative Commons license and indicate if changes were made.

The images or other third party material in this chapter are included in the chapter's Creative Commons license, unless indicated otherwise in a credit line to the material. If material is not included in the chapter's Creative Commons license and your intended use is not permitted by statutory regulation or exceeds the permitted use, you will need to obtain permission directly from the copyright holder.

CHAPTER 20

AI-Based Framework for Supporting Micro and Small Hosting Service Providers on the Report and Removal of Online Terrorist Content

*George Kalpakis, Caterina Paternoster, Marina Mancuso,
Denitsa Kozhuharova, Theoni Spathi,
Theodoros Semertzidis, Theodora Tsikrika,
and Stefanos Vrochidis*

G. Kalpakis (✉) • T. Spathi • T. Semertzidis • T. Tsikrika • S. Vrochidis
Information Technologies Institute, Center for Research and Technology Hellas,
Thessaloniki, Greece
e-mail: kalpakis@iti.gr

C. Paternoster • M. Mancuso
Transcrime Università Cattolica del Sacro Cuore, Milan, Italy

D. Kozhuharova
Law and Internet Foundation, Sofia, Bulgaria

© The Author(s) 2025   249
I. Gkotsis et al. (eds.), *Paradigms on Technology Development
for Security Practitioners*, Security Informatics and Law
Enforcement, https://doi.org/10.1007/978-3-031-62083-6_20

Introduction

Despite the huge efforts put by Member States (MS) to suppress terrorist content online (TCO), terrorist groups have proven versatile and adaptable in finding new ways to distribute such content. Terrorist organisations and extremist groups leverage a large ecosystem of Internet-based platforms, varying from social media to hosting/cloud services and file-sharing systems, cutting across the Surface and the Dark Web, in order to accomplish their subversive intentions and support their goals and operations [1]. A huge amount of terrorism-related multimedia content is spread every day in several languages through such channels to reach a large audience for violent radicalisation and recruitment purposes, as well as for training and financing terrorist acts, which are not restricted by the different locations of the parties involved. Furthermore, terrorist and extremist groups often use fake digital identities to post links of relevant sources on darknets, to sell illegal goods and services, and organise terrorist attacks and criminal activities.

To address this challenge, Directive 2017/541 [2] urged MS to collaborate among themselves and also with third countries to remove such content or block it from the EU territory. This legal provision has also invoked the Internet industry to support MS by preventing such misuse of their services, albeit on a voluntary basis [2]. A step further has been taken by the EU by adopting Regulation (EU) 2021/784 [3] (also known as the TCO Regulation) urging the implementation of specific protective measures, requiring online hosting service providers (HSPs) to remove any illegal terrorist content within one hour after receiving an official removal order issued by a European competent authority (e.g. EUROPOL and MS LEAs). The obligations stemming from the Regulation are enforced to any HSP, irrespective of their size or popularity, creating significant operational, financial, and technical overhead, mainly impacting micro and small HSPs, given their limited capacity and human resources to apply effective response measures in a timely manner.

In this context, the ALLIES[1] project (https://www.alliesproject.com/) aims to assist micro and small HSPs in adhering to the TCO Regulation and protecting their platforms from terrorist abuse in an efficient and cost-effective way, as well as to enhance their capability to respond in a timely

---

[1] The ALLIES project full name is: 'AI-based framework for supporting micro and small Host Service Providers on the report and removal of online terrorist content'.

manner to removal orders issued by MS competent authorities. In particular, ALLIES proposes a framework that equips HSPs with a powerful suite of AI tools permitting the automated detection and the proactive removal of TCO on their end in a short period after the content upload and the prevention of its re-upload. In addition, the framework encompasses a distributed infrastructure allowing HSPs to leverage federated learning techniques for (re-)training, validating, and testing the underlying AI models utilising an ecosystem of annotated datasets hosted in private HSP storage spaces, ensuring that the raw data are inaccessible by any third party. Furthermore, the framework comprises a secure centralised online environment where multimodal hash representations of TCO are shared among HSPs to further facilitate the automated removal of subversive content. Finally, the ALLIES framework integrates a unified reporting mechanism to support the submission of removal orders issued by competent authorities, as well as the reporting of TCO removed by HSPs when necessary. The framework is being developed in a user-driven manner based on the requirements of HSP and law enforcement agency (LEA) personnel, while its development is also guided by the outputs of desktop and empirical research on online radicalisation, extremism, and terrorism, providing up-to-date information on these criminal activities.

TCO Regulation

The Regulation (EU) 2021/784 [3] of the European Parliament and of the Council of 29 April 2021 on addressing the dissemination of TCO also known as the TCO Regulation is the fruit of a four-year work based upon EU-led efforts of cooperating with HSPs, so that the dissemination of TCO is eradicated [4]. The TCO Regulation became applicable as of 7 June 2022, as per its Art. 24.

The main idea behind the TCO Regulation is to make sure that the digital single market of the EU is functioning seamlessly, and that public security is upheld by providing guarantees that hosting services available in the EU are not abused for terrorism-related purposes [3]. To this end, the TCO Regulation contains provisions requiring EU MS to introduce measures for the prompt removal of TCO, while also safeguarding fundamental rights, namely freedom of expression and freedom of information—two central EU values embedded in its core, being promulgated by Art. 12 of the Charter of Fundamental Rights of the EU. Additionally, as with other EU legal instruments in the security domain, a second requirement to the measures to be put into place by MS concerns the cooperation within the EU tackling this threat.

The main set of obligations addresses two types of entities: (i) HSPs—any entity storing information provided by and at the request of a content provider (the regulation applies regardless of the HSPs' location, as long as they provide their services in the EU) [3], and (ii) competent authorities—authorities specially designated by an EU MS to issue removal orders, scrutinise removal orders issued by competent authorities of other MS, oversee the implementation of specific measures, and be competent to impose penalties for the non-compliance with the TCO Regulation [3].

The main instrument established in Art. 3 of the TCO Regulation is the 'removal order'. The removal order prescribes either the complete removal or deactivation of access to the piece of content which has been identified as terrorism-related. The latter is understood as 'material that incites or solicits an individual or a group of people to commit a terrorist act, or that provides instruction on making weapons or on other methods or techniques for use in a terrorist attack' [3]. The removal order should be predeceased by information related to the applicable procedures and deadlines at least 12 hours prior its issue, in case this is the very first removal order issued in view of a respective HSP, and provided the case at hand does not constitute an emergency. Once the removal order is placed, the HSP has only one hour to remove the identified content. The removal order should be issued using a template provided by the TCO Regulation itself in its Annex 1. As outlined above, the TCO Regulation requires that a balance is stuck between fundamental rights and public security. This is quite visible in the said removal order template, which prescribes that information on the available redress mechanism is likewise noted. Being an EU instrument, the TCO Regulation foresees how cross-border cases should be handled. In case the HSP, whose content is identified as TCO, is based in another MS, a copy of the removal order should be sent to the competent authority of the said MS.

Last but not least, the TCO Regulation imposes obligations to HSPs which have been exposed to terrorist content—in case the competent authority has issued a relevant decision based on objective facts, such as the receipt of two or more removal orders in the last 12 months. In such a case, the HSP must take measures to protect itself and the provided services from further dissemination of TCO. The TCO Regulation does not prescribe such measures but leaves them to the discretion of the respective HSP; it only enumerates several examples including technical and organisational measures, such as allocating sufficient human resources to monitor the uploaded content or enabling users to report suspicious cases. The measures taken should meet the following requirements imposed by the TCO Regulation: (i) they must be effective in mitigating the level of exposure of the HSP services to TCO; (ii) they should be targeted and proportionate considering the TCO exposure level of the HSP services, as well as the technical and operational capabilities, financial strength, the number of users, and the amount of content they provide; (iii) they should be applied in a manner that takes full account of the rights and legitimate interest of the users, in particular the users' fundamental rights on freedom of expression and information, respect for private life and protection of personal data; and (iv) they should be applied in a diligent and non-discriminatory manner [3].

Online Radicalisation, Extremism, and Terrorism Landscape

Traditionally, radicalisation has relied on direct social interactions in physical settings, like prisons, places of worship, and other places of community. In the last decades, the advent of the Internet has revolutionised how individuals communicate and spend their time, providing new opportunities for the spread of violent extremist ideologies [5]. Indeed, in recent years the expansion towards the online environment has facilitated the rapid dissemination of extremist ideologies, allowing to reach larger audiences with greater efficiency and guaranteeing more anonymity to the recruiters. Despite not being the only factor, the presence of TCO has been demonstrated to be a catalyst for radicalisation, which can lead individuals to commit acts of violence and terrorist attacks [6, 7]. Tackling TCO has therefore become a matter of urgency in recent years, within and beyond the EU.

Drawing on both desktop and empirical research (semi-structured interviews with 11 experts, namely scholars and representatives of LEAs), ALLIES has conducted a comprehensive overview of the strategies employed by jihadists, specifically the Islamic State (IS), and far-right supporters to disseminate content online.

The analysis highlights that the actors involved in the dissemination of TCO adapt their narratives to current events and regional dynamics to engage and keep the audience engaged, yet often drawing from traditional themes and ideologies. In particular, for what concerns the IS the main narratives are focused on: double-salvation, victimhood, and oppression [8, 9], while for far-right groups, movements, and supporters are: white supremacy, antisemitism, and male supremacy [10]. The theme of martyrdom and glorification of attackers is spreading in the far-right environment, showing similarities to Salafi Jihadism [11]. In various online platforms, far-right supporters, especially within the white supremacy movements, are increasingly sanctifying perpetrators for the purpose of propaganda and proselytism [1].

The strategic objectives pursued by the IS and far-right actors through the dissemination of these narratives partially overlap: radicalisation, recruitment, fundraising, and mobilisation are primary aims shared by both movements [9]. The target audience varies, but experts highlight that is increasingly younger, especially within the far-right environment.

The channels employed to disseminate TCO are multiple, and the strategy varies comparing IS and far-right supporters. The IS and its supporters have actively strived to establish a presence across various platforms, mainly through the massive dissemination in mainstream platforms of hyperlinks to file-sharing repositories and terrorist-operated websites, which guarantee a null or weaker content moderation and removal, frequently using bots [12]. The focus is therefore on the swiftness of dissemination rather than on the content or specific channels. Online violent far-right actors also adopt a multiplatform strategy for spreading TCO, but they differ from IS supporters in their dissemination methods. Instead of drawing on extensive lists of out-links, far-right actors primarily rely on loosely connected content creators to spread their messages. This reflects the nature of the extreme right environment, which, unlike the IS one, consists of a dynamic and fragmented network of individuals, groups, and movements, embracing various ideologies [1].

Both the IS and far-right extremists disseminate a variety of media outputs and content online, adapting them to the target and channel used, and adopting several visual strategies to make them more appealing [13, 14]. Aesthetics play a central role in conveying extremist narratives, making ideologies more accessible and attractive allowing users to familiarise with violence [14]. In this respect, humour and memes are strategic means used within the far-right extremist movements to disguise hate speech and make it more accessible for non-radicalised individuals [15]. The use of gameplay characteristics, known as gamification, is another visual strategy employed by both far-right and IS supporters: the gaming aesthetic makes the content instantly recognisable by a large audience, being familiar to millions of people, mainly young male, who are the primary target for both IS and far-right extremists [16, 17].

The activity of IS and far-right supporters shows a complex and rapidly mutating nature, driven by current events and new opportunities. As the Internet continues to play a pivotal role in the process of radicalisation and recruitment, it is necessary to further investigate and understand how these strategies adapt to new challenges and opportunities.

THE ALLIES FRAMEWORK

The technological solutions being developed as part of the ALLIES framework will assist micro and small HSPs towards implementing the TCO Regulation and protecting their online spaces from terrorist abuse, while facilitating the seamless communication of competent authorities and HSPs in the context of the submission and monitoring of removal orders. ALLIES aspires to build an ecosystem of modular interconnected AI solutions providing HSPs with the necessary means towards detecting and removing TCO in an efficient and cost-effective manner, as well as to promote their collaboration through a federated learning infrastructure and the sharing of existing knowledge in terms of TCO removal through a centralised hash repository. ALLIES will also support the speedy and trustworthy reporting between competent authorities and HSPs through a secure unified reporting mechanism.

By enhancing their arsenal with advanced AI tools supporting the continuous automated monitoring, processing, and analysis of large amounts of data uploaded by numerous users of their platforms, the HSPs will be in a position to proactively detect and remove TCO in a more efficient manner. The federated infrastructure will contribute towards improving the performance of the AI tools and the underlying models, leveraging multilingual and multimodal annotated datasets hosted in the private HSP spaces, ensuring that the raw data are not accessible by any third party. Additionally, any TCO detected and validated by an HSP will be fingerprinted based on multimodal hashing techniques and will be stored in a shared hash repository, enabling the prevention of re-uploading identical or highly similar content to any HSP platform. Finally, a unified reporting mechanism will facilitate the submission of removal orders by competent authorities to HSPs, as well as the notification of the authorities for TCO content that has been proactively removed by an HSP.

Figure 20.1 depicts an overview of the ALLIES framework. At its core, the powerful multimodal AI analysis tools enable HSPs to continuously monitor any content uploaded to their end by their users, and auto-detect any potentially violating material. The analysis tools process all modalities of interest including text, audio, and video, as well as their multimodal combinations thereof.

In particular, the framework is equipped with AI-based natural language processing (NLP) tools providing useful insights by analysing the multilingual online textual content and extracting, disambiguating, linking, and semantically enhancing concepts and named entities [18] associated with terrorism- and extremism-related content. These tools are further enhanced by machine translation and transliteration capabilities for multilingual support, targeting languages of interest for the ALLIES stakeholders. An automatic speech recognition (ASR) tool processes audio and visual files to produce accurate transcriptions for languages of interest that can be subsequently analysed by NLP technologies [19]. To this end, ALLIES is assessing and fine-tuning pretrained state-of-the-art

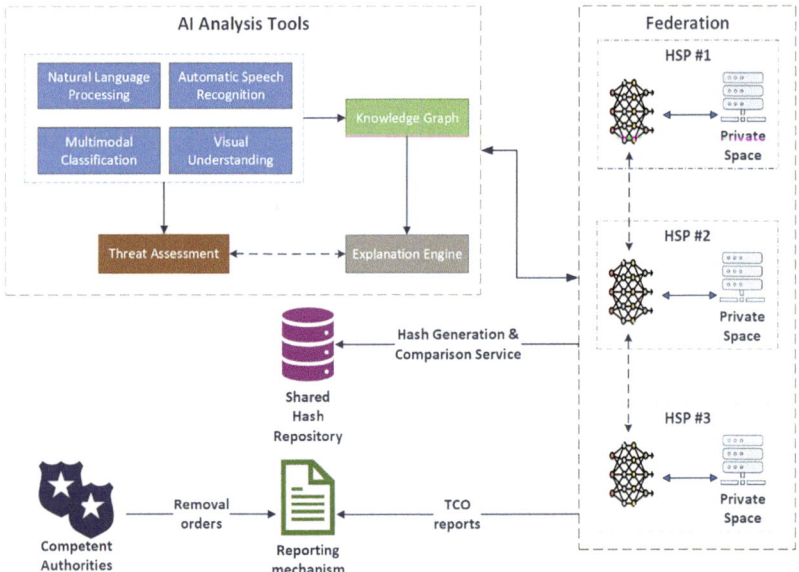

Fig. 20.1   The ALLIES framework

open-source neural sequence-to-sequence ASR models. In addition, a multimedia visual understanding tool based on deep learning approaches and algorithms analyses videos and images to detect and recognise objects, concepts, human behaviour, and activities [20] relevant to the domain of terrorism. In particular, state-of-the-art deep neural network methods for robust object detection and recognition on multimedia content (i.e. images and videos), as well as for event and activity detection on videos, will be deployed based on supervised and semi-supervised techniques.

The indicators produced by the separate analysis of textual, audio, and visual modalities are leveraged by the classifiers that automatically categorise the online data into predefined categories relevant to the terrorism domain based on deep learning multimodal fusion techniques combined with neural representation methods [21].

The multimodal AI analysis tools serve as a first step towards the automated detection of TCO on the online spaces of the HSPs. Their outputs are introduced into a knowledge graph which meaningfully integrates knowledge extracted from multiple and heterogeneous data silos and different HSPs. The relevant content and semantics are leveraged by a powerful AI-boosted explanation generator engine [22] supporting human interpretation and inference, providing HSP moderators with reasoning and explanation on the results delivered by the AI solutions. The engine is accompanied by a set of analytics capabilities and visualisation dashboards, enabling an informative and user-friendly overview of the associated processes. The explanation engine together with the threat assessment tool, providing insights regarding the severity of threats posted online [23], helps the HSP moderators making informed decisions related to the removal of abusive content from their online spaces in an efficient manner, in the sense that this process eliminates the need for human inspection on all data uploaded to a hosting platform, and requires validating only the pieces of content flagged as TCO by the AI analysis tools and the explanation engine.

The framework is built upon a distributed infrastructure that enables the application of federated machine/deep learning techniques [24] for generating and training the underlying AI models of the analysis tools. In this context, each HSP platform is considered as a private space, inaccessible by third parties. The federated infrastructure allows (re-)training, validating, and testing AI models in a distributed way, leveraging annotated datasets included in the separate private spaces, prohibiting the access on the actual data and permitting the exchange of the relevant federated analysis results. The federated approach is respectful of social, ethical, legal, and privacy rules and constraints, adopting a privacy-by-design architecture that ensures personal data protection, and integrity and robustness of the produced models.

To further enhance the capability of micro and small HSPs in terms of preventing the terrorist abuse of their spaces, ALLIES introduces a centralised shared repository with multimodal hashes of TCO. In particular, the hash repository will be populated by HSPs with hashes of TCO detected on their end, whereas HSPs will also be equipped with hash-comparison services to quickly identify already flagged material on the space of other HSPs, and thus prevent its (re-)upload to their space. The shared hash repository permits the inexpensive cross-provider collaboration for the identification and removal of similar TCO that is potentially stored on multiple HSP platforms, without the need to share the actual content among multiple partners. To this end, ALLIES develops a multimodal data hashing service generating multimodal hash representations of the content hosted by the HSPs aiming to facilitate the inexpensive detection of identical or highly similar material posted on the same or different providers. Hashes will be generated for all modalities of interest, including text, audio, and image/video, as well as their multimodal combinations thereof, based on supervised and/or unsupervised AI approaches [25].

Furthermore, ALLIES will facilitate the submission of removal orders by competent authorities to micro and small HSPs according to the TCO Regulation, through a unified reporting mechanism towards the efficient and effective progress of the removal order, enhanced with real-time monitoring of the order status. The goal of the framework is to enhance the capacity of micro and small HSPs to respond to removal orders issued by competent authorities within the timeframe set by the one-hour rule. Finally, the reporting mechanism will also be used by HSPs when it is mandatory according to the TCO Regulation to notify competent authorities for TCO that has been proactively removed.

CONCLUSIONS

This work proposed a framework that integrates several AI tools aiming to assist micro and small HSPs in adhering to the TCO Regulation and protecting their platforms from terrorist abuse. It is envisaged that within such a framework, the tools developed not only will enhance the capacity of micro and small HSPs in terms of proactively removing TCO and implementing the TCO Regulation with the limited resources available but will also establish a model of communication among such enterprises so that they share best practices and experiences. The development of this framework is currently ongoing, and the integrated tools and technologies will be further adapted, improved, tested, and evaluated, in accordance with the end user and domain expert requirements in the context of the activities of the ALLIES project.

Acknowledgements This work was supported by the ALLIES project funded by the European Security Funds under the topic ISFP-2021-AG-TCO from the EU Home Affairs—Grant Agreement No 101080090. Views and opinions expressed are, however, those of the author(s) only and do not necessarily reflect those of the European Union. The European Union cannot be held responsible for them.

REFERENCES

1. EUROPOL. *European Union Terrorism Situation and Trend report 2023* (TE-SAT). Publications Office of the European Union.
2. European Parliament and the Council of the European Union. (2017). *DIRECTIVE (EU) 2017/541 OF THE EUROPEAN PARLIAMENT AND OF THE COUNCIL of 15 March 2017 on combating terrorism and replacing Council Framework Decision 2002/475/JHA and amending Council Decision 2005/671/JHA.* Official Journal of the European Union.
3. European Parliament and the Council of the European Union. (2021). *Regulation (EU) 2021/784 of the European Parliament and of the Council of 29 April 2021 on addressing the dissemination of terrorist content online (Text with EEA relevance).* Official Journal of the European Union.
4. European Commission. (2018). *Proposal for a REGULATION OF THE EUROPEAN PARLIAMENT AND OF THE COUNCIL on preventing the dissemination of terrorist content online, Brussels.*
5. Mølmen, G. N., & Ravndal, J. A. (2021). Mechanisms of online radicalisation: How the internet affects the radicalisation of extreme-right lone actor terrorists. *Behavioral Sciences of Terrorism and Political Aggression*, 1–25.
6. Gaudette, T., Scrivens, R., & Venkatesh, V. (2022). The role of the internet in facilitating violent extremism: Insights from former right-wing extremists. *Terrorism and Political Violence, 34*(7), 1339–1356.
7. Meleagrou-Hitchens, A., & Kaderbhai, N. (2017). Research perspectives on online radicalisation: A literature review 2016–2106. *International Centre for the Study of Radicalisation, 19*, 1–98.
8. Katon, A., Brugh, C. S., Desmarais, S. L., Simons-Rudolph, J., & Zottola, S. A. (2020). A qualitative analysis of drivers among military-affiliated and

civilian lone actor terrorists inspired by jihadism. *Studies in Conflict & Terrorism, 44*(2), 138–155.

9. Winter, C. (2015). *The virtual 'Caliphate': Understanding Islamic State's propaganda strategy* (Vol. 25). Quilliam.

10. Farinelli, F. (2021). *Conspiracy theories and right-wing extremism: Insights and recommendations for P/CVE.* Publications Office of the European Union.

11. Am, A. B., & Weimann, G. (2020). Fabricated martyrs. *Perspectives on Terrorism, 14*(5), 130–147.

12. Alrhmoun, A., Winter, C., & Kertész, J. (2023). Automating terror: The role and impact of telegram bots in the Islamic State's online ecosystem. *Terrorism and Political Violence,* 1–16.

13. Winter, C. (2022). *The terrorist image: Decoding the Islamic State's photopropaganda.* Hurst Publishers.

14. Bogerts, L., & Fielitz, M. (2020). *The visual culture of far-right terrorism.* Global Network on Extremism & Technology.

15. EUROPOL. (2020). *European Union terrorism situation and trend report* (TE-SAT_2020).

16. Gleeson, C. (2023). Countering terrorist exploitation of online gaming. In *Conference presentation at the third annual GNET conference.*

17. Kingdon, A. (2023). From ancient Greece to modern day Montana: Gaming aesthetics in white supremacist propaganda. In *Conference presentation at the third annual GNET conference.*

18. Liu, T., Jiang, Y., Monath, N., Cotterell, R., & Sachan, M. (2022). Autoregressive structured prediction with language models. *arXiv preprint arXiv, 2210.14698.*

19. Veselý, K., Ghoshal, A., Burget, L., & Povey, D. (2013). Sequence-discriminative training of deep neural networks. *Interspeech,* 2345–2349.

20. Touska, D., Gkountakos, K., Tsikrika, T., Ioannidis, K., Vrochidis, S., & Kompatsiaris, I. (2023). Graph-based data association in multiple object tracking: A survey. In *International conference on multimedia modeling.* Springer.

21. Stylianou, N., Chatzakou, D., Tsikrika, T., Vrochidis, S., & Kompatsiaris, I. (2023). Domain-aligned data augmentation for low-resource and imbalanced text classification. In *European conference on information retrieval* (pp. 172–187). Springer.

22. Gunning, D., Stefik, M., Choi, J., Miller, T., Stumpf, S., & Yang, G. Z. (2019). XAI – Explainable artificial intelligence. *Science robotics, 4*(37).

23. Theodosiadou, O., Chatzakou, D., Tsikrika, T., Vrochidis, S., & Kompatsiaris, I. (2023). Real-time threat assessment based on hidden Markov models. *Risk Analysis.*

24. Zhang, C., Xie, Y., Bai, H., Yu, B., Li, W., & Gao, Y. (2021). A survey on federated learning. *Knowledge-Based Systems, 216.*

25. Song, J., Zhang, H., Li, X., Gao, L., Wang, M., & Hong, R. (2018). Self-supervised video hashing with hierarchical binary auto-encoder. *IEEE Transactions on Image Processing, 27*(7), 3210–3221.

Open Access   This chapter is licensed under the terms of the Creative Commons Attribution 4.0 International License (http://creativecommons.org/licenses/by/4.0/), which permits use, sharing, adaptation, distribution and reproduction in any medium or format, as long as you give appropriate credit to the original author(s) and the source, provide a link to the Creative Commons license and indicate if changes were made.

The images or other third party material in this chapter are included in the chapter's Creative Commons license, unless indicated otherwise in a credit line to the material. If material is not included in the chapter's Creative Commons license and your intended use is not permitted by statutory regulation or exceeds the permitted use, you will need to obtain permission directly from the copyright holder.

CHAPTER 21

Kriptosare: Behavior Analysis in Cryptocurrency Transactions

Francesco Zola, Jon Elduayen, Igor Pallin, and Raúl Orduna-Urrutia

INTRODUCTION

Undoubtedly, the cryptocurrency industry is experiencing rapid innovation and constant evolution derived from its power and utility. Despite the promises of security, immutability, and complete transparency offered by blockchain technology, certain cryptocurrencies, particularly Bitcoin, have been utilized in both legal and illegal activities such as trading, buying goods, money laundering, scams, terrorism financing, and ransomware payments. In this sense, tackling terrorist financing through investigation, prosecution, and prevention has become a worldwide issue that extends

F. Zola (✉) • I. Pallin • R. Orduna-Urrutia
Vicomtech Foundation, Basque Research and Technology Alliance (BRTA), San Sebastian, Spain
e-mail: fzola@vicomtech.org; ipallin@vicomtech.org; roduna@vicomtech.org

J. Elduayen
European Anti-Cybercrime Technology Development Association (EACTDA), San Sebastian, Spain
e-mail: jon.elduayen@eactda.eu

© The Author(s) 2025
I. Gkotsis et al. (eds.), *Paradigms on Technology Development for Security Practitioners*, Security Informatics and Law Enforcement, https://doi.org/10.1007/978-3-031-62083-6_21

beyond Europe. Every day, terrorists find new mediums to communicate, campaign, and finance their activities. For example, as reported by EUROPOL in the IOCTA report [1], two main trends are related to crowdfunding campaigns and generating revenue in markets. However, in both cases, to maintain anonymity, they often employ a combination of cryptocurrencies and dark market technologies [2].

Consequently, law enforcement officers (LEOs) face a critical challenge in analyzing these crypto transactions and identifying the responsible parties, especially due to properties like the (pseudo) anonymity of the network, the absence of regulatory oversight, the utilization of anonymizer mechanisms, the changing behavior of entities, and the emergence of new dynamics all contribute to the complexity of this task. Additionally, the sheer volume of information that needs to be examined can lead to a significant waste of time and resources, thereby impeding the progress of investigations.

To tackle these needs, and combat cybercrime, new paradigms, such as artificial intelligence (AI) and big data, can be used alongside conventional systems to create novel investigation tools. In particular, in this work, we present Kriptosare, a tool able to classify entity behaviors belonging to three main cryptocurrencies: Bitcoin (BTC), Bitcoin Cash (BCH), and Litecoin (LTC). Kriptosare is able to extract behaviors (or classes) from interactions and dynamics of different known entities involved in the transactions and then predicts the behaviors of new unseen entities. Predefined ML models are provided for a first classification, although users can train new ones using always new information and so they can reclassify the whole blockchains. For this task, the blockchain information is combined with open-source external data containing information about crypto addresses and real-world entity names detected over the years. This additional information facilitates the behavior definition following the taxonomy[1] provided by Interpol (Exchange, Mixer, Miner Pool, Marketplace, etc.) and represents a *ground-truth* for the ML training. However, these external data show uneven distribution, i.e., several entity behaviors are more represented than others introducing a class imbalance problem [3]. The imbalance problem is very critical since it can strongly affect ML performance, leading the model to learn skewed scenarios. Furthermore, addressing this issue is even more challenging in cryptocurrency applications, where detecting and collecting new observation data is complex and

---

[1] https://interpol-innovation-centre.github.io/DW-VA-Taxonomy/taxonomies/entities

expensive in terms of resources and costs. Indeed, it is easier to find labeled behaviors of entities related to licit transactions rather than those involved in illicit activities, which are the most interesting from an investigation point of view. For this reason, Kriptosare also includes a synthetic data generator module, i.e., a crypto simulator able to create and manage a private Bitcoin, Bitcoin Cash, or Litecoin network. The control of this crypto environment allows users to replicate real behaviors generating synthetic data and then use them to address the imbalance problem introduced by external sources. More specifically, for creating their private network, users have two options, (a) deploy standard wallets, i.e., traditional and behavioral-free entities, or (b) pre-defined behavioral entities, i.e., *intelligent* wallets able to replicate real specific behavior assigned. In this way, on the one hand, it is possible to enhance the performance of the Kritposare.class reducing the costs. On the other hand, LEOs can study behaviors in captivity, i.e., in an isolated and controlled environment, to improve their knowledge about them.

In summary, Kriptosare allows users to manage both the classification and the generator modules in an easy way, through an intuitive and user-friendly interface (frontend). To the best of our knowledge, the presented tool can be used by LEOs to search and highlight the most important red flag indicators that could suggest criminal behavior, for example, a divergence between real labels obtained from external sources and the Kriptosare.class predictions, or the usage of specific entities that are usually involved in illicit activities, such as anonymizer or tumblers. These results can also be used for supporting LEOs' analysis and optimizing their investigation resources by focusing their effort just on the most relevant behaviors, excluding the ones that are completely unregulated and which would require longer analysis times.

RELATED WORK

The capability for non-transparent transactions and the absence of robust regulatory measures have spread the usage of cryptocurrency, in both legal and illegal/criminal activities. The most striking case is represented by Bitcoin [4]. In fact, over the years, the number of transactions involved in activities such as money laundering, selling illegal goods, ransomware, and Ponzi schemes has abruptly increased. This trend is confirmed in the

"2023 Crypto Crime Report" [5] released by Chainalysis[2] [5], in which they count that in **2022**, $20.6B were moved by illicit addresses. Consequently, the task of reducing anonymity within the network and categorizing crypto entities has become challenging and essential for law enforcement agencies (LEAs) [6].

For these reasons, many studies [7–9] have tried to address this task by using new paradigms like artificial intelligence (AI) and machine learning (ML). However, the majority of them, although valid from an academic point of view, are not used and validated in an operative context (investigation) by an end user. On the other hand, the most common tools like Chainalysis, Graphsense [10], BlockSci [11], Blockchair,[3] and Ciphertrace[4] [12] are mainly focused on detecting entity behavior by gathering tags, labels, and information from the clear and dark web, rather than using AI and ML algorithms for forecasting them. In that sense, they need to be continuously fed with new external information (tags/labels) that is not always easy—and cheap—to find.

For this reason, in this study, we try to merge the two needs by introducing Kriptosare, a tool able to predict entity behaviors within cryptocurrencies using ML techniques. The tool analyzes interactions and dynamics of entities engaged in transactions, and from a few known (tagged/labeled), it is able to generalize their behaviors for detecting similar behaviors across the blockchain. Furthermore, Kriptosare allows the generation of synthetic data in a private and isolated environment. In this way, it is possible to reduce the issues related to the acquisition of external information.

GENERAL ARCHITECTURE

As shown in Fig. 21.1, Kriptosare includes a central database (DB) and five units interconnected among them: four of them representing the backend *(kripto_data, kripto_brain, kripto_API, and kripto_twins)* and one *(kripto_viz)* the frontend.

In particular, the backend is based on the following technologies:

 – Python-Flask: Microweb service AP.
 – Swagger: Python-Flask API development.

---

[2] https://www.chainalysis.com/
[3] https://blockchair.com/
[4] https://ciphertrace.com/

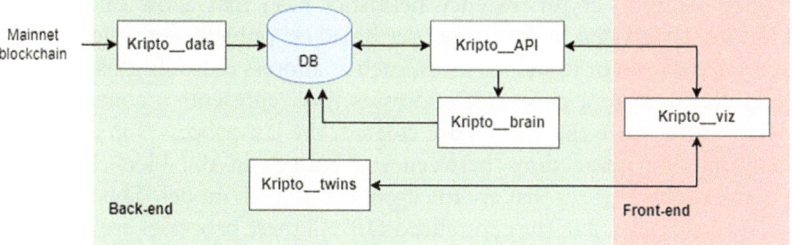

Fig. 21.1 Kriptosare architecture

- Python Scikit Learn: Python library for ML application.
- Cassandra: Database DB.
- Litecoin and Bitcoin Core: daemon for running real wallet.

Whereas the frontend:

- Vue.js: JavaScript framework.

As already described, the backend is composed of four units. The first one is the *kripto_data* unit, which is in charge of the data collection. More specifically, this unit allows Kriptosare to download all the available blockchains (BTC, BCH, LTC) until the current date. This operation is done by running a blockchain daemon with a specific configuration inside a docker container (one container for each cryptocurrency). Once these containers are created and linked to the real network (Mainnet), they start to synchronize themselves and so download the data. At the same time, during this synchronization phase, a specific task is in charge of copying the blockchain data into the centralized DB so that the information can be further consumed by other units. This unit constitutes a safeguard for the data in the networks that are created and used over the tool's lifetime.

The Second Unit That Composes the Backend Is the kripto_brain
This unit represents the CORE of Kriptosare. In fact, it is in charge of three main processes which are: *data preprocessing, entity creation,* and *feature extraction.* More specifically, in the first process, blockchain data are analyzed and processed in order to extract direct relations between input and output addresses. This information is a key aspect of the LEOs' investigations as well as for applying the follow-the-money approach [13]. As the data are preprocessed, the entity creation process is run. This script

applies common cryptocurrency heuristics [14] that allow one to link addresses controlled by the same user based on publicly available transaction information or users' mistakes, such as address reuse. In this way, it is possible to create a cluster of addresses that represents a concrete user [15]. Finally, once the entities are created, the last process is in charge of analyzing the interactions between the entities in the blockchain and extracting the features that are the inputs of the ML model. This information is finally stored in the centralized DB. All these processes are executed as Python scripts that operate uninterrupted in the background so that new information is continuously preprocessed and updated.

The primary objectives of the third unit (*kripto_API*) are twofold. Firstly, it serves as a conventional API, i.e., the contact point between the user interface and the data. In fact, it allows users to consult the stored information and get classification results, statistics, and so on. Secondly, kripto_API also executes the scripts that control the training of the ML models and the (re-)classification task. More specifically, the ML model used by Kriptosare recalls the cascading machine learning approach presented in [16]. This ML strategy already showed to reach very promising performance in scientific investigations. Again, all the predictions and the new models are stored in the centralized DB.

Finally, the last module that composes the backend is the *kripto_twins* (or *simulator*). This unit allows controlling and generating private cryptocurrency networks (or Regtest) of BTC, BCH, or LTC. This simulator is implemented following the instruction released in [17], where Docker containers are used for simulating the different crypto wallets. In each of these containers, the appropriate crypto daemon is run, and then, remote procedure call (RPC) commands are used to control these nodes for creating connections, transactions, mining blocks, and simulating complex behaviors.

The Frontend Is Based on the Kripto_viz Module

This unit serves as a bridge between the user and the underlying functionalities of the tool, enabling a seamless and user-friendly experience. In fact, it promotes *interactivity, helping users to retrieve the classification information* and the complete parametrization of synthetic networks, specifying values such as the number of wallets or their behavior (within a pre-defined set of available types of behavior). It also improves the operations' *efficiency*, allowing users to run complex tasks and workflows in a few steps, and at the same time, it provides meaningful *error messages* for recovering mistakes and preventing critical errors. Finally, this unit allows the users to *interpret and visually understand* the results through a series of graphs and tables.

As shown in Fig. 21.2, Kriptosare main page is composed of 4 main sections, each one highlighted with a different color. On the right-hand side (red section), there is a menu that allows the user to navigate through the tool functionalities: *Classifier*, for getting classification results; *Model Management*, dedicated to modify or train new machine learning models; *Network Management*, facilitating the creation, deletion, and status retrieval of private blockchains; and lastly *Behavior Simulator*, designed for generating transactions, mining blocks, and simulating intricate behaviors within the simulator. The central section (green area) represents the

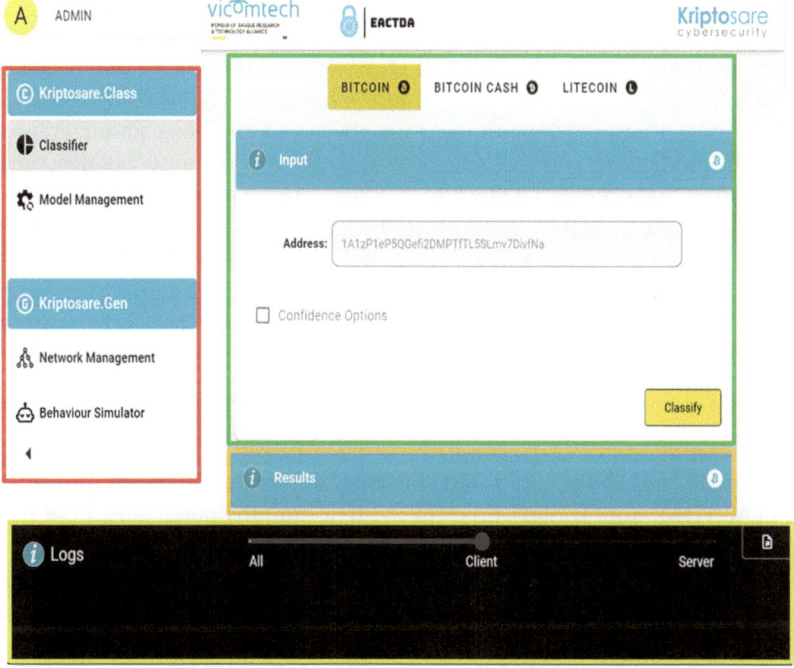

Fig. 21.2 Kriptosare interface (main page)

command area where the user can select the desired cryptocurrency and configure relevant parameters according to the chosen functionality. In the specific instance presented in Fig. 21.2, when users navigate on to the *Classifier* menu, they must input a valid cryptocurrency address. Situated at the lower part of the interface (highlighted in yellow), a log area is designated to display all messages related to ongoing operations, their real-time statuses, and their ultimate outcomes. This setup ensures that users remain well-informed about the progress of all the operations, particularly considering that certain tasks may demand considerable time, such as training a new model or generating a substantial volume of blocks. Concluding the layout, the orange segment is dedicated to showcasing the results attained following the function's execution. An example of the results that can be obtained using the *Classifier* function is reported in Fig. 21.3. As it is possible to see, Kriptosare shows statistical information related to the searched address as well as the behavior prediction (provided by the ML model) of the entity that controls or can control the address, using a very intuitive view.

All the functionalities provided by Kriptosare are functionable to all users without any prior knowledge. However, it is possible to differentiate two different groups of users: *basic* and *ML experts*. The first one includes basic users that use the interface for their investigations about crypto address predictions and the generator to create private networks that validate their hypotheses (*Classifier, Network Management,* and *Behavior Simulator* menu). The second group includes users who know how beneficial could be to train a new ML model and reclassify the whole blockchain, as well as they know how to include the synthetic data in the loop for improving the model's abilities. In this sense, the ML experts fully exploit the model management features.

VALIDATION AND CONCLUSIONS

Kriptosare has been evaluated in two different European projects: TITANIUM[5] and Tools4LEAs.[6] In the first one, the initial version of the tool (a prototype) was made available to the project stakeholders (mainly LEAs from Germany, Spain, Finland, and Interpol) during two events called Field Labs. These events were Capture-The-Flag (CTF) exercises,

[5] https://www.titanium-project.eu/
[6] https://eactda.eu/projects/Tools4LEAs/home.html

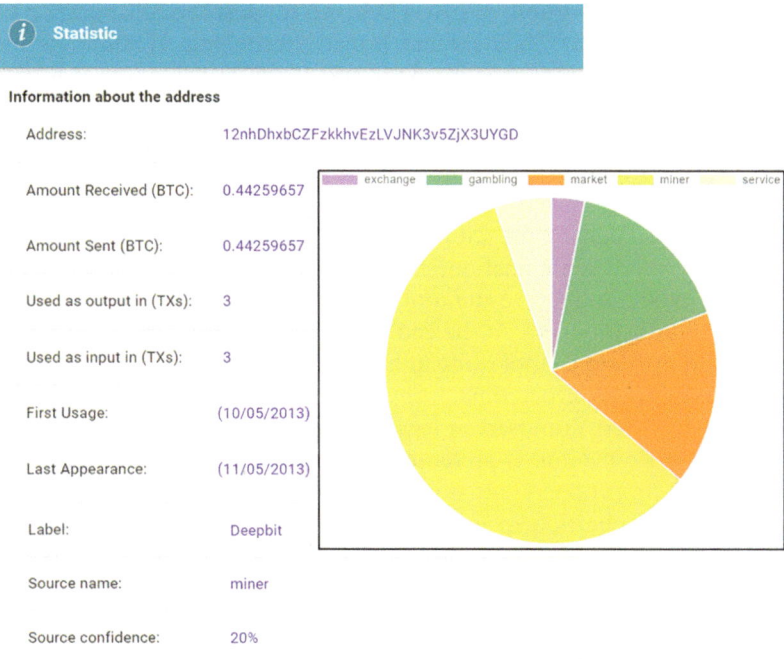

Fig. 21.3 Example of address statistics and classification results extracted and visualized in Kriptosare

in which LEAs used Kriptosare and other project tools to tackle challenges associated with criminal investigations and terrorist activities involving virtual currencies and underground markets in the darknet. This approach provided valuable insights from end-users regarding the tool's relevance to their investigations and day-to-day responsibilities. Additionally, it allowed us to gather feedback on how to improve the tool, i.e., include new functionalities, increase the interoperability of the tool, and improve usability and user experience.

Thus, the tool was improved, and its maturity level was enhanced thanks to the Tools4LEAs project. In this second project, the tool was again evaluated by domain experts selected by the EUROPEAN ANTI-CYBERCRIME TECHNOLOGY DEVELOPMENT ASSOCIATION

(EACTDA).[7] The purpose of EACTDA is to support the collaboration of multiple essential stakeholders and provide technological solutions for European Law Enforcement Agencies and Forensic Laboratories to use them in their fight against crime. In this second validation, domain experts had the chance to read all about the tool (installation and user guide), and then, they tested it freely, using only the provided materials as a guide. After that, they evaluated the tool according to the eight software characteristics, as defined by the ISO/IEC25010:2011[8] standard. Finally, the experts answered some final questions such as the type of enhancements that they would suggest to the tool, if they considered the tools valuable in their investigation, etc. At the end of this process, Kriptosare was a fully tested and operational tool ready to be used by EU public security organizations for fighting cybercrime.

In this chapter, Kriptosare, a tool for cryptocurrency entity behavioral analysis and simulation, is presented. Some preliminary results gathered from LEAs, practitioners, and domain experts proved the potential of this tool and its application in use case investigation. However, on its first deployment, the tool takes a long time to have all the blockchains up to date (depending on the physical resources). In fact, each time a new instance of Kriptosare is run, it needs months to download, preprocess, train, and classify all the data of the three blockchains, considering that just the Bitcoin blockchain has about 866 M transactions and more than 1000 M of addresses generated in 14 years (until the publication date).

As a product of the Tools4LEAs project, Kriptosare is now accessible to EU public security organizations, practitioners, and customers. To gain access to the tool, interested parties may reach out to EACTDA at info@eactda.eu.

REFERENCES

1. Europol. (2021). *Internet Organised Crime Threat Assessment (IOCTA)*. Publications Office of the European Union.
2. Europol. (2022). *European Union terrorism situation and trend report*. Publications Office of the European Union.

---

[7] https://eactda.eu/Contact.html
[8] https://www.iso.org/standard/35733.html

3. Zola, F., et al. (2022). Attacking bitcoin anonymity: Generative adversarial networks for improving Bitcoin entity classification. *Applied Intelligence, 52*(15), 17289–17314.
4. Dhali, M., et al. (2023). Cryptocurrency in the Darknet: Sustainability of the current national legislation. *International Journal of Law and Management.*
5. The 2023 crypto crime report, Chainalysis, 2023.
6. Zola, F., et al. (2019). Bitcoin and cybersecurity: Temporal dissection of blockchain data to unveil changes in entity behavioral patterns. *Applied Sciences, 9*, 23.
7. Mujlid, H. (2023). A survey on machine learning approaches in cryptocurrency: Challenges and opportunities. In *4th international conference on computing, Mathematics and Engineering Technologies (iCoMET)* (p. 2023). IEEE.
8. Turner, A. B., McCombie, S., & Uhlmann, A. J. (2020). Analysis techniques for illicit bitcoin transactions. *Frontiers in Computer Science, 2.*
9. Lorenz, J., et al. (2020). Machine learning methods to detect money laundering in the bitcoin blockchain in the presence of label scarcity. In *Proceedings of the first ACM international conference on AI in finance.*
10. Haslhofer, B., et al. (2021). GraphSense: A general-purpose cryptoasset analytics platform. *arXiv preprint arXiv, 2102.13613.*
11. Kalodner, et al. (2020). BlockSci: Design and applications of a blockchain analysis platform. In *29th USENIX Security Symposium* (pp. 2721–2738).
12. Srivasthav, D. P., Maddali, L. P., & Vigneswaran, R. (2021). Study of blockchain forensics and analytics tools. In *2021 3rd conference on Blockchain Research & Applications for Innovative Networks and Services (BRAINS).* IEEE.
13. Dearden, T. E., & Tucker, S. E. (2023). Follow the money: Analyzing Darknet activity using cryptocurrency and the bitcoin Blockchain. *Journal of Contemporary Criminal Justice, 39*(2), 257–275.
14. Zhang, Y., Wang, J., & Luo, J. (2020). Heuristic-based address clustering in bitcoin. *IEEE Access, 8*, 210582–210591.
15. Androulaki, E., et al. (2013). Evaluating user privacy in bitcoin. In *Financial cryptography and data security: 17th international conference, FC 2013, Okinawa, Japan, April 1–5, 2013* (Revised Selected Papers 17). Springer.
16. Zola, F., Eguimendia, M., Bruse, J. L., & Urrutia, R. O. (2019). Cascading machine learning to attack bitcoin anonymity. In *2019 IEEE international conference on Blockchain (Blockchain)* (pp. 10–17). IEEE.
17. Zola, F., Pérez-Solà, C., Zubia, J. E., Eguimendia, M., & Herrera-Joancomartí, J. (2019). Kriptosare. gen, a dockerized bitcoin testbed: Analysis of server performance. In *2019 10th IFIP international conference on new technologies, mobility and security (NTMS)* (pp. 1–5). IEEE.

Open Access   This chapter is licensed under the terms of the Creative Commons Attribution 4.0 International License (http://creativecommons.org/licenses/by/4.0/), which permits use, sharing, adaptation, distribution and reproduction in any medium or format, as long as you give appropriate credit to the original author(s) and the source, provide a link to the Creative Commons license and indicate if changes were made.

The images or other third party material in this chapter are included in the chapter's Creative Commons license, unless indicated otherwise in a credit line to the material. If material is not included in the chapter's Creative Commons license and your intended use is not permitted by statutory regulation or exceeds the permitted use, you will need to obtain permission directly from the copyright holder.

CHAPTER 22

CTC Project: Advancing the Fight against Terrorist Financing with AI Technologies

Theoni Spathi, Maria Jofre, Mateo Djouadi,
Taxiarchis Skouras, Nikolaos Stylianou, Sotirios Raptis,
Andreas Kosmatopoulos, Ourania Theodosiadou,
Alexandros Koufakis, Theodoros Semertzidis,
Gisela Sanchez, Giorgio Da Bormida, Efstathios Skarlatos,
Nikolaos Lykousas, Clément Pavué, Dimitris Kavallieros,
Theodora Tsikrika, and Stefanos Vrochidis

T. Spathi (✉) • T. Skouras • N. Stylianou • S. Raptis • A. Kosmatopoulos •
O. Theodosiadou • A. Koufakis • T. Semertzidis • D. Kavallieros • T. Tsikrika •
S. Vrochidis
Information Technologies Institute, Center for Research and Technology Hellas,
Thessaloniki, Greece
e-mail: tspathi@iti.gr

M. Jofre
Università Cattolica del Sacro Cuore and Transcrime, Milan, Italy

M. Djouadi • G. Sanchez • G. Da Bormida
Finance Innovation, Paris, France

© The Author(s) 2025   275
I. Gkotsis et al. (eds.), *Paradigms on Technology Development*
for Security Practitioners, Security Informatics and Law
Enforcement, https://doi.org/10.1007/978-3-031-62083-6_22

INTRODUCTION: AIMS AND OBJECTIVES
OF THE CTC PROJECT

Terrorist financing (TF) poses a significant threat to the security and stability of the EU and its member states. The prevention and disruption of TF require international cooperation and advanced investigative techniques based on new technologies. The CTC project[1] capitalizes on an end user-driven methodology, having as its main objective to clearly depict the users' operational, procedural, and organizational needs, integrating the final outcomes to the developed training and technical solutions. The whole project has been built on four main phases: *Phase 1* includes the identification and analysis of use cases and end user requirements, along with the domain trend identification; *Phase 2* focuses on the design and development of the CTC training framework with focus on cross-border collaboration and the public-private synergies; *Phase 3* outlines (a) the acquisition, ingestion, and processing of financial and financial-related data for detecting malicious and unusual patterns, (b) the development of tools and solutions for the detection and analysis of suspicious patterns in financial transactions based on the data acquired, and (c) the development of agile solutions for sharing of evidence-related information in a secure and timely manner; *Phase 4* includes the establishment of a wide stakeholder community of law enforcement agencies (LEAs), financial authorities, private organizations, and academia, along with the information exchange mechanism based on decentralized technologies, to ensure secure and leveled information and intelligence sharing.

[1] https://ctc-project.eu/

E. Skarlatos
Center for Security Studies (KEMEA), Athens, Greece

N. Lykousas
Data Centric, Bucharest, Romania

C. Pavué
Scorechain, Esch-Belval, Esch-sur-Alzette, Luxembourg

BACKGROUND

Understanding threats and trends in the modus operandi of TF networks is crucial for developing effective countermeasures and mitigating associated risks. In doing so, a comprehensive assessment was conducted that encompassed a thorough review of diverse sources, including public- and private-sector intelligence, police and judicial reports, institutional publications and press releases, EU legislation, academic literature, civil society reports, and media articles. The outcomes of this assessment revealed three primary categories of TF threats and trends. The *first category* encompasses traditional systems and technologies commonly employed by terrorists to finance their activities. Terrorist financers demonstrate a high level of confidence in using the formal banking system to conduct multiple small-amount deposits, often across different countries and involving cross-border transactions [1]. Bank transfers are sometimes accompanied by hand-transferred cash or integrated into schemes involving shell companies and money transfer services [2]. Hawala networks, both simple and complex, also play a prevalent role in facilitating the transfer of unlawfully obtained funds (e.g., through fraud, extortion, or kidnapping) via various schemes [3] [4]. Furthermore, evidence suggests the use of high-value commodities such as gold and diamonds, often obtained through robberies or smuggled from conflict areas controlled by terrorist groups [5]. The *second category* revolves around emerging payment systems and obfuscation techniques. Terrorist organizations have increasingly adopted cryptocurrencies, with Bitcoin being the most commonly used, posing a significant risk due to their pseudonymity, high negotiability, and real-time transaction capabilities [6–8]. Additionally, terrorists have embraced alternative crypto-based payment methods, including crypto debit and credit cards, Bitcoin ATMs, and local trades [9]. To evade detection, terrorist financers employ various obfuscation techniques such as crowdfunding, mixers, digital wallets, chain hopping, NFTs, and engaging in gambling activities within the Metaverse. Financers of terrorism also exploit other financial technologies, such as Internet-based payment systems like PayPal and e-commerce platforms like eBay, either individually or in combination [10]. The *third and final category* of TF threats pertains to Internet-based communication platforms and social media. Platforms such as Twitter, Telegram, Facebook, Wickr, YouTube, video games, personal blogs, and chat rooms play a crucial role in terrorist activities. These platforms enable terrorists to leverage the digital multiplier effect and reach a large audience with their messages. While fundraising is one aspect, terrorists also use these platforms for other purposes, such as recruiting new members, spreading propaganda messages, and disseminating technical knowledge related to their activities [11].

In parallel to the desk research, the CTC project garnered valuable insights into observed trends in financial transactions by *conducting interviews with subject matter experts* in anti-money laundering (AML) and counter-terrorist financing (CTF). These interviews adhered strictly to GDPR guidelines to ensure compliance and data protection. To obtain a holistic comprehension of AML/CTF, interviews were conducted with professionals possessing diverse backgrounds and roles, including (a) financial intelligence unit (FIU) experts, (b) AML/CTF analysts, (c) financial experts, (d) policy officers, and (e) AML factory (RegTech/fintech entity). Table 22.1 outlines the main outcomes, which provided the CTC consortium with a comprehensive understanding of the prevailing scenario in this domain, granting access to information not readily available in reports, analyses, policies, or academic papers.

The CTC Solution

Having identified the current landscape, gaps, and existing challenges, CTC aimed to function as a well-rounded response to the needs of the relevant end users, who are interested in the most updated information related to TF. In particular, to be able to (a) identify relevant suspicious transactions (fiat and cryptocurrency), label, classify, and properly annotate them, (b) identify potential networks of interest, (c) detect anomalous patterns and correlate the relevant findings, and (e) merge all relevant information in a suitable and useful format. To this end, the project proposes a combination of tools that aim to track and analyze transactions involving cryptocurrencies as well as traditional financial systems, while providing actionable insights for counter actions.

1. *Cryptocurrency Value Transfer Analysis Platform (CVTA)*

To strengthen the capacity of AML/CFT tools, CTC focuses on cryptocurrency transactions, which are typically used by malicious actors to finance their activities. The CVTA tool aims to perform investigations on multiple blockchains and visualize the interactions between different elements regrouping addresses, entities, lists, and types. It analyzes and tracks

Table 22.1 Main lessons learned from the interviews with experts

Lack of financial transaction data sharing	Financial institutions often refrain from communicating and sharing their transactional data with other banks, resulting in complexities while monitoring financial transactions, even within a single country
Human involvement in transactional monitoring	Machine algorithms and automated systems detect only approximately half of the anomalies reported to the FIU, with the rest requiring human intervention and analysis
Thresholds for flagging transactions	Different jurisdictions have varying thresholds for minimum and maximum transaction amounts that trigger red flags. Moreover, certain businesses automatically raise flags for large sums, contingent upon the type of commercial activity
Flagged transactions and true cases	Roughly 5% to 7% of all transactions are flagged, and within this subset, out of which 50% to 60% correspond to actual AML/CTF cases. Among these, 98% pertain to ML and 2% to TF
Challenges in identifying TF cases	The identification of TF cases proves exceedingly challenging without the crucial know your customer (KYC) information sourced from external entities like LEAs
Complexity introduced by crypto wallets	The advent of cryptocurrency wallets has significantly complicated the tracking of financial transactions. Mixing services, for instance, can be construed as a basic ML tactic, transitioning funds from fiat currency to cryptocurrencies
Emerging money mulling phenomenon	A novel trend involves hiring young individuals to open accounts for illicit purposes, further complicating AML/CTF efforts
Primary contributors to ML	Corruption and tax evasion emerge as the two primary contributors to money laundering (ML) activities
High-Risk countries for ML/TF	Approximately 60% of countries worldwide are categorized as high risk with respect to ML/TF
Traditional risk assessments	Risk assessments conventionally focus on three core areas: Customer risk, transaction risk, and geographical risk

cryptocurrency transactions, through detecting anomalous patterns and providing actionable insights to counter TF in the digital currency domain. The end users will be provided with an in-depth analysis that delves into identifying behavioral and transactional patterns, as well as exploring relationship and interactions within the blockchain network. This analysis will guide them in effectively addressing high or critical risk alerts based on AML/CFT rules and to contribute to solving criminal cases. The outcomes of this analysis will be integrated into subsequent CTC tools such as the multilingual text analysis module (MTAM), the social network analysis module (SNAM), the AI-based pattern recognition of terrorist and criminal activities (AIBPRT), and the cross modal correlation module

(CMCM), enabling significant correlations and the recognition of intricate patterns.

2. *Infrastructure for Traditional Currencies Transaction (ITCT)*

In parallel to the cryptocurrency transaction analysis, focus has been given to money flows and financial transactions across the financial services supply chain, including cross-border transactions, processing data, and identifying red flags, through an early warning system. The ITCT tool aims to monitor, analyze, and detect suspicious transactions in traditional financial systems, providing comprehensive coverage of both digital and conventional value transfers related to TF. Same as before, the outcomes of this analysis will be integrated into the CTC architecture.

3. *Content Acquisition Tool (CAT)*

One of the most crucial issues during a financial investigation in the online domain is for investigators to be able to monitor both the surface and the dark web to discover and identify relevant suspicious links to darknets, and hidden services to online marketplaces. For that reason, both web and social media crawlers have been developed and widely utilized by counter-terrorism units and other relevant actors. As part of the CTC pipeline, the CAT aims to facilitate the discovery and extraction of content relevant to the CTC domain from social media channels, surface websites (also to detect links related to darknets websites), and the dark web. CAT consists of one Web Crawler and one social media crawler. The Web Crawler is responsible for extracting text-based content from surface, deep, and dark web resources. End users provide web entry points of interest (i.e., URLs) and then additional sources with content related to terrorist financing are being identified. Parallel to that, the Social Media Crawler is responsible for the extraction of data (text based) from social networks of interest on behalf of the end users, thus for the discovery of TF-related content. CAT framework has been built to be privacy aware and GDPR compliant, since the controller applies techniques to pseudonymize any personal data found before storing it in the Data Store.

4. *Multilingual Text Analysis Module (MTAM)*

After having identified and listed all the suspicious financial activities from the previous CTC components, a thorough analysis of the textual findings is of interest to the relevant end users. The MTAM has been developed with its main objective to extract key information nuggets from a continuously updated collection of resources that originate from social media, surface, deep, and dark web, thus being used to identify suspicious activities and events including indications of illicit trade and funding. This module consists of three subcomponents: multilingual information extraction (MIE), automatic topic modeling (TM), and user-defined topic classification (TC). The MIE component processes the data collection by automatically extracting fine-grained information within multilingual documents, relevant to illicit activities or funding. It consists of two subcomponents, a multilingual named entity recognition (NER) deep learning model and a heuristic component that together extract entities of interest from a given text and it enables the extraction of Named Entities in 11 languages, including code-mixing, extraction of social identifiers, and extraction of blockchain addresses from four different cryptocurrencies. The TM component is an unsupervised procedure that is based on statistical methods deployed on the set of words in the corpus, which automatically clusters documents of similar semantic meanings and provides insights to the users in terms of similar documents and the topics described in them. Finally, the TC component classifies the instances into user-defined topics. The MTAM uses a combination of natural language processing (NLP) and machine-learning algorithms, trained on multilingual collections of texts, and a set of heuristics and linguistic rules, such as regular expressions, to support the systems' capabilities. The components of the module are built utilizing publicly available datasets as well as data collected for purposes of the project, hence representing secondary use data, without the inclusion of personal data.

5. *Social Network Analysis Module (SNAM)*

The identification of potential networks of interest and, in particular, the community detection based on user interactions along with the key actor identification based on the influence exerted by specific users in each community can play a pivotal role in the investigation of TF activities. The social structure of financial and crypto transactions, as well as the network structures of digital currencies, can be revealed by applying social network analysis (SNA) techniques. These techniques can enhance the understanding of how funds flow and how patterns and relationships emerge in financial transactions. Some studies [12, 13] have used SNA techniques to analyze financial transaction data, while others [14] have focused on specific cryptocurrencies and their network structures. Overall, the use of SNA in the context of financial transactions and crypto transactions has the potential to provide valuable insight into the behavior of market participants and the underlying structure of financial systems. The CTC project aims to analyze online social networks and their associated graphs using the SNAM, which provides a systematic method for uncovering specific information inside a network. To begin with, user groups are being identified based on the frequency and commonality of their interactions (community detection). The community detection feature is made possible by the deployment of innovative AI algorithms that employ graph embedding methods. More specifically, the SNAM tool uses the graph embedding methods DeepWalk [15] and node2vec [16] to produce a low-dimensional vector representation for each graph node given an input social network and its accompanying graph. Then, the tool focuses on each of the detected communities to identify its significant individuals or, in other words, those key actors who exercise an influence on other community members, either indirectly or directly, because of the content they contribute or their frequent interactions with one another. The key actor identification utilizes sophisticated centrality measures (e.g., betweenness centrality and PageRank centrality), which estimate the level of influence a community member has as a "bridge" node or as a node that exerts impact beyond their local neighbors. Similarly, the degree centrality identifies users that share many interactions with other users in the same community. The final outcomes of the analysis are depicted in a user-friendly format, to ameliorate and further advance the investigative endeavors.

6. *AI-Based Pattern Recognition of Terrorist and Criminal Activities (AIBPRT)*

The AIBPRT component focuses on pattern detection in traditional finance and cryptocurrency transactions with emphasis on terrorist financing. The aim is to leverage the rich data acquired from various sources and tools of the CTC solution and identify paths of transactions across different platforms. The component delivers an advanced model that utilizes machine-learning research to detect and identify patterns of suspicious events using time-evolving graph neural network architecture and time-series analysis. The model is designed to handle incoming data by integrating information from different sources, such as traditional financing, cryptocurrency transaction data, and the monitoring of surface web and social media and others. It incorporates the results of CTC's advanced techniques for data analysis, entity extraction, and multilingual text analysis to identify specific entities and information related to suspicious activities, including indications of illicit trade and funding. As all data are entering a single large graph, the analysis is able to reveal information beyond the single data source.

7. *Cross-Modal Correlation Module (CMCM)*

The financing of terrorist acts forms an interconnected network of interactions that is composed of different modalities (i.e., traditional banking services and cryptocurrencies). For that reason, it is important not only to unveil potential suspicious patterns, but also to correlate and combine information of different types and compose a more thorough overview of a situation. The CTC project has developed the CMCM, which provides a framework that allows for the combination of different modalities with the target to assist the investigation efforts by pinpointing time instances that exhibit irregular transaction activity. The proposed CMCM framework consists of three steps: *feature extraction, feature selection, and change point analysis.* In the first step, the transaction activity of an entity is provided as input to the module and the feature extraction mechanism generates features in the form of time series that represent multiple facets of the entity's activity. The next step entails grouping the relevant time series into clusters in an unsupervised manner to perform feature selection and eliminate overlapping information among the several extracted features. Finally, change point detection is applied to the multivariate time series that is formed by the medoids of the formulated clusters to estimate time locations of statistically significant changes. This analysis enables the identification of potential relationships between time instances and event incidents that could have triggered the changes observed in the transaction activity. Overall, the developed framework can be used on both traditional banking and cryptocurrency data, serving as a digital forensics tool in two ways: first, it provides a comprehensive overview of the transaction activity by extracting several time series features of interest, and then, it can indicate time instances linked to event occurrences that could be further investigated to identify possible trends and patterns potentially related to illicit actions.

CONCLUSIONS AND FUTURE WORK

To fight TF and prevent ML in the EU, the CTC project aims to make an important step forward. The capacity of the EU to detect, analyze, and predict TF activities is strengthened in terms of speed and accuracy through the use of innovative AI-based tools, while enhancing public-private partnerships (PPP). CTC supports the maintenance, as well as the improvement of the EU's security against this persistent threat by harnessing the power of AI, collaboration, and innovation. In this way, the identification and prevention of TF threats by authorities will be improved in the following three pillars:

1. *Facilitation of efficient cooperation and information sharing*: In combating TF, the project seeks to foster collaboration among governments and private actors, pulling together intelligence and resources, in order to lead to more comprehensive observations and a consistent response to TF networks and activities.
2. *Increase of understanding on the way terrorists finance themselves*: To analyze financial data and gain a clearer understanding of TF methods and trends, the CTC project employs advanced AI algorithms, which provide information essential to support evidence-based policies and strategic decision-making, helping the EU to stay abreast of the ever-changing tactics employed by terrorist organizations.
3. *Awareness raising and development of an innovative culture*: Raising awareness on the importance of TF and its implications for the EU is one of the objectives of the project, while highlighting and promoting the culture of innovation and collaboration among all interested parties will lead to a sustained commitment, assistance, and financing for the fight against terrorism.

The CTC project will also integrate a blockchain-based chain of custody that reinforces secure and transparent intelligence sharing among stakeholders. This chain of custody, embedded within the project's infrastructure, serves as a pivotal tool for secure, transparent, and auditable exchange of intelligence. Through the integration of blockchain technology, the CTC project facilitates the precise creation and access control of file objects that store potential evidence, improving the traceability and verifiability of shared intelligence. This blockchain-enabled evidence-sharing system presents a credible, immutable record of transactions that can be substantiated in a court of law [17]. As such, it strengthens the chain of evidence, having potentially critical implications for the successful prosecution of terrorism financing cases. All tools and modules will be embedded in a user-friendly interface, while their impact will be closely monitored and assessed to ensure their effectiveness and relevance to stakeholders and end users, supporting an effective fight against TF.

Acknowledgments This project was funded by the European Union's Internal Security Fund—Police under Grant Agreement No. 101036276. The content of this article represents the views of the authors only and is their sole responsibility. The European Commission does not accept any responsibility for use that may be made of the information it contains.

REFERENCES

1. Europol. (2023). *European Union terrorism situation and trend report* (TE-SAT).
2. Teichmann, F. (2019). Recent trends in money laundering and terrorism financing. *Journal of Financial Regulation and Compliance, 27*(1), 2–12.
3. Jamwal, N. (2002). Hawala-the invisible financing system of terrorism. *Strategic Analysis, 26*(2), 181–198.
4. Looney, R. (2003). Hawala: The terrorist's informal financial mechanism. *Middle East Policy, 10*(1), 164–167.
5. Freeman, M., & Ruehsen, M. (2013). Terrorism financing methods: An overview. *Perspectives on Terrorism, 7*(4), 5–26.
6. Chain Analysis. (2023). *The 2023 Crypto crime report*.
7. Dion-Schwarz, C., Manheim, D., & Johnston, P. B. (2019). *Terrorist use of cryptocurrencies: Technical and organizational barriers and future threats*. RAND Corporation.
8. Choo, K.-K. (2015). Cryptocurrency and virtual currency: Corruption and money laundering/terrorism financing risks? In *Handbook of digital currency* (pp. 283–307). Academic.
9. Choo, K.-K. (2009). Money laundering and terrorism financing risks of prepaid cards instruments? *Asian Journal of Criminology, 4*(1), 11–30.

10. Laksmi, S. (2017). Terrorism financing and the risk of internet-based payment Services in Indonesia. *Counter Terrorist Trends and Analysis, 9*(2), 21–25.
11. Rudner, M. (2017). Electronic Jihad': The internet as Al Qaeda's catalyst for global terror. *Studies in Conflict & Terrorism, 40*(1), 10–13.
12. Kondor, D., Posfai, M., Csabai, I., & Vattay, G. (2014). Do the rich get richer? An empirical analysis of the bitcoin transaction network. *PLoS One, 9*(5).
13. Wu, J., Liu, J., Zhao, Y., & Zheng, Z. (2021). Analysis of cryptocurrency transactions from a network perspective: An overview. *Journal of Network and Computer Applications, 190,* 103139.
14. Ao, Z., Horvath, G., & Zhang, L. (2023). *Is decentralized finance actually decentralized? A social network analysis of the Aave protocol on the Ethereum blockchain.* Revised.
15. Perozzi, B., Al-Rfou, R., & Skiena, S. (2014). DeepWalk: Online learning of social representations. *arXiv, 1403.6652.*
16. Grover, A., & Leskovec, J. (2016). node2vec: Scalable feature learning for networks. *arXiv, 1607.00653.*
17. Bacon, J., Michels, J. D., Millard, C., & Singh, J. (2018). Blockchain demystified: A technical and legal introduction to distributed and centralized ledgers. *Richmond Journal of Law and Technology, 25*(1).

Open Access   This chapter is licensed under the terms of the Creative Commons Attribution 4.0 International License (http://creativecommons.org/licenses/by/4.0/), which permits use, sharing, adaptation, distribution and reproduction in any medium or format, as long as you give appropriate credit to the original author(s) and the source, provide a link to the Creative Commons license and indicate if changes were made.

The images or other third party material in this chapter are included in the chapter's Creative Commons license, unless indicated otherwise in a credit line to the material. If material is not included in the chapter's Creative Commons license and your intended use is not permitted by statutory regulation or exceeds the permitted use, you will need to obtain permission directly from the copyright holder.

CHAPTER 23

LAW-GAME: Elevating Experiential Training Through Gamification Technologies

Katerina Margariti, Pantelis Velanas,
Christos Malliarakis, John Soldatos,
and Vassileios Roussakis

INTRODUCTION

The challenges posed by evolving criminal activities, as well as the need for skilled, adaptable officers, have never been more pressing in the realm of law enforcement. Law Enforcement Agencies (LEAs) throughout the European Union are constantly confronted with issues ranging from increasing criminal sophistication to the growing threat of terrorism. Traditional approaches to preparing police officers for the multifaceted demands of modern policing have limitations.

K. Margariti (✉) • P. Velanas • C. Malliarakis
European University Cyprus, Nicosia, Cyprus
e-mail: k.margariti@research.euc.ac.cy

J. Soldatos
INNOV-ACTS Ltd., Nicosia, Cyprus

V. Roussakis
Center for Security Studies, Athens, Greece

© The Author(s) 2025   287
I. Gkotsis et al. (eds.), *Paradigms on Technology Development for Security Practitioners*, Security Informatics and Law Enforcement, https://doi.org/10.1007/978-3-031-62083-6_23

The problem at hand is twofold. First and foremost, traditional training methods frequently fail to bridge the gap between theoretical knowledge and real-world practice, leaving officers unprepared to deal with dynamic, high-pressure situations. Second, the criminal landscape is constantly changing, with criminals leveraging technology and new tactics, making law enforcement difficult to keep up.

The need for immersive, experiential learning that replicates real-world scenarios, fostering critical thinking and decision-making skills, is one of the challenges in law enforcement training. Furthermore, the sheer variety of situations that officers may face, from crime scene investigations to negotiations with suspects, necessitates extensive training programs.

While law enforcement training methods such as simulators and e-learning platforms have advanced, many of these approaches remain fragmented and fail to provide a cohesive, integrated training experience. Existing solutions frequently lack the realism and adaptability required to fully prepare officers for the complexities of their roles.

The LAW-GAME project aims to transform law enforcement training by delivering an integrated, immersive, and adaptable learning platform that addresses the identified challenges. Our strategy entails the creation of four distinct "mini games" aimed at training officers in critical competencies. These mini games cover a wide range of scenarios, from crime scene investigations to counterterrorism strategies, and allow officers to gain practical experience in a safe virtual environment.

LAW-GAME creates realistic training scenarios by utilizing cutting-edge technologies such as virtual reality (VR) devices and AI-assisted procedures. The platform provides real-time feedback, allowing officers to learn from their mistakes and continuously improve their skills. LAW-GAME aims to bridge the gap between theory and practice by combining these elements, providing officers with the tools and knowledge they need to excel in their roles.

To summarize, the LAW-GAME training platform is a game-changing solution to the pressing challenges that Law Enforcement Agencies (LEAs) face. LAW-GAME equips officers with the competencies needed to combat modern criminal activities effectively and ensure the safety and security of the European Union by providing immersive, adaptable, and realistic training experiences.

An Outline of Law-Game Project

The aim of our project is to train police officers on the procedure, enhancing the transition between the theory and real-life practice through gamification technologies in a safe and controlled virtual environment. Essential tasks during the creation of LAW-GAME serious games are to virtualize and accurately recreate the real world, by realistically simulating and analyzing aspects of real-world situations. LAW-GAME introduce an attractive approach to the development of core competencies required for performing intelligence analysis, through a series of AI-assisted procedures for crime analysis and prediction of illegal acts, all within the LAW-GAME game realm.

Building upon an in-depth analysis of police officers' learning needs and inspired by a multitude of disciplines, LAW-GAME develops an advanced learning experience, embedded into four comprehensive "mini games" dedicated to train police officers and measure their proficiency in:

1. Conducting forensic examination, through a one-player or multiplayer cooperative gaming scenario, played through the role of a forensics expert.
2. Effective questioning, threatening, cajoling, persuasion, or negotiation.
3. Recognizing and mitigating potential terrorist attacks, where the trainees impersonate an intelligence analyst tasked with preventing an impending terrorist attack under a didactic and exciting "bad and good" multiplayer game.
4. Car accident forensic analysis: Scene investigation, witness testimony, safety protocols.

The proposed learning experience focuses on the development of the key competences needed for successfully operating in diverse and distributed teams, as required by several cross-organizational and international cooperation situations that police officers face.

LAW-GAME project is an innovative training platform that combines state-of-the-art technology with immersive gameplay to provide a comprehensive training experience for law enforcement professionals.

The project is divided into various modules that are integrated into the final platform. These modules include human emotion modeling, stance modeling, dialogue engine, AI narrator, scenario configurator, crime scene reconstruction, object and human detection, and ballistics analysis.

The human emotion and stance modeling modules utilize physiological signals, facial expressions, and game video, audio, and player video feeds to accurately predict the players' emotional state and stance, respectively. The dialogue engine enables smooth interactions between human players and non-player characters (NPCs), while the AI narrator provides context-aware hints and instructions to guide players through the game. Finally, the scenario configurator generates a wide range of mini game scenarios, each designed to challenge players.

In addition to the mini game scenarios, the crime scene reconstruction, object and human detection, and ballistics analysis modules form a stand-alone tool for crime scene analysis. This tool provides a highly realistic simulation of a crime scene, including the physical environment, evidence placement, and sequence of events. This allows trainees to practice their investigation skills in a controlled environment and receive immediate feedback on their decision-making and problem-solving abilities.

LAW-GAME Mini Games

The project consists of four distinct highly immersive and appealing games that are designed and implemented to provide various types of training to police officers.

Crime Scene Investigation (CSI) Mini Game

The learning methodology consists of both theoretical and practical training, taking place in immersive virtual environments. This involves the basic knowledge on the meaning and scope of CSI process, the general principles of the crime scene, the preparatory steps, and the basic actions before starting the exploration, the basic obligations of first arrived officers, scene surveillance, and recording of witnesses. The training framework concludes with the evidence collection and subsequent analysis principles including the subjects of discrimination of indications, the traces and persuasive, the detection of biological, the selecting evidence from a crime scene, and contamination of evidence.

The game is played through the role of a forensic expert. In this mode, the trainees are able to do virtual forensic examinations on both real and hypothetical scenarios. The expert walks into the crime scene, restricted by the physical laws in normal mode. The most important part is the increased interaction with the environment and objects that enable the ability to catalogue the evidence found. All the tools that the expert has in the real world also exist in the virtual one, for the execution of all the measurements and recordings that are needed.

Police Interview Mini Game
In this game, two types of scenarios are performed, the interrogation and negotiation scenarios. The theoretical training in the interrogation scenario consists of planning and preparing the examination. The initial stage focuses on the suitable tactics, techniques, and procedures (TTPs) to assess and approach the suspect. The practical training mostly deals with the escalation of interrogation, confession, and the interpreted interrogation. The complete training also involves the methodology and system to evaluate the officer's performance of an interrogation. The game creates a VR police interrogation room and suspect's 3D avatar. Here, the trainees try, using their cognitive background, to persuade the suspect to cooperate, changing the tone of their voice or the way they ask questions depending on the suspect's reactions. The trainees also observe body language and any other signs indicating the suspect's psychological state. The trainee's emotional state gives the avatar the same capability, humanizes it, and sets the level of difficulty much higher.

The negotiation training module is in coherence with the interrogation one as, in both, police officers shall interview with citizens for the resolution of critical incidents and crimes. The training module is used to train inexperienced law enforcement personnel in negotiation strategies, consisting of both theoretical and VR practical training. The module introduces the meaning, the purpose, and the issues addressing the general context of negotiations. The specialized texts begin with the presentation of negotiation's main categories and the separations of hostage from non-hostage incidents by the referring real cases that have taken place around the world. Planning and preparation of a negotiation case follow, informing the reader about the stages of information exchange, validation, bargaining, and the Behavioral Change Stairway Model (BCSM), and continue with the evaluation methods used in police negotiations. The theoretical training concludes with semantic negotiation incidents such as the "Stockholm Syndrome," the phenomenon of "Enforced Action," the distinction of negotiable vs non-negotiable and satisfiable vs non-satisfiable incidents and suicidal persons.

The practical course utilizes more complex immersive virtual reality (VR) environments and human-agent negotiation settings. The trainee negotiation skills are evaluated based on the negotiation outcome, negotiation knowledge, and emotional intelligence.

Terrorist Attack Mini Game

The third training module focuses on the best police tactics, techniques, and procedures (TTPs) for preventing terrorist attacks, drawing insights from lessons learned, and strategies employed by the EU and Europol in countering terrorist threats. This module undertakes a foundational review of existing theories and practices, guiding Law Enforcement Agencies (LEAs) on the identification of potential terrorist actors through their movements, international cooperation, and detection of preparatory actions and high-risk targets, including critical infrastructure and public spaces.

Within the LAW-GAME Terrorist Attack mini game, officers are immersed in a novel VR experience designed to enhance their understanding of and preparedness for actions commonly associated with acts of terrorism. Additionally, the game leverages AI modules to elevate its intelligence, particularly in assessing risks related to terrorist activities. The mini game functions as a data generator module, providing essential data for training machine-learning algorithms. It is essential to note that ongoing development is underway, with a subset of the gaming engine modules already implemented.

Car Accident Analysis Mini Game

The final training module of LAW-GAME is dedicated to car accident scene analysis. This module introduces trainees to the general aspects, objectives, and protocols that Law Enforcement Agencies (LEAs) must follow when examining car accidents, including several types, factors, and causes associated with traffic collisions. The training system thoroughly analyzes the sequence of actions that LEA officers must take, both before and upon arriving at the accident scene.

Within the fully immersive 3D gamified training system, trainees engage in a comprehensive exploration of the scene investigation process, aided by state-of-the-art artificial intelligence tools. Special emphasis is placed on training officers to prevent evidence contamination and effectively conduct interviews with involved individuals. The module provides best practices and tactics, techniques, and procedures (TTPs) for evidence collection, measurement, and photography of the scene, incorporating the latest advancements in experiential learning and theory.

Technical Outline

The architecture of the LAW-GAME project is described in detail in the following section. Figure 23.1 illustrates a high-level architecture of the LAW-GAME platform and its various components, providing an in-depth understanding of how the various elements interact to create a seamless, engaging, and effective user training experience. Upon further examination of the architecture, one will acquire a more profound comprehension of the project's inventive approaches to law enforcement training and its overall structure.

There are three primary pillars that make up the architecture of the server that hosts the LAW-GAME applications. Each of these pillars contributes to the overall structure of the architecture. The first pillar consists of VR devices communicating in real time with the Photon Engine. This real-time interconnection provides load balancing and network synchronization, allowing us to provide a fluid, real-time gaming experience for all players.

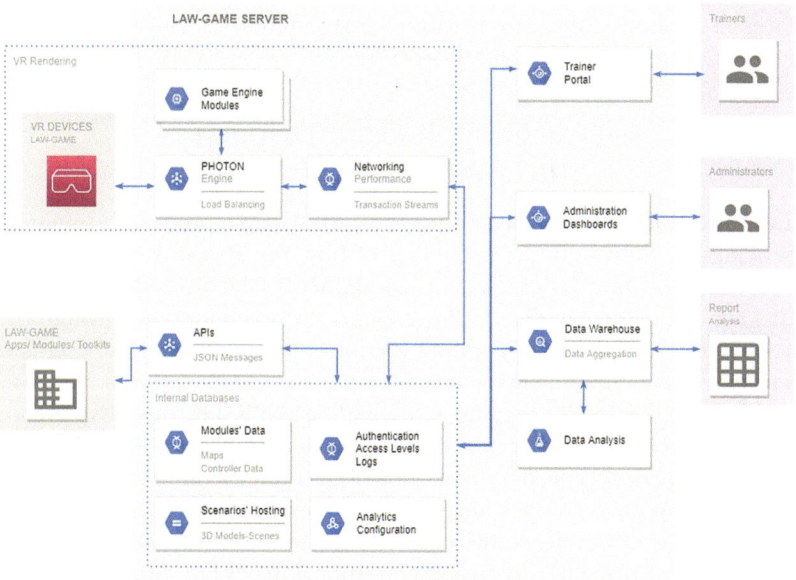

Fig. 23.1 LAW-GAME platform system architecture

In addition to this, the server is in charge of hosting and storing the game engine modules that play a vital role in ensuring the smooth and effective functioning of the Photon Engine. These modules are essential to the game's ability to function. In addition to the Photon Engine, there are additional methods for monitoring the network's operation and ensuring that the game is played uninterruptedly.

The user authentication credentials of users, the relevant access levels, and all server logs are all stored in the internal databases that make up the second pillar of the architecture. In addition, the relevant access levels are also stored in the external databases. The databases are where the information for the user interfaces, maps, and scenarios in three dimensions of the game, as well as the information for any other necessary components, is stored. Application programming interfaces (APIs) and JSON messages, both of which are saved on the server, are used to transmit this data to the apps, modules, and toolkits.

The interfaces that the administrators make use of to evaluate the overall level of quality of the server are the focus of the third and final pillar. Trainers are able to control game sessions, trainees, and other training-related data through other graphical user interface instances. These interfaces also include customized reports (data warehouse) generated from an analysis of the data stored on the server.

The User's Journey Through the LAW-GAME Ecosystem

In this section, the user journey for the LAW-GAME project is examined, providing a thorough understanding of how the system is navigated by users and how interactions with its components take place. The sequence diagram as depicted in Fig. 23.2 represents the user's journey through the LAW-GAME project, highlighting the interactions between the user and various system components.

The user begins by logging into the Social VR Learning Experience Portal, which also provides access to collaboration tools for enhanced learning experiences. From there, the portal connects the user to Moodle, an e-learning platform that hosts training content. Moodle is designed to be highly customizable, allowing for the creation of personalized learning environments that cater to diverse educational needs [1]. Moodle delivers the necessary training content to the user, preparing them for the immersive learning experience.

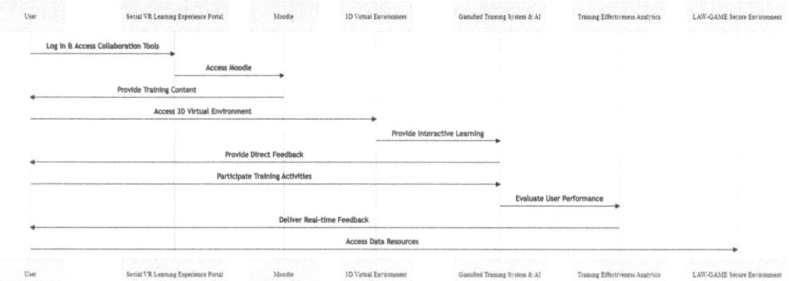

Fig. 23.2 High-level LAW-GAME sequence diagram—user interactions with system components

Once the user enters the 3D virtual environment, they are exposed to the realistic simulations and scenarios that facilitate interactive learning through the mini serious games that are described in section "Introduction". The gamified training system, integrated within the 3D virtual environment, offers a range of interactive learning activities and real-time feedback to the user. The user engages in various training activities, such as simulations and challenges, to develop their understanding of LAW-GAME concepts and improve their skills.

During this process, the gamified training environment works in conjunction with the analytics component provided by Moodle to evaluate the user's performance. The analytics component provides real-time feedback to the user, allowing them to understand their progress and identify areas for improvement. It is important to mention that the user can access the LAW-GAME secure environment, which offers, apart from training, a range of data resources to support and enhance their learning experience. Overall, the sequence diagram illustrates a comprehensive, effective, and interactive learning experience offered by the LAW-GAME project.

Conclusion

The findings of the research and the facts point in the general direction of the conclusion that illegal activities are detrimental to the welfare of the European Union [2]. As a direct result of organized crime groups in the EU, law enforcement officials are forced to deal with a wide variety of illegal activities daily. These activities include intentional homicide and assault, as well as the trafficking of drugs, humans, and illegal firearms.

The education of those responsible for the procedure is the LAW-GAME project's main objective. This is accomplished by easing the transition between theory and practice by applying gamification technologies within a secure and strictly governed virtual setting.

To accomplish this, the project presents an intriguing strategy for assisting in the development of core competencies that are required to carry out intelligence analysis. This is accomplished through a series of AI-assisted procedures for the analysis of criminal behavior and the prediction of future illegal acts.

Acknowledgments This project has received funding from the European Union's Horizon 2020 research and innovation program under grant agreement No. 101021714. This article reflects only the authors' views and the Research Executive Agency, and the European Commission are not responsible for any use that may be made of the information it contains.

References

1. Al-Azawei, A., Parslow, P., & Lundqvist, K. (2017). Investigating the effect of learning styles in a blended e-learning system: An extension of the technology acceptance model (TAM). *Australasian Journal of Educational Technology, 33*(2), 1–23. https://doi.org/10.14742/ajet.2885
2. Europol. (2020). *European Union Serious and Organised Crime Threat Assessment (SOCTA) 2021*. Retrieved from https://www.europol.europa.eu/activities-services/main-reports/european-union-serious-and-organised-crime-threat-assessment

Open Access   This chapter is licensed under the terms of the Creative Commons Attribution 4.0 International License (http://creativecommons.org/licenses/by/4.0/), which permits use, sharing, adaptation, distribution and reproduction in any medium or format, as long as you give appropriate credit to the original author(s) and the source, provide a link to the Creative Commons license and indicate if changes were made.

The images or other third party material in this chapter are included in the chapter's Creative Commons license, unless indicated otherwise in a credit line to the material. If material is not included in the chapter's Creative Commons license and your intended use is not permitted by statutory regulation or exceeds the permitted use, you will need to obtain permission directly from the copyright holder.

CHAPTER 24

Learning Domain-Invariant Spatio-Temporal Visual Cues for Video-Based Crowd Panic Detection

Javier Calle ⓘ, *Luis Unzueta* ⓘ, *Peter Leskovsky* ⓘ, *and Jorge García* ⓘ

INTRODUCTION

Public spaces such as shopping malls, transportation hubs, and entertainment venues are often crowded environments in which it can be challenging for security personnel to monitor the safety of everyone present. CCTV cameras play a crucial role in monitoring the safety of people in crowded public spaces. Artificial intelligence (AI) can enhance the effectiveness of CCTV cameras by using advanced algorithms and machine-learning techniques to provide real-time footage analysis. However, this task is challenging due to the complexity and diversity of scenarios, occlusion and motion of people, and noise and quality of video streams.

J. Calle (✉) • L. Unzueta • P. Leskovsky • J. García
Fundación Vicomtech, Basque Research and Technology Alliance (BRTA), Donostia-San Sebastián, Spain
e-mail: jcalle@vicomtech.org; lunzueta@vicomtech.org; pleskovsky@vicomtech.org; jgarciac@vicomtech.org

© The Author(s) 2025
I. Gkotsis et al. (eds.), *Paradigms on Technology Development for Security Practitioners*, Security Informatics and Law Enforcement, https://doi.org/10.1007/978-3-031-62083-6_24

Obtaining real-world data that meet these requirements can be difficult specifically for two principal reasons: data collection process should comply with privacy-related regulations (e.g. EU's GDPR [1]), implementing data protection principles such as anonymization; and it demands considerable time and cost, given the substantial quantity, variety, and the need for a balanced representation of labelled visual appearances to contain all possibilities during training. A common technique to overcome the challenge of data scarcity is using synthetic data to simulate real-world scenarios. This provides AI algorithms with a rich and diverse training dataset, improving the model's performance. However, challenges related to the domain gap must be considered, such as (i) ensuring the synthetic data accurately represents real-world scenarios, (ii) balancing the proportion of real and synthetic data in the training set, and (iii) avoiding potential biases introduced by the data generation process. In this paper, the main contributions include:

- Presenting a methodology to train a model with real and synthetic data, avoiding the domain gap issues. Discussing different strategies for incorporating synthetic data when real data are scarce.
- Explaining a novel approach to visual-based panic detection by learning domain-invariant spatio-temporal visual cues.
- Delving into the specifics of the data used to train and test the model. Giving details about how the model can learn characteristics to avoid domain gap issues.
- Comparing the results of different experiments to identify the most effective approach and configuration of hyperparameters.
- Benchmarking the results against the current state of the art.

Related Work

Panic Detection Methods

Due to the limited number of studies that focus specifically on panic detection, the state of the art is based on anomaly detection strategies and their application to panic detection. Anomaly detection involves two different strategies:

1. Hand-crafted features: used for a long time to detect anomalies, with several approaches available. One approach is analysing the group behaviour, where there are different phenomena to analyse

such as collectiveness, stability, or uniformity [2]. Another option is to analyse the crowd density, as proposed by [3] where they use the crowd density and motion of individuals within the crowd as a feature. Other strategies are based on spatio-temporal analysis using the gradient sum of the frame difference as a feature [4]. One of the most widely used strategies is the use of optical flow to extract features. In a recent study [5], they propose an optical flow framework based on a GAN and use transfer learning to detect behavioural abnormalities in large-scale crowd scenes. Other option proposed by [6] is to use entropy-based methods where experimental results shows that panic crowd motion states have higher entropy, while normal crowd states have lower entropy.

2. Automatic features: can be extracted using deep neural networks (DNNs). One widely used method for detecting anomalies is autoencoders [7]. Another approach is the use of convolutional neural networks (CNNs) [8].

Training with Synthetic and Real Data

There are some strategies to train a model with real and synthetic data:

1. A simple method is to simultaneously train the model with synthetic and real data. It consists of building batches with images from both domains (synthetic and real). When defining a ratio, the real images should dominate the distribution [9].

2. Another approach is to pre-train the model on synthetic data and then fine-tune it on real data [10]. Allowing the model to learn general patterns with the synthetic data and then adapt to the real world with the fine-tuning with real data. This approach and the last one described highly depend on the quality and amount of real data.

3. Another method is to add a domain classifier that predicts if the image belongs to the real or synthetic domain, aiming to learn useful features for both domains, in a way that the domain classifier cannot distinguish, achieving the extraction of domain-invariant features from the data [11]. They usually follow two steps: (i) learn features that minimize the loss of the target task and (ii) learn features that maximize the loss of a domain classifier.

4. A completely different approach is to use image-to-image translation techniques. These use generative adversarial networks (GANs)

[12]. The idea is to make the synthetic images look more realistic so the model can learn from both domains without confusion. These methods are making remarkable advances, but they still tend to introduce artefacts or distortions in the translated images, which can affect the model's performance. Therefore, this strategy was not considered in this study.

Our proposed strategy combines the extracted domain-invariant features with spatio-temporal cues to represent the direction of motion of the crowd.

TRAINING METHODOLOGY

Multi-task learning has shown that dedicating some networks to classify specific features related to the principal task can help improve the model's performance, as shown in [13]. Our training strategy uses this principle and adds classifiers that identify the direction of the crowd runs during panic, along with a domain classifier, to outperform a conventional panic detector model. The proposed training strategy extracts domain-invariant features from the data and captures both the input sequence's spatio-temporal dynamics and the crowd motion information. The architecture consists of three different components (see Fig. 24.1):

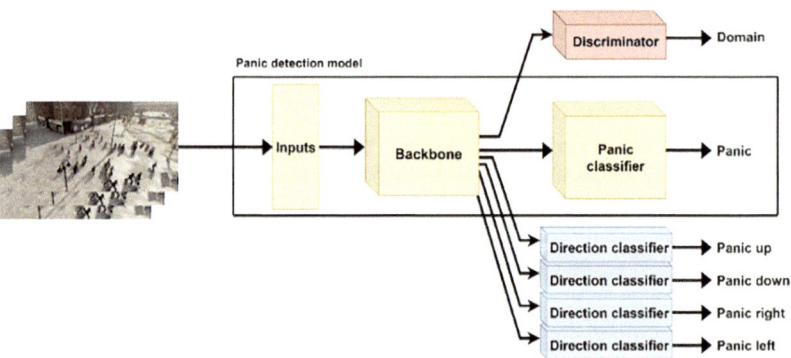

Fig. 24.1 Proposed training architecture. Including a panic detection model, a discriminator, and four direction classifiers

1. Panic detection model: a standard model with an input layer that receives frame sequences, a backbone network, and a classifier layer.
2. Discriminator: is a domain classifier that identifies the input sample to its corresponding domain (real or synthetic) using a fully connected layer, and a two-class output layer with a softmax activation. The purpose of adding a discriminator is to help the backbone learn domain-invariant features from the data. This is achieved by taking the loss of the discriminator and applying it to the backbone in a reverse manner using a gradient reversal layer. In the end, the discriminator will not be able to accurately classify the domain of the input due to the information given to the backbone that helped reduce the extraction of domain-specific features.
3. Direction classifiers: these are classifiers that consist of a fully connected layer, and a two-class output layer with a softmax activation. Each classifier is attached to the backbone of the panic detection model. Their purpose is to determine the direction of the panic runs. Each discriminator activates the panic output when the panic happens in their direction, otherwise its value is not panic. While there is no panic, in any direction, all classifiers should predict as no panic.

During the testing phase, the discriminator and direction classifiers are removed, using only the panic detection model.

During training, we create batches that include sequences from both domains, each with six labels: panic, domain, and four panic directions (top, bottom, right, and left). This approach narrows the domain gap by learning domain-invariant features from the data, allowing us to incorporate more synthetic data and have more flexibility in choosing the ratio of distribution of data from both domains. However, a highly imbalanced dataset could yield worse results, and the threshold ratio that determines the trade-off between performance and data balance is still an open question.

Experimental Results

Datasets

Real datasets. There are few real datasets for panic detection events using the CCTV camera's point of view (see Fig. 24.2). The available datasets are:

Fig. 24.2 Samples from real datasets: (left) UMN, (centre) MED, (right) PETS2009

- UMN [14]: a common benchmark created by the University of Minnesota with ten sequences of three scenes, recorded with a fixed camera at 30 fps and 480 × 640 resolution. Is used for testing abnormal event detection in crowds. They walk with no regular motion pattern and then suddenly run away in different directions.
- MED (Motion Emotion Dataset) [15]: has videos of people walking on individual walkways, filmed from above with a stationary camera at 30 fps, with labels for the emotions and behaviours of the crowd. The videos begin with normal scenes and end with abnormal events (panic, fights, fallen people, abandoned objects, etc).
- PETS2009 [16]: has multiple cameras, actors, and some abnormal events where people run away, but has a low frame rate of 8 fps. It is good to estimate the number and density of people, track individuals, and detect different flows and events.

Synthetic dataset. We use a synthetic dataset [17] composed of recordings showing the behaviour of crowds in normal and panic situations. Each recording shows a group of pedestrians walking randomly with different density levels (low, mid, and high). The dataset contains sequences generated in six places (see Fig. 24.3), using different camera configurations (angle and position), weather conditions, and pedestrian locations for each simulation. Each simulation has at least four cameras placed in different positions and angles to observe the scene from different perspectives. The resulting dataset contains videos of 375 frames each. The recordings include 6 different places, 31 different CCTV cameras in total (with a slight variation in each simulation), obtaining 320 different videos with various weather conditions.

Training dataset. Our proposed architecture uses spatio-temporal data to make predictions, so the frame rate of the training dataset and testing

Fig. 24.3 Synthetic dataset samples showing the six scenarios

dataset must match. We discarded the usage of the PETS2009 dataset due to its low frame rate. The UMN dataset is a single video composed of different panic scenes, which we have separated into independent videos. The labels are printed over the frames when the panic begins. To avoid the detection of the text box as panic, we have cropped all the frames removing the top part from the image The MED dataset contains a variety of anomalous situations, so we have selected only the sequences containing panic (1, 2, 3, 4, 5, 8, 9, 10, and 11) for our training and testing. After this process, the UMN offers 11 videos and the MED dataset offers 9 videos, for a total of 20 available videos. As there are insufficient data to train our model, we will use our synthetic dataset.

We annotated the three filtered datasets considering the crowd directions during the panic and created different versions of each training dataset by extracting sequences of varying length (30, 15, or 10 frames) and frame rate (30 or 15 fps) from the filtered video datasets. Any sequence containing both normal and panic frames is excluded; we also avoid any overlap between the sequences. After this process, we obtained 164 normal and 17 panic sequences from the UMN dataset; 1343 normal and 31 panic sequences from the MED dataset; and 11,888 normal and 5944 panic sequences from the synthetic dataset.

Implementation Details

Model selection. The proposed model for action classification is MoviNet-A0 [18], a state-of-the-art model designed for efficiency and accuracy in video recognition tasks. MoviNet is based on 3D CNNs and can handle challenges such as varying camera angles, lighting conditions or background changes. By using MoviNet, we expect high accuracy and efficiency in detecting panic and normal events, as we believe it is suitable for our problem due to its ability to extract spatio-temporal features.

Training parameters. The selection of sequence length and frame rate is crucial in our approach. A consistent frame rate for both training and inference is essential. The "loss weight" is a key parameter for training our proposed methodology, which determines the contribution of each branch to the overall model learning. It can be adjusted to balance the learning of the different branches and prevent one branch from dominating the others. It is important to monitor the training process and evaluate the performance of each classifier, ensuring that all classifiers, except the discriminator, learn to classify the sequences correctly. The discriminator should have an accuracy close to 0.5, indicating that it cannot differentiate between real and fake sequences.

We experimented with different frame rates and sequence lengths to find the optimal dataset for testing our methodology. Using 15 frames at 15 fps produced comparable results to using 30 frames at 30 fps, but with lower computational cost. Using frame rates lower than 15 fps or sequences shorter than 1 s resulted in worse performance. We also tested different image sizes and learning rates, selecting 256×256 pixels and 0.001 for the learning rate.

Testing method. In the testing phase, we use a real dataset that was not used in the training phase to evaluate the performance of the model on unseen data. We use a sliding window with a step of 1 and respect the frame rate. We add one every two frames of the video, which has a frame rate of 15 fps.

A sequence is considered panic when all the frames are panic. The panic ground truth labels have been moved 30 frames (1 s, the covered time of a sequence). To evaluate the performance of our model, we use the area under the curve (AUC) metric. It is reliable for evaluating our model's performance, as it considers both true positive and false positive rates. It measures how well it can distinguish between the two classes, making it useful when dealing with unbalanced datasets.

Analysis

Importance of the distribution of data. Several tests were conducted using the configuration explained in the Implementation Details section, one for determine the data distribution (between real and synthetic data). The model was trained with the panic detector and discriminator to verify that the selected number of synthetic sequences was appropriate for training. The desired behaviour was that the discriminator did not improve its performance and the panic detector learned to identify panic properly. It was

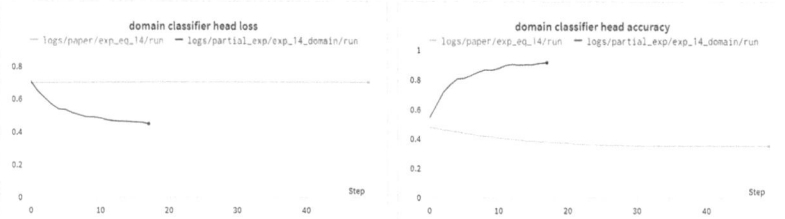

Fig. 24.4 Domain classifier training metrics. It shows the loss (left) and the accuracy (right) of models trained with different data distribution

found that the difference between the number of synthetic and real sequences was too large. The value was reduced until the training worked as expected. The original synthetic dataset had 17,832 sequences, while the filtered dataset had only 3155. The red experiment (see Fig. 24.4), which used the entire synthetic dataset, learned to identify the domain due to the increasing accuracy and the decreasing loss. The orange experiment showed that the discriminator could not learn the domain, lowering its accuracy to approximately 0.38. As a result, the reduced synthetic dataset was selected for the rest of the test.

Loss weight. To evaluate the impact of the loss weight variable on the performance of the model, a series of tests were conducted. The model was trained using a combination of one real dataset and the reduced synthetic dataset and tested on the non-used-for-training real dataset. This process was repeated, alternating both the UMN and MED datasets. The complete proposed architecture, including the panic model, the discriminator, and the direction classifiers, was utilized for this experiment.

The results shown in Fig. 24.5 correspond to a model trained with the MED dataset mixed with the reduced synthetic dataset and tested on the UMN dataset. Four different models were trained, each with a different set of loss weight values per branch: one for the discriminator, one for the direction classifiers (all sharing the same loss weight value), and one for the panic classifier.

The influence of the loss weight is significant. After evaluating the models trained with both MED and UMN datasets (combined with the synthetic dataset), it was found that the best loss weight configuration was to assign 0.5 for the panic detection classifier, 0.0001 for the discriminator, and 0.4 for the direction classifiers.

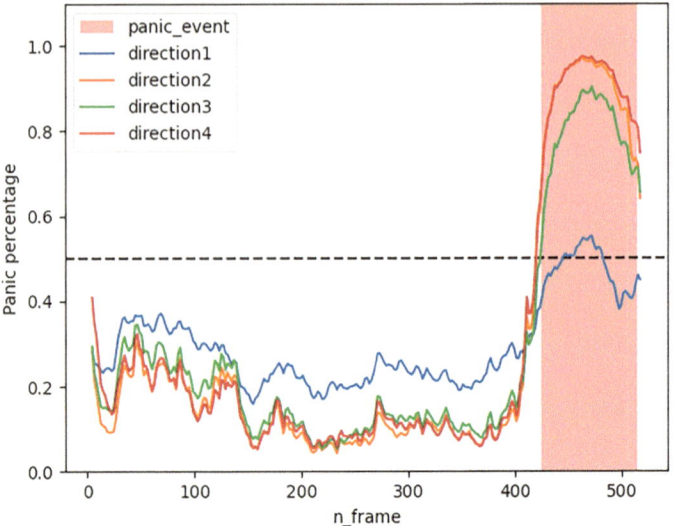

Fig. 24.5 Comparison of model performance with different loss weight values

COMPARATIVE OF METHODS

In this section, we compare the different strategies. These are referred to as: "Public": trained only with real data; "Mixed": trained with real and synthetic data; "Finetuned": trained first with synthetic data and finetuned with real data; "Domain": the panic model with the discriminator; and finally, "Direction": our proposed method.

After testing all the models with both public datasets (see Fig. 24.6), it can be observed that our proposed method ("Direction," light blue coloured) outperforms the other alternatives when the model is trained with the MED, followed by the "Domain" proposal (yellow). Analysing the models trained with UMN, it can be seen that the performance of both the "Domain" and "Direction" methods is superior, but in this case, the "Direction" model performs slightly better. Although the selected value of the discriminator loss weight may seem low, when comparing the results of the "Mixed" method (orange), that does not have a discriminator, with the "Domain" method, that has one, it is clear that the performance of the model improves, demonstrating that the addition of a discriminator works. The average AUC results of each model for each

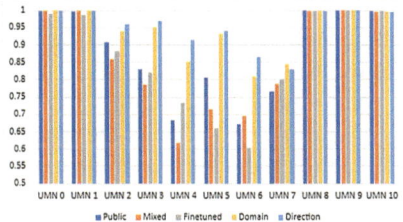 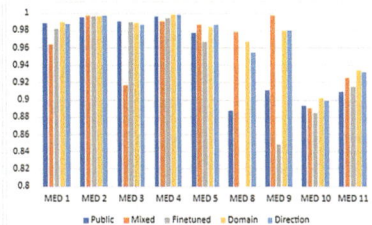

Fig. 24.6 Models AUC by video. (Left) Models trained with MED, prediction over UMN dataset. (Right) Models trained with UMN, prediction over MED dataset

Table 24.1 Methods AUC metric result

Test dataset	Public	Mixed	Finetuned	Domain	Direction
UMN	0.8776	0.8586	0.8605	0.9378	**0.9515**
MED	0.9303	0.9495	0.9578	**0.9706**	0.9689
Average	0.9040	0.9041	0.9092	0.9542	**0.9602**

Table 24.2 Comparison of our proposed method with the state-of-the-art methods

Method	UMN	MED
Ammar et al. [3]	0.991	0.9170
Zang et al. [6]	0.97	–
Ilyas et al. [4]	**0.9955**	–
Alafif et al. [5]	0.9810	–
Our proposal	0.9515	**0.9689**

AUC performance of anomaly detection in the datasets UMN and MED

dataset can be seen in Table 24.1, along with the average metric of both datasets. Our proposed method is the best option among the tested models.

COMPARISON WITH STATE OF THE ART

As can be seen in Table 24.2, our method offers acceptable results when tested on the UMN dataset but is surpassed by the rest of the SOTA methods. Regarding the MED dataset, only one of the methods offers results. In this case, our method improves it, becoming the SOTA.

We aim to compare our solution with other available methods. However, it is important to note that a direct comparison may not be entirely fair, as each state-of-the-art model was designed to solve different tasks, with some focusing on anomaly detection. Additionally, these models may not have been developed under the same conditions or to address the specific challenges we are tackling in this work, such as reducing the domain gap. It is worth highlighting the robustness of our method, as it offers good results on both datasets, demonstrating that it is capable of generalizing and functioning in different environments and situations.

CONCLUSION

We have presented a novel training methodology to train a model with real and synthetic data avoiding the domain gap issues demonstrating that it is the most effective option for training models combining synthetic and real data. Our approach has successfully shown the feasibility of utilizing synthetic data, even in situations where real data are limited and have achieved results that are competitive with the current state of the art. Additionally, we have shown that the use of multi-task learning approaches during training can enhance the performance of the primary task. By implementing a domain classifier equipped with a reversal gradient layer and properly configured hyperparameters, we can extract domain-invariant features during training, effectively addressing domain gap issues present in the datasets used. These findings underscore the potential and efficacy of our methodology in overcoming the challenges associated with combining synthetic and real data in model training.

Acknowledgements This project has received funding from the European Union's Horizon 2020 research and innovation programme under grant agreement no. 101021981, APPRAISE project.

REFERENCES

1. European Parliament, & Council of the EU. Regulation (EU) 2016/679 (2016) Official Journal of the European Union, L 119(1). (GDPR).
2. Afiq, A., Zakariya, M., Saad, M., et al. (2019). A review on classifying abnormal behavior in crowd scene. *Journal of Visual Communication and Image Representation*.

3. Ammar, H., & Cherif, A. (2021). DeepROD: A deep learning approach for real-time and online detection of panic behavior in human crowds. *Machine Vision and Applications.*

4. Ilyas, Z., Aziz, Z., Qasim, T., et al. (2021). A hybrid deep network based approach for crowd anomaly detection. *Multimedia Tools and Applications, 80,* 24053–24067.

5. Alafif, T., Alzahrani, B., Cao, Y., et al. (2022). Generative adversarial network based abnormal behavior detection in massive crowd videos: A hajj case study. *Journal of Ambient Intelligence and Humanized Computing.* https://doi.org/10.1007/s12652-021-03323-5

6. Zhang, X., Shu, X., & He, Z. (2019). *Crowd panic state detection using entropy of the distribution of enthalpy* (Physica A: Statistical Mechanics and Its Applications). Elsevier.

7. Xu, M., Yu, X., Chen, D., Wu, C., & Jiang, Y. (2019). An efficient anomaly detection system for crowded scenes using variational autoencoders. *Applied Sciences, 9*(16), 33–37.

8. Singh, K., Rajora, S., Vishwakarma, D. K., et al. (2020). Crowd anomaly detection using aggregation of ensembles of fine-tuned convnets. *Neurocomputing, 371,* 188–198.

9. Ros, G., Sellart, L., Materzynska, J., et al. (2016). The SYNTHIA dataset: A large collection of synthetic images for semantic segmentation of urban scenes. In *Proceedings IEEE CVPR* (pp. 3234–3243).

10. Shafaei, A., Little, J. J., & Schmidt, M. (2016). Play and learn: Using video games to train computer vision models. In *Proceedings BMVC.*

11. Tonutti, M., Ruffaldi, E., Cattaneo, A., & Avizzano, C. A. Robust and subject-independent driving manoeuvre anticipation through domain adversarial recurrent neural networks. *Robotics and Autonomous Systems, 115,* 162–173. 201.

12. Zhu, J.-Y., Park, T., Isola, P., & Efros, A. A. (2017). Unpaired image-to-image translation using cycle-consistent adversarial networks. In *IEEE international conference on CV.*

13. Rodriguez, A. M., Unzueta, L., Geradts, et al. (2023). Multi-task explainable quality networks for large-scale forensic facial recognition. *IEEE JSTSP, 17*(3), 612–623.

14. MultiMedia LLC. (n.d.). *Unusual crowd activity dataset of University of Minnesota.* Retrieved January, from http://mha.cs.umn.edu

15. Rabiee, H., Haddadnia, J., Mousavi, H., Kalantarzadeh, M., Nabi, M., & Murino, V. (2016). Novel dataset for fine-grained abnormal behavior understanding in crowd. In *IEEE international conference on advanced video and signal based surveillance.*

16. Ferryman, J., & Shahrokni, A. (2009). PETS2009: Dataset and challenge. In *IEEE international workshop on performance evaluation of tracking and surveillance*.
17. Calle, J., Leskovsky, P., Garcia, J., & Sanchez, M. (2023). *Synthetic dataset for panic detection in human crowded scenes*. Eurographics 2023 – Posters.
18. Kondratyuk, D., Yuan, L., Li, Y., Zhang, L., Brown, M., & Gong, B. (2021). *MoViNets: Mobile video networks for efficient video recognition*.

Open Access   This chapter is licensed under the terms of the Creative Commons Attribution 4.0 International License (http://creativecommons.org/licenses/by/4.0/), which permits use, sharing, adaptation, distribution and reproduction in any medium or format, as long as you give appropriate credit to the original author(s) and the source, provide a link to the Creative Commons license and indicate if changes were made.

The images or other third party material in this chapter are included in the chapter's Creative Commons license, unless indicated otherwise in a credit line to the material. If material is not included in the chapter's Creative Commons license and your intended use is not permitted by statutory regulation or exceeds the permitted use, you will need to obtain permission directly from the copyright holder.

CHAPTER 25

A Generative Adversarial Network (GAN) Solution for Synthetically Generated Botnet Attack Data Samples

Nikolaos Peppes, Theodoros Alexakis,
Emmanouil Daskalakis, Evgenia Adamopoulou,
and Konstantinos Demestichas

INTRODUCTION

The widespread adoption of digital services in people's daily lives has resulted in an increased demand for cybersecurity. With the proliferation of new software and hardware, detecting known botnets or other types of

N. Peppes (✉) • T. Alexakis • E. Daskalakis • E. Adamopoulou
Institute of Communication and Computer Systems, School of Electrical and Computer Engineering, National Technical University of Athens, Athens, Greece
e-mail: npeppes@cn.ntua.gr; talexakis@cn.ntua.gr; edaskalakis@cn.ntua.gr; eadam@cn.ntua.gr

K. Demestichas
Department of Agricultural Economics and Rural Development, Agricultural University of Athens, Athens, Greece
e-mail: cdemest@aua.gr

© The Author(s) 2025   311
I. Gkotsis et al. (eds.), *Paradigms on Technology Development for Security Practitioners*, Security Informatics and Law Enforcement, https://doi.org/10.1007/978-3-031-62083-6_25

attacks has become a daunting task for cybersecurity professionals. Botnets as one type of cyberattack can have disastrous consequences [1, 2], as they allow attackers to remotely control infected machines, since they have the potential to impact numerous devices in parallel, particularly within IoT networks, due to a large number of devices interconnected.

Cybersecurity incidents are predominantly addressed reactively, subsequently to the occurrence of an attack, necessitating the engagement of cybersecurity professionals to respond and mitigate resultant damage. To combat these infections, cybersecurity experts are developing proactive systems that utilize machine-learning and deep learning (ML & DL) technologies. Consequently, the primary dataset for cybersecurity analysis predominantly comprises historical attack data. This essentially implies that nearly all cybersecurity systems are developed based on historical attack patterns, rendering them susceptible to emerging variants. Nonetheless, many organizations refrain from sharing their attack data, resulting in a scarcity of such information, consequently hindering the effective training of ML or DL models and the development of such systems.

The current study proposes a methodology for generating botnet-type data in a tabular format. This methodology employs an 8-layer generative adversarial network (GAN) model [3] to evaluate its effectiveness in generating synthetic data with high precision while minimizing computational expenses The generated samples will be assessed using a wide range of graphical data quality indicators, such cumulative sums, absolute log mean and STD diagrams, correlation matrices, and heatmaps.

The remainder of this study contains Section "Related Works" that investigates related research on botnet attack generation techniques, Section "BNGAN: A Proposed Solution For Addressing the Data Issue" that provides a more in-depth explanation of the GAN model design methodology, whereas Section "Data Generation Results Evaluation" focuses on evaluating the synthetic dataset's significance and Section "Conclusion and Future Work" examines the revealed discoveries.

Related Works

The escalating damage inflicted on computer systems by botnet attacks has underscored the imperative need to delve deeper into detection methods. Consequently, a plethora of studies in this domain can be found in the existing literature. Within the context of our study's core elements outlined earlier, the works discussed in this section primarily concentrate on two key aspects: the generation and classification of botnet attack datasets.

Yin et al. [4] concentrated on augmenting botnet detection. Their research introduced a GAN designed to generate nearly lifelike botnet attack samples, enhancing the training of machine-learning classifiers. The Bot-GAN consistently supplied "synthetic" data to the discriminator, which classified these samples using a softmax function. This approach resulted in improved accuracy and precision when compared to pretrained models utilizing the original imbalanced dataset. Pursuing a similar route to mitigate the challenges posed by imbalanced datasets, Song et al. [5] introduced the GAN-efficient lifelong learning algorithm (ELLA) solution. Their methodology demonstrated that dataset expansion through a GAN architecture not only boosted the performance of traditional ML solutions for botnet identification but also enhanced the lifelong learning approach of the ELLA algorithm.

Tram Truong-Huu and his team [6] investigated the application of GANs in network anomaly detection. They employed multiple datasets to assess GANs' performance in comparison with other network anomaly detection methods. Their experiments revealed significant improvements over existing deep learning techniques, indicating promise in detecting unknown anomalous behavior and zero-day attacks focusing on botnet traffic.

Zhong et al. [7] introduced MalFox, a solution designed to demonstrate the limitations of existing black box detectors. MalFox employs a convolutional GAN and adopts a confrontational strategy to create perturbation paths. These paths incorporate up to three methods (Obfusmal, Stealmal, and Hollowmal) to generate adversarial malware examples. Their results showed promising performance, with an accuracy of approximately 99%, while the detection rate of the generated samples was at a lower percentage, around 45%.

The significance of GANs for data augmentation, especially in the cybersecurity realm, was underscored also by Habibi et al.'s Conditional Tabular GAN (CTGAN) model [8]. They experimented with various CTGAN versions and parameters to identify the most effective one. The outcomes demonstrated CTGAN's ability to preserve the structure of both continuous and discrete data. This provided a solution for ML classifiers or detectors, addressing dataset imbalances and training these algorithms for novel threats, given that GAN-generated data are novel and unseen.

Lingam et al. [9] conducted a study on imbalanced data concerning bot identification. Their objective was to tackle the issue of imbalanced data for ML classifiers by employing a GAN with a gated recurrent unit (GRU). This enabled them to generate synthetic data closely resembling real data, effectively balancing benign user and bot classes. Results indicated that their approach outperformed ML methods trained solely on the original Twitter dataset, achieving an average accuracy of approximately 91% with the GAN-generated dataset.

BNGAN: A PROPOSED SOLUTION FOR ADDRESSING THE DATA ISSUE

Generative adversarial networks (GANs) [1] utilize an architecture that generates new data based on input data and random noise. GANs consist of two components: the generator and discriminator. The generator uses random noise to create realistic data, while the discriminator classifies input samples as either real or fake. Both components are optimized based on the discriminator's ability to accurately classify real and fake data.

Hence, there is significant importance in conducting experiments involving various GAN architectures and adjusting their hyperparameters to discover the most suitable model tailored to a particular dataset and objective. Approaches like hyperparameter optimization and architectural exploration serve as valuable tools in pinpointing the ideal GAN structure and hyperparameters tailored to a specific task.

This study aims to evaluate the effectiveness of a proposed 8-layer GAN architecture called BNGAN in generating synthetic data that accurately represent malicious cyber-attacks, specifically botnet attacks [3]. To accomplish this, the study evaluates the performance of the proposed 8-layer GAN model [10] for both the generator and discriminator, using the CTU-13 dataset [11] from the Stratosphere IPS. This dataset includes captures of diverse malware samples and normal traffic, with 32 million packets. The training dataset has 216,352 records, with 140,849 marked as "0" for malware and 75,503 labeled as "1" for legitimate. The evaluation dataset has 88,258 records without any labels.

The study utilizes the BNGAN model architecture, which is designed to generate 1D synthetic data from the input dataset. The model was implemented using Tensorflow 2.0 and Keras API. The proposed BNGAN architecture utilizes the sequential API to stack the different layers of the deep neural network. The generator model is built using the sequential API and consists of an input layer for accepting appropriately scaled, randomly generated noise with the intended size. This input is then processed through six subsequent hidden layers utilizing the "ReLU" activation function, ultimately leading to an output layer. This output layer employs a "linear" activation function, aligning its dimension with that of the preprocessed dataset.

The discriminator, by itself, takes the form of a sequential model, composed also of eight dense layers. In the initial seven layers, the "ReLU" activation function is utilized, while the last layer employs the "sigmoid" function to classify input samples as either authentic (genuine) or counterfeit (malware). To bolster the model's precision, a 20% dropout rate is applied to both the visible (input) layer and the six concealed layers within the discriminator model. The ultimate choice of this dropout rate was reached through a series of iterative experiments, considering its influence on preventing overfitting while ensuring the model's capability to capture pertinent data patterns.

After detailing the generator and discriminator models, the proposed BNGAN model is characterized as a sequential model that integrates these components in an adversarial manner. Figure 25.1 illustrates how the BNGAN model uses (preprocessed) botnet data samples to generate synthetic, tabular data.

DATA GENERATION RESULT EVALUATION

In the previous chapter, the generator and discriminator models were established, combining them to form the comprehensive BNGAN model. Subsequently, the training process was initiated to facilitate the generation of datasets mirroring the originals. The training process encompassed a total of 1000 epochs, with each epoch involving batch training of a predefined size for both the generator and discriminator networks. In this process, the discriminator received as input a predetermined batch of data samples from the original dataset as well as the generated output data sample from the generator. For each (data) batch, the discriminator computed the loss for both the genuine and the generated data. The losses computed (by the discriminator) served to refine the predictions made by the discriminator model, subsequently enabling the computation of generator losses and gradients via backpropagation techniques. In this iterative process, the generator persists in enhancing the quality of the

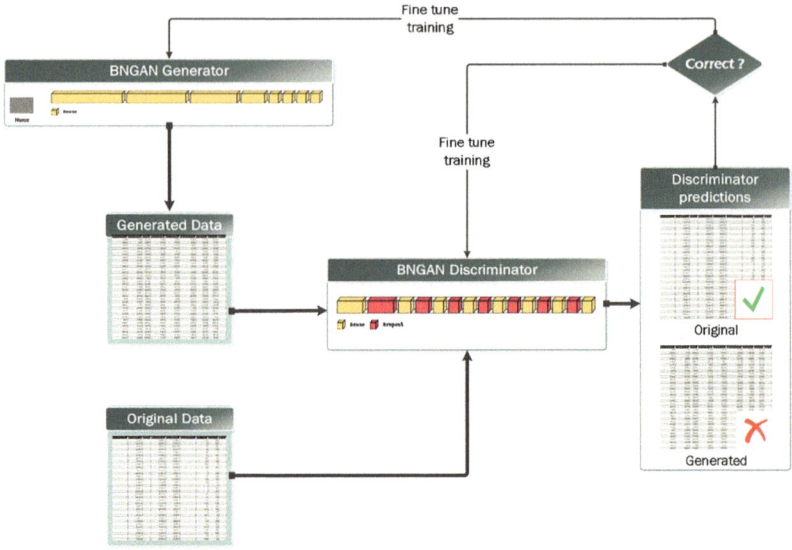

Fig. 25.1   BNGAN Model Implementation

synthetically generated data samples by constantly adjusting its weights based on these gradients.

Visual representations, such as diagrams, prove to be an effective way of assessing and illustrating the similarity between datasets (synthetically) generated by a GAN model and real data. These similarity estimation methods offer valuable insights into the fidelity and precision of the generated dataset, aiding researchers in pinpointing areas where enhancements to the GAN model might be necessary to produce synthetic data that closely mirrors real data. Furthermore, the GAN model's performance in creating synthetic data that closely resembles the real data can be determined. The choice of diagram types is contingent upon the nature of the analyzed data and the particularly considered objectives of the research. The current study includes the following diagrams to evaluate the generated data: correlation matrices with heatmaps, highlighting clusters illustrating distinctions between the real and generated datasets, cumulative sum (cumsum) diagrams for visualizing the accumulation over time of the original and the generated data and STD diagrams to compare the (similarity) scores between the original and generated datasets from the GAN

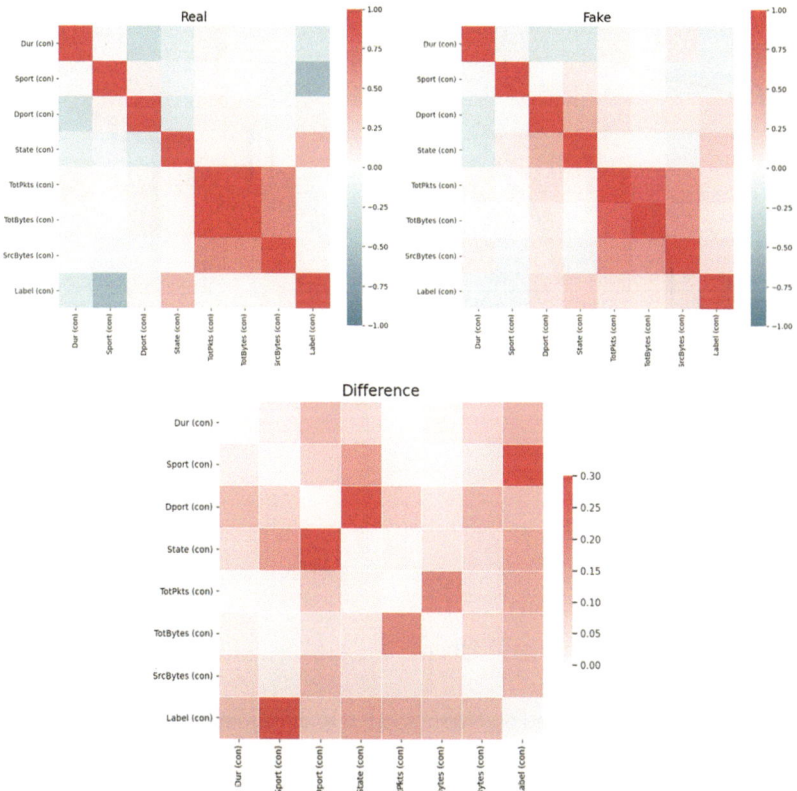

Fig. 25.2 Correlation Matrices with Heatmap

model. Figures 25.2, 25.3, and 25.4 visualize the comparison results between the original and the generated data from the GAN model.

Based on the results depicted previously, the cumulative sum diagrams reveal notable insights regarding the similarity between real and generated datasets for eight variables. Five of these variables (Dur, TotPkts, TotBytes, SrcBytes, and Label) exhibit a consistent, steadily increasing similarity score in both datasets, suggesting a continuous pattern. In contrast, the remaining three variables (Sport, Dport, and State) display a fluctuating pattern with abrupt spikes and drops in the similarity score, indicating deviations in the synthetic dataset. These fluctuations suggest certain data

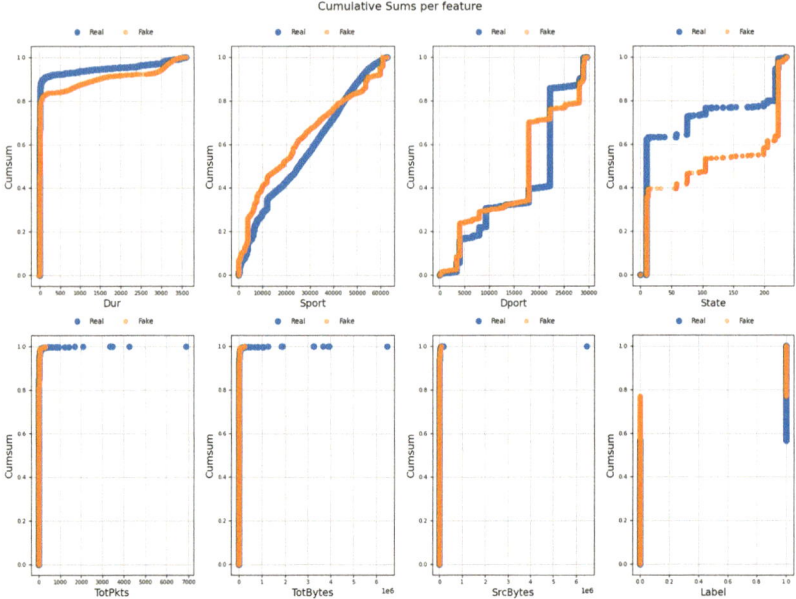

Fig. 25.3 Cumulative Sum Diagrams

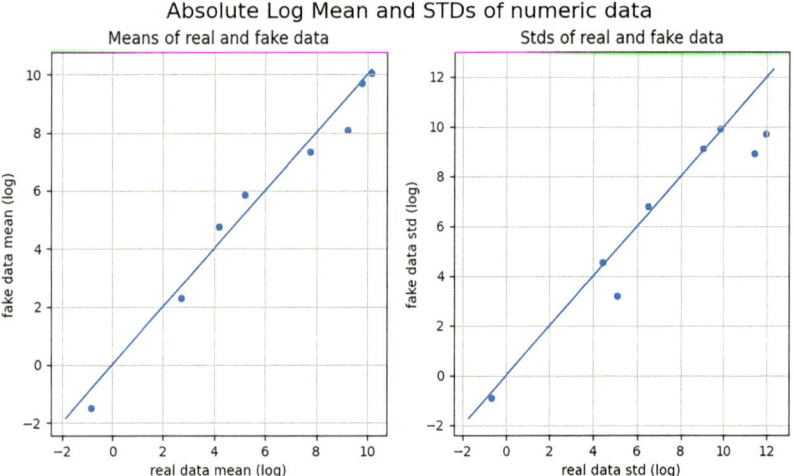

Fig. 25.4 STD Diagrams

points significantly diverge from the overall pattern, contributing to lower overall similarity scores for these variables. Furthermore, the cumulative sum diagrams suggest that the GAN model might require more training epochs to produce a synthetic dataset closely in order to resemble the real one. Moving to the correlation matrix diagrams, the real dataset illustrates a strong positive correlation among its variables. However, in the generated synthetic dataset, the positive correlations are weaker, and no significant negative correlations emerge. Additionally, a strong positive correlation arises between the various features in the "Difference" section, signifying that as the epochs progress, the generated data faithfully replicates patterns and characteristics from the real dataset in a realistic manner. Finally, examining the absolute mean and standard deviation diagrams reveals that the synthetic dataset contains higher values for certain features compared to the real dataset. This variance may suggest that the generated data for some features could not precisely mirror the real data, at least initially. However, as the number of training epochs increases, the synthetic dataset progressively aligns more closely with the real dataset provided.

CONCLUSION AND FUTURE WORK

As digital tools continue to evolve and become more prevalent, the need for effective cybersecurity measures has become increasingly critical. The primary objective of this study is to outline a comprehensive methodology for generating synthetic data for botnet attacks using generative adversarial networks (BNGAN). The generation process utilizes an open-source dataset, the CTU-13 dataset, provided by Stratosphere IPS, which is a collection of network traffic captures that have been widely used in the field of cybersecurity research. This tabular format data is used as input for the suggested BNGAN architecture [11]. The BNGAN model generates over 200,000 new botnet data samples that closely resemble the original data. Subsequently, the generated botnet data samples are evaluated using a wide range of graphical data quality indicators, including cumulative sums, absolute log mean and STD diagrams and correlation matrices with heatmaps, to assess the quality of the generated data. Overall, this proposed methodology provides a promising approach to improving botnet attack detection and prevention. The future prospect of this research involves expanding data categories and domains into various fields, encompassing diverse data formats and addressing a broader range of cyberthreats. Furthermore, an important avenue of exploration is the integration of lifelong learning techniques, both for data generation and the zero-day detection and classification of such attacks.

REFERENCES

1. Shinan, K., Alsubhi, K., Alzahrani, A., & Ashraf, M. U. (2021). Machine learning-based Botnet detection in software-defined network: A systematic review. *Symmetry, 13*. https://doi.org/10.3390/sym13050866
2. Check point check point research reports a 38% increase in 2022 global cyberattacks Available online: https://blog.checkpoint.com/2023/01/05/38-increase-in-2022-global-cyberattacks/. Accessed on 22 July 2023.
3. Randhawa, R. H., Aslam, N., Alauthman, M., Rafiq, H., & Comeau, F. (2021). Security hardening of Botnet detectors using generative adversarial networks. *IEEE Access, 9*, 78276–78292. https://doi.org/10.1109/ACCESS.2021.3083421
4. Yin, C., Zhu, Y., Liu, S., Fei, J., & Zhang, H. (2018). An Enhancing framework for botnet detection using generative adversarial networks. In *Proceedings of the 2018 international conference on artificial intelligence and Big Data (ICAIBD)* (pp. 228–234).
5. Song, C., Wushouer, M., & Tuerho, G. (2022). Botnet detection based on generative adversarial network and efficient lifelong learning algorithm. In *Proceedings of the 2022 international conference on Big Data, information and computer network (BDICN)* (pp. 48–54).
6. Truong-Huu, T., Dheenadhayalan, N., Pratim Kundu, P., Ramnath, V., Liao, J., Teo, S. G., & Praveen Kadiyala, S. (2020). An empirical study on unsupervised network anomaly detection using generative adversarial networks. In *Proceedings of the proceedings of the 1st ACM workshop on security and privacy on artificial intelligence* (pp. 20–29). Association for Computing Machinery: New York, NY, USA.
7. Zhong, F., Cheng, X., Yu, D., Gong, B., Song, S., & Yu, J. (2023). MalFox: Camouflaged adversarial malware example generation based on Conv-GANs against Black-Box detectors. *IEEE Transactions on Computers*, 1–14. https://doi.org/10.1109/TC.2023.3236901
8. Habibi, O., Chemmakha, M., & Lazaar, M. (2023). Imbalanced tabular data modelization using CTGAN and machine learning to improve IoT Botnet attacks detection. *Engineering Applications of Artificial Intelligence, 118*, 105669. https://doi.org/10.1016/j.engappai.2022.105669
9. Lingam, G., Yasaswini, B., Jagadamba, P. V. S. L., & Kolliboyana, N. (2022). An improved Bot identification with imbalanced data using GG-XGBoost. In

Proceedings of the 2022 2nd international conference on intelligent technologies (CONIT) (pp. 1–6).

10. Peppes, N., Alexakis, T., Adamopoulou, E., & Demestichas, K. (2023). The effectiveness of zero-day attacks data samples generated via GANs on deep learning classifiers. *Sensors, 23.* https://doi.org/10.3390/s23020900

11. García, S., Grill, M., Stiborek, J., & Zunino, A. (2014). An empirical comparison of Botnet detection methods. *Computers & Security, 45,* 100–123. https://doi.org/10.1016/j.cose.2014.05.011

Open Access This chapter is licensed under the terms of the Creative Commons Attribution 4.0 International License (http://creativecommons.org/licenses/by/4.0/), which permits use, sharing, adaptation, distribution and reproduction in any medium or format, as long as you give appropriate credit to the original author(s) and the source, provide a link to the Creative Commons license and indicate if changes were made.

The images or other third party material in this chapter are included in the chapter's Creative Commons license, unless indicated otherwise in a credit line to the material. If material is not included in the chapter's Creative Commons license and your intended use is not permitted by statutory regulation or exceeds the permitted use, you will need to obtain permission directly from the copyright holder.

CHAPTER 26

Geo-temporal Crime Forecasting Using a Deep Learning Attention-Based Model

Fabio Caffaro, Lorenzo Bongiovanni, and Claudio Rossi

INTRODUCTION

Crime forecasting plays a vital role in law enforcement agencies' efforts to prevent and address criminal activities. Accurate predictions regarding the spatial and temporal patterns of crime can assist in resource deployment, proactive intervention, and effective crime prevention strategies [1]. In this sense, it is essential to identify the possible crime hotspots within narrow regions spatially as general predictions on larger areas, such as the city or district level, do not allow to design and implement strategies to combat crimes effectively [2]. A substantial amount of previous research has been performed on the application of machine learning (ML) for the task of crime predictions [3]. In this chapter, we present a deep learning (DL) attention-based approach to geo-temporal crime forecasting.

Our research focuses on developing a transformer-based model specifically designed for crime forecasting. The model consists of an encoder that

F. Caffaro (✉) • L. Bongiovanni • C. Rossi
LINKS Foundation, AI, Data & Space (ADS), Torino, Italy
e-mail: fabio.caffaro@linksfoundation.com;
lorenzo.bongiovanni@linksfoundation.com; claudio.rossi@linksfoundation.com

© The Author(s) 2025   323
I. Gkotsis et al. (eds.), *Paradigms on Technology Development
for Security Practitioners*, Security Informatics and Law
Enforcement, https://doi.org/10.1007/978-3-031-62083-6_26

takes as input the crimes that occurred during a given context window of n days and a decoder that generates the forecasts for the next m days based on the input fed by the encoder. By leveraging the power of DL techniques, we aim to capture and utilize the intricate relationships between crime occurrences over time and their spatial context. The model leverages the power of transformers and attention mechanisms to capture the spatial and temporal correlations of crime occurrences. Our experimental results highlight the superior performance of our model and its potential to contribute significantly to the field of crime prevention and law enforcement efforts.

RELATED WORK

Numerous studies have tackled the challenging task of geo-temporal crime forecasting, aiming to provide accurate predictions and assist law enforcement agencies (LEAs) in combating crime. In this section, we discuss relevant works that have explored different approaches and techniques in this domain.

Traditional statistical models, such as linear regression [4] and random forest [5], have been widely used for crime prediction and to identify possible crime hotspots. These models often rely on historical crime patterns and spatial information to identify correlations and forecast future crime occurrences. However, their limitations in capturing complex spatial and temporal relationships restrict their predictive capabilities.

Other ML approaches have also been employed for crime forecasting. Clustering algorithms, for instance, have been utilized to identify crime hotspots and spatial patterns [6]. These methods leverage spatial analysis to detect areas with high crime rates and predict future criminal activities. However, the absence of temporal dynamics may hinder their forecasting accuracy.

In recent years, transformer-based models have gained attention across various domains, including natural language processing and computer vision [7]. Their ability to capture long-range dependencies and model interactions across different input elements makes them suitable also for crime forecasting tasks. By incorporating attention mechanisms, transformers are able to effectively consider both spatial and temporal contexts, improving the predictive performance for geo-temporal crime forecasting.

Our work focuses on developing a novel attention-based model tailored specifically for the task of geo-temporal crime forecasting, aiming to overcome the limitations of previous approaches and achieve enhanced accuracy in predicting crime occurrences within a given area.

Model and Data

In this section, we outline the methodology adopted and the data used for the development of our geo-temporal crime forecasting model.

Model Architecture

We developed a transformer-based model (see Fig. 26.1) adopting an encoder–decoder architecture [7] that consists of multiple layers of self-attention and feedforward networks, which allows the model to capture long-term dependencies in the sequential data.

The encoder receives as input a context window, containing the crime occurrences from the previous n days, while the decoder generates the forecasts for the next m days based on the input provided by the encoder. The attention mechanism within the model facilitates the consideration of both spatial and temporal correlations, enabling effective crime prediction.

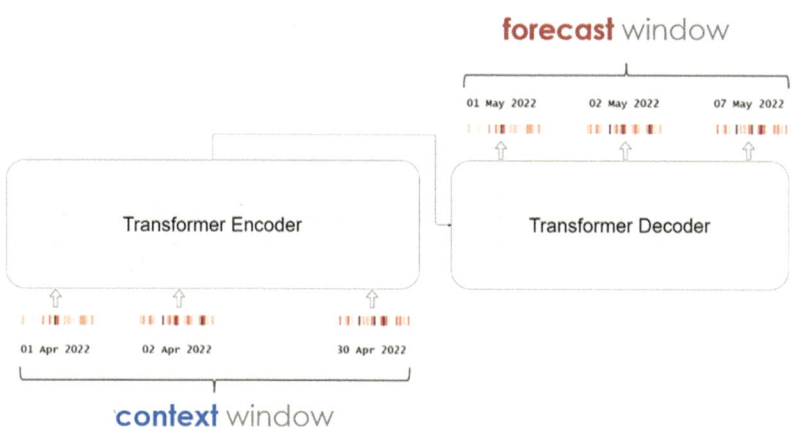

Fig. 26.1 Depiction of the encoder–decoder architecture adopted. Each input token of the model represents the daily distribution of crimes

Data Source

To test the model, we utilized the public dataset "Boston Incident Crime Report" published by the Boston Police Department[1]. This comprehensive dataset covers crimes that occurred from August 2015 to December 2022 on incidents such as larceny, burglary, and robbery. A total of 4,68,208 crimes are reported in the dataset, with an average of 5202 crimes per month. Each crime is geo-localized (with latitude and longitude coordinates) and time-stamped.

Spatial Grid and Input Data Representation

To perform crime forecasting on a fine-grained spatial level, we adopted a grid-based approach. The grid consists of cells with dimensions of 1 km by 1 km. By dividing the area of interest into these cells, we can effectively capture localized crime patterns and predict crime occurrences at a granular level. The grid-based approach enables us to assess crime trends and forecast crime hotspots within each cell. The resulting grid is composed of 122 cells (see Fig. 26.2).

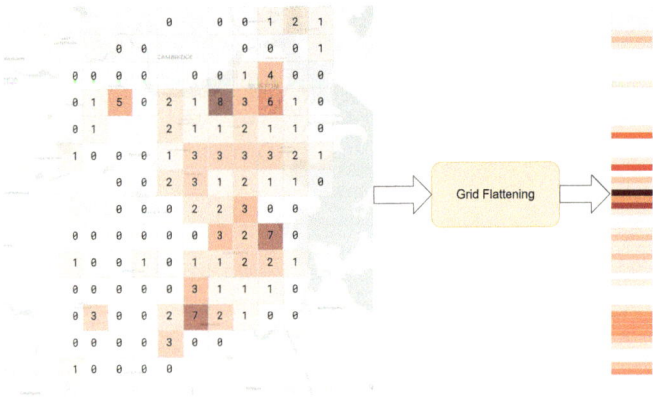

Grid   Features embedding vector

Fig. 26.2 Depiction of the crime distribution for a given day and the corresponding feature embedding vector obtained by flattening the 2D-grid representation

Starting from the grid division, we constructed the input tokens of the model for each day. The distribution of crimes over the grid was flattened, resulting in a feature embedding vector of dimension 1 times the number of cells in the grid. Each element of the feature vector represents the total number of crimes within a specific cell on a given day. Consequently, the input tokens capture the spatial distribution of crimes for each day, facilitating the learning of spatial correlations by the model.

Training and Testing

For model training, we used the data from 2015 up to the end of 2021, encompassing several years' worth of crime incidents. To evaluate the model's performance and assess its generalization capability, we tested it on the data of 2022.

We considered a context window composed of the crimes that happened during the previous 30 days and a forecast window of the following 7 days.

We implemented this work on Google Colaboratory Pro+ with Python 3.10.11, using Pytorch 2.0 for the implementation of the transformer model (i.e., nn.TransformerEncoder and nn.TransformerDecoder) and scikit-learn for the baseline models (i.e., RandomForestRegressor and LinearRegression). We set the transformer model with a hidden size equal to 32, 2 layers, a dropout equal to 0.1, and a learning rate of 1e-4, while for the random forest model, we use 100 trees and a maximum depth of 4.

RESULTS

We evaluated the model's performance by measuring the mean average error (MAE) and mean squared error (MSE) of each cell's predicted daily number of crimes. The dataset was split, considering as a training set all the crimes that happened before January 1, 2022, and as a test set all the remaining ones. Our experimental results show that the proposed model outperforms traditional machine learning models, such as the linear regression model [8] and random forest [9] for crime forecasting. As it is possible to observe from Table 26.1, the transformer model proposed provides a substantial improvement with respect to standard machine learning models. In particular, the model obtains a score of 1.674 in MSE, achieving a reduction of 68% and about 18% compared to the linear regression and random forest models, respectively.

Table 26.1 The
obtained MAE and MSE
for different models on
the test set

Model	MAE	MSE
Linear regression	1.319	5.276
Random forest	0.797	2.041
Transformer	**0.791**	**1.674**

The baseline models (linear regression
and random forest) are compared with the
proposed model

CONCLUSIONS

Accurate crime predictions can assist law enforcement agencies in allocating resources to effectively address crime in specific areas, thereby improving public safety. In this chapter, we proposed a deep learning model based on an encoder–decoder transformer architecture for geo-temporal crime forecasting. The model demonstrated its ability to capture crime incidents' spatial and temporal dependencies and forecast localized crime patterns, improving the prediction accuracy against baseline models proposed in previous studies. In future work, we plan to extend our model by incorporating additional features (e.g., weather forecasts, point of interest, and land use) to make the model spatially agnostic and scalable to different cities.

Acknowledgments This project received EU funding through the STARLIGHT project (grant agreement no. 101021797), the APPRAISE project (grant agreement no. 101021981), and the LAGO project (grant agreement no. 101073951).

REFERENCES

1. Benbouzid, B. (2019). To predict and to manage. Predictive policing in the United States. *Big Data & Society, 6*(1). https://doi.org/10.1177/2053951719861703
2. Weisburd, D., Bernasco, W., & Bruinsma, G. J. N. (2009). *Putting crime in its place: Units of analysis in geographic criminology.* Springer.
3. Jenga, K., Catal, C., & Kar, G. (2023). Machine learning in crime prediction. *Journal of Ambient Intelligence and Humanized Computing, 14,* 2887–2913. https://doi.org/10.1007/s12652-023-04530-y
4. Cavadas, B., Branco, P., & Pereira, S. (2015). Crime prediction using regression and resources optimization. In F. Pereira, P. Machado, E. Costa, & A. Cardoso (Eds.), *Progress in artificial intelligence. EPIA 2015. Lecture notes in computer science()* (Vol. 9273). Springer. https://doi.org/10.1007/978-3-319-23485-4_51

5. Yao, S., et al. (2020). *Prediction of crime hotspots based on spatial factors of random forest*. 2020 15th international conference on computer science & education (ICCSE), Delft, Netherlands, pp. 811–815, https://doi.org/10.1109/ICCSE49874.2020.9201899.
6. Cesario, E., Lindia, P., & Vinci, A. (2023). Detecting multi-density urban hotspots in a smart city: Approaches, challenges and applications. *Big Data and Cognitive Computing, 7*(1), 29. https://doi.org/10.3390/bdcc7010029
7. Vaswani, A., et al. (2017). Attention is all you need. *Advances in Neural Information Processing Systems, 30*, 5998.
8. Nelder, J. A., & Wedderburn, R. W. M. (1972). Generalized linear models. *Journal of the Royal Statistical Society. Series A (General), 135*(3), 370–84. JSTOR.
9. Breiman, L. (2001). Random forests. *Machine Learning, 45*, 5–32.

Open Access This chapter is licensed under the terms of the Creative Commons Attribution 4.0 International License (http://creativecommons.org/licenses/by/4.0/), which permits use, sharing, adaptation, distribution and reproduction in any medium or format, as long as you give appropriate credit to the original author(s) and the source, provide a link to the Creative Commons license and indicate if changes were made.

The images or other third party material in this chapter are included in the chapter's Creative Commons license, unless indicated otherwise in a credit line to the material. If material is not included in the chapter's Creative Commons license and your intended use is not permitted by statutory regulation or exceeds the permitted use, you will need to obtain permission directly from the copyright holder.

CHAPTER 27

Leveraging Continuous Learning for Fighting Misinformation

Evgenia Adamopoulou, Theodoros Alexakis, Nikolaos Peppes, Emmanouil Daskalakis, and Konstantinos Demestichas

INTRODUCTION

The eruption of digitization and the establishment of social media as a major content production and reproduction means have led to new paradigms of journalism and news spreading. The rapid changes that took place in the last 20 years led to an environment of pluralism without borders, where also many threats are lurking. One of these threats is the rapid

E. Adamopoulou (✉) • T. Alexakis • N. Peppes • E. Daskalakis
Institute of Communication and Computer Systems, School of Electrical and
Computer Engineering, National Technical University of Athens, Athens, Greece
e-mail: eadam@cn.ntua.gr; talexakis@cn.ntua.gr; npeppes@cn.ntua.gr;
edaskalakis@cn.ntua.gr

K. Demestichas
Department of Agricultural Economics and Rural Development, Agricultural
University of Athens, Athens, Greece
e-mail: cdemest@aua.gr

© The Author(s) 2025   331
I. Gkotsis et al. (eds.), *Paradigms on Technology Development
for Security Practitioners*, Security Informatics and Law
Enforcement, https://doi.org/10.1007/978-3-031-62083-6_27

spreading of misinformation and disinformation. It has been reported that fake news is spreading even six times faster than credible information [1]. This phenomenon represents a major concern, firstly, for media organizations and professionals, and, secondly, for law enforcement agencies (LEAs) due to the fact that the rapid spread of disinformation can severely threaten several aspects of society. According to the European Commission, the spread of both disinformation and misinformation can have a range of harmful consequences, such as the threatening of our democracies, the polarization of debates, and the setting of the health, security, and environment of EU citizens at risk [2].

As the practices of misinformation and disinformation evolve, it is of utmost importance to design, develop, and engage innovative technologies and solutions in order to tackle such phenomena. In this light, numerous approaches have emerged taking advantage of machine learning (ML) in order to address this problem from different viewpoints. Even though, from a technical perspective, different solutions for fake news detection and identification of misinformation exist, such as transfer learning, multitask learning, reinforcement learning, and online learning, no universal solution that can address all the aspects of the issue has been developed so far. Almost each and every single solution aims to address the problem in a specific topic or narrow domain and based on a limited dataset.

The purpose of this study is to present an approach that combines and evaluates the results of different machine learning prediction models into a common environment named "Meta-Detection Toolset." This solution relies on the calculation of a meta-score by using weights-based voting among different "prediction models," which are referred to herein as "verification services." The weights of the verification services are constantly updated by the end users of the toolset based on an annotation procedure. This leverages the current solution into a lifelong learning approach that is future-proof and adaptable as the machine learning models improve or deteriorate, through the course of time, and might perform better or worse for different topics or styles of writing.

The remainder of this study is structured as follows: Section "Related Work" contains related works concerning natural language processing applications (e.g., topic selection and language modeling) and lifelong learning studies. Section "Meta-Detection Toolset" presents in detail the proposed Meta-Detection Toolset, and, finally, Section "Conclusion: Future Work" concludes the article and paves the path for future updates of the presented toolset.

RELATED WORK

Lifelong learning (LL) or continuous learning (CL) is an emerging trend in computer science as well as in artificial intelligence. Thus, in the last few years, there has been an upward trend for studies focused on producing systems and solutions based on the concept of LL. The vast majority of dis/misinformation fighting tools are based on machine learning and deep learning algorithms. A comparative analysis of six available state-of-the-art fake news detection tools was made by Giełczyk et al. [3]. This comparison was feasible due to the fact that the datasets used were labeled and this is something rare in real-life conditions. The use of LL comes as an answer to minimize the need of expensive and scarce labeled data. In the domain of dis/misinformation and the trustworthiness of news and articles, LL is at a nascent stage and, thus, this section presents notable LL studies relevant to other fields of text analysis and natural language processing.

Topic identification is an application that can be enabled by LL approaches. More specifically, ML-based models called "topic models" extract hidden structures and correlations from a collection of documents in order to classify similar documents under a common topic. Each topic contains sets of common or contextually related words or characteristics [4]. In this light, Chen et al. [5] proposed a lifelong topic model called non-negative matrix factorization (NMF)—lifelong topic model (LTM). The method of Chen et al. [5] showed better performance compared to other methods after extensive experiments on public corpora. In the same direction, Xu et al. [6] proposed the lifelong learning topic (LLT) model that tries to lift the limitations when there are limited co-occurrences in a dataset. The LLT model is based on the notion of lifelong learning and expands the topic knowledge discovered by learning new word-embeddings based on the topics generated in previous iterations. Another interesting approach to topic modeling and learning was made by Zhang et al. [7]. More specifically, Zhang et al. combined generative adversarial network (GAN) with lifelong learning in their solution named lifelong knowledge-enhanced adversarial neural topic model (LKATM). LKATM discovers topics in documents by using a knowledge extractor that utilizes knowledge distillation and data augmentation in order to transfer prior topic knowledge.

Apart from topic identification, language modeling is another task where LL is exploited to offer state-of-the-art solutions. In this context, Sun et al. [8] proposed the language modeling for lifelong language modeling (LAMOL) framework. Specifically, LAMOL is a language model that learns to solve tasks and, at the same time, generates training samples. The dynamic representations for imbalanced lifelong learning (DRILL) solution is presented by Ahrens et al. [9] and mainly focuses on addressing dataset limitations. In particular, DRILL is defined as a novel lifelong learning architecture for open-domain sequence classification. DRILL is a hybrid architectural and rehearsal-based continuous learning method that utilizes meta-learning and a self-organizing neural architecture in order to adapt to new unseen data while trying to avoid catastrophic forgetting.

Chen et al. [10] proposed an LL solution for sentiment classification based on product reviews. Their approach focused on negative or positive reviews for various products. Each product represented a different task for the LL learner they proposed. He et al. [11] studied and proposed the applicability of an LL model in Weibo rumor detection. Their approach aimed to mitigate the rapid changes happening in online news and rumors as well as the limited availability of data.

To the best of our knowledge and at the time of writing the current study, there is no dedicated LL method or approach to the trustworthiness of news or articles. Thus, the related works presented in this section constitute LL solutions and approaches concerning applications relevant to text and language processing in other domains, such as rumor detection, sentiment detection, language modeling, and topic detection.

META-DETECTION TOOLSET

The proposed solution of the Meta-Detection Toolset engages different verification services. These diverse verification services serve as predictors of credibility for a given piece of content (typically, an article provided in the form of a URL or text). Based on the integration and implementation of a weighted majority algorithm [12], equal weights are initially assigned to each verification service. During the continuous training process, the weight assigned to a verification service is automatically adjusted, according to the accuracy of its predictions. Verification services with more correct predictions during the training phase are provided with higher weights, thus playing a more significant role when the MDT is calculating the credibility of a certain article. The verification services that the MDT can host may vary, ranging, for instance, from BERT-based models up to sentiment and stylometric analysis models (Fig. 27.1). Also, the input that

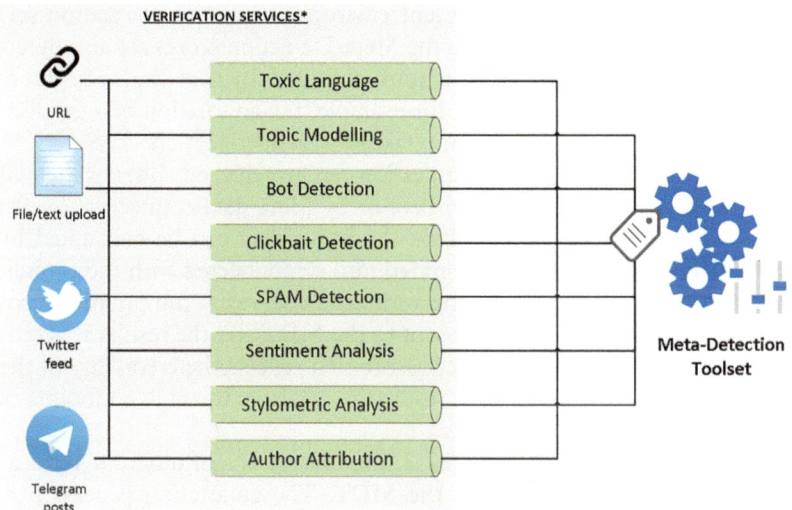

Fig. 27.1 Potential verification services that can be hosted in the MDT

the MDT can process might come from diverse sources, such news sites, or social media posts (Twitter, Telegram, etc.), as shown in Fig. 27.1.

End users, for example, fact-checkers, also play an active role in the training process. More specifically, end users can insert their credibility evaluation of specific articles (i.e., indicating whether a specific piece of content represents legitimate or fake news). The aforementioned users' evaluations are provided in the form of ground-truth labels (legitimate/fake), stored in a database, and utilized during the continuous training phase for updating the weights assigned to each verification service. Thus, a growing number of these annotations lead to improved verification results of the Meta-Detection Toolset.

The accumulated experience of the toolset leads to the generation of a model that extensively utilizes contemporary AI technologies for combating the spread of dis/misinformation on the web or in social media. This model is comprised of multiple specialized verification services and has the ability to combine them, aiming to evaluate the truth based on a complex scoring mechanism. This AI-based process is called Meta-Detection and achieves continuous improvement established by annotation processes performed by specialized end-users. In the context of the Meta-Detection

Toolset, an integrated management environment of the verification services has been developed, where the Meta-Detection scores are also determined according to the annotations provided by fact-checkers. More specifically, for a specific article, for example, the annotation of a ground-truth label is provided (legitimate/fake) by certified fact-checkers.

As shown in Fig. 27.2, data ingestion can be achieved either at the end users' side over the HTTPS protocol or by using data connectors (Kafka topics and/or REST APIs). Then, the input data can be consumed by various verification services integrated into or connected with the toolset. Following the completion of the verification services' computation processes, the prediction results are sent to the MDT and the results are combined in order to compute a meta-score that reflects the credibility of the digital content. The meta-score results are available through endpoints of REST APIs and/or Kafka topics.

As shown in Figs. 27.2 and 27.3, the evaluations of different verification services are combined by the MDT. The annotation process performed by the fact-checkers helps to identify which verification services perform better compared to the rest. These annotations are provided through the Meta-Detection Toolset user interface and better-performing verification services are provided with higher weights. In this way, the MDT enables the knowledge retention from previous evaluations and is

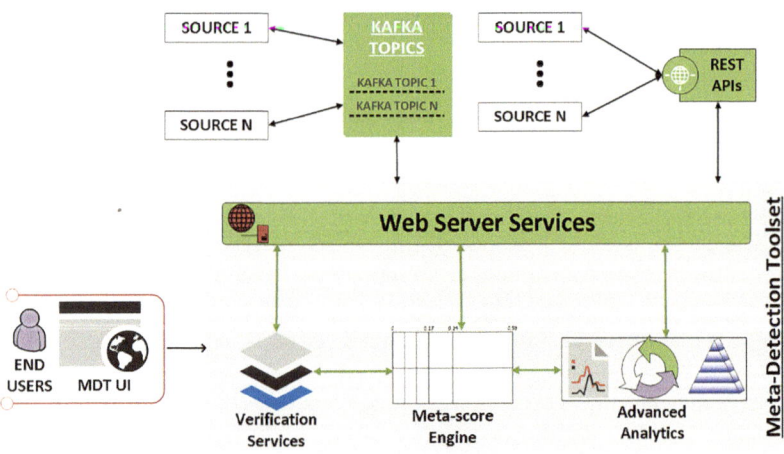

Fig. 27.2 High-level architecture of the Meta-Detection Toolset

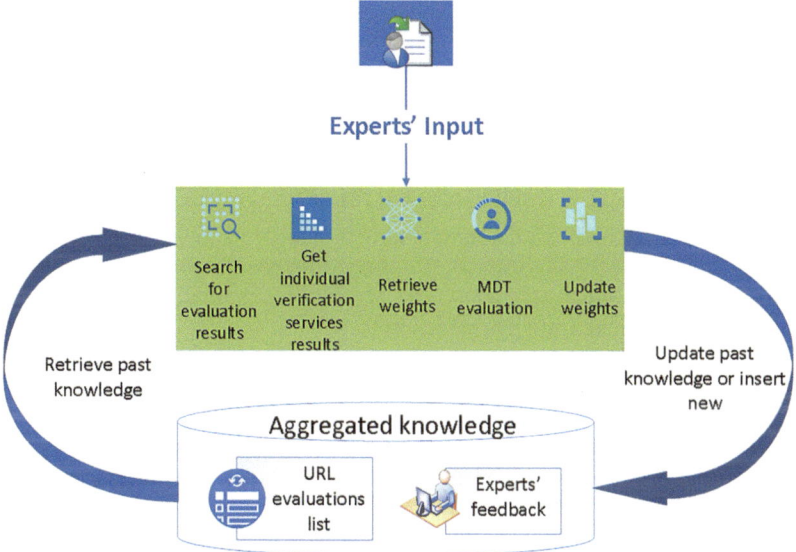

Fig. 27.3 Continuous learning process of MDT

capable of updating the weights in a continuous way, leading to a continuous learning paradigm. Last but not least, the system is constantly expanding by evaluating new pieces of content and recalculating the weights based on the experts' feedback. This entire process is depicted in Fig. 27.3.

Conclusion: Future Work

The work presented in this study (Section "Meta-Detection Toolset") combines the prediction results from various dis/misinformation prediction models and computes a meta-score that reflects the credibility of the digital content, aiming to achieve continuous improvement based on annotation processes. Through this solution, an aggregation of different ML prediction models is implemented in order to provide more trustful insights on content credibility for news articles. Thus, end users are provided with a reliable indicative score about the credibility of the content under evaluation.

The future steps involve the expansion of the Meta-Detection Toolset so as to integrate more verification services that could work on different data formats, such as pictures, video, voice, and other types as well, in order to assess their credibility. In addition to the legitimate/fake annotations, the future steps of the Meta-Detection Toolset focus on enabling end users to also provide annotations of the type of news included in a URL (e.g., political news and sports) or even insert a news category annotation of their own choice. For each category of news (including the user-defined categories), a distinct set of verification services' weights will be calculated. This aims to improve the predictions of the MDT, as more annotations arrive over time. It will also enhance the proposed solution with the ability of learning new tasks (credibility evaluation of additional content categories), initially unknown to it.

REFERENCES

1. Vosoughi, S., Roy, D., & Aral, S. (2018). The spread of true and false news online. *Science, 359*, 1146–1151. https://doi.org/10.1126/science.aap9559
2. European Commission Tackling Online Disinformation Available online: https://digital-strategy.ec.europa.eu/en/policies/online-disinformation. Accessed on 18 April 2023.
3. Giełczyk, A., Wawrzyniak, R., & Choraundefined, M. (2019). Evaluation of the existing tools for fake news detection. In *Proceedings of the computer information systems and industrial management: 18th international conference, CISIM 2019, Belgrade, Serbia, September 19–21, 2019, proceedings* (pp. 144–151). Springer-Verlag.
4. Khan, M. T., & Khalid, S. (2018). Trends and challenges in lifelong machine learning topic models. In *Proceedings of the 2018 international conference on computing, electronic and electrical engineering (ICE Cube)* (pp. 1–6).
5. Chen, Y., Wu, J., Lin, J., Liu, R., Zhang, H., & Ye, Z. (2020). Affinity regularized non-negative matrix factorization for lifelong topic modeling. *IEEE Transactions on Knowledge and Data Engineering, 32*, 1249–1262. https://doi.org/10.1109/TKDE.2019.2904687
6. Xu, M., Yang, R., Harenberg, S., & Samatova, N. F. (2017). A lifelong learning topic model structured using latent embeddings. In *Proceedings of the 2017 IEEE 11th international conference on semantic computing (ICSC)* (pp. 260–261).
7. Zhang, X., Rao, Y., & Li, Q. (2022). Lifelong topic modeling with knowledge-enhanced adversarial network. *World Wide Web, 25*, 219–238. https://doi.org/10.1007/s11280-021-00984-2
8. Sun, F.-K., Ho, C.-H., & Lee, H.-Y. (2019). LAMOL: Language modeling for lifelong language learning. *arXiv preprint arXiv:1909.03329.*

9. Ahrens, K., Abawi, F., & Wermter, S. (2021). DRILL: Dynamic representations for imbalanced lifelong learning. In *CoRR. abs/2105.08445.*

10. Chen, Z., Ma, N., & Liu, B. (2018). Lifelong learning for sentiment classification. In *CoRR. abs/1801.02808.*

11. He, X., Tuerhong, G., Wushouer, M., & Xin, D. (2022). Rumors detection based on lifelong machine learning. *IEEE Access, 10,* 25605–25620. https://doi.org/10.1109/ACCESS.2022.3152842

12. Littlestone, N., & Warmuth, M. K. (1994). The weighted majority algorithm. *Information and Computation, 108,* 212–261. https://doi.org/10.1006/inco.1994.1009

Open Access   This chapter is licensed under the terms of the Creative Commons Attribution 4.0 International License (http://creativecommons.org/licenses/by/4.0/), which permits use, sharing, adaptation, distribution and reproduction in any medium or format, as long as you give appropriate credit to the original author(s) and the source, provide a link to the Creative Commons license and indicate if changes were made.

The images or other third party material in this chapter are included in the chapter's Creative Commons license, unless indicated otherwise in a credit line to the material. If material is not included in the chapter's Creative Commons license and your intended use is not permitted by statutory regulation or exceeds the permitted use, you will need to obtain permission directly from the copyright holder.

PART IV

Protection of Public Spaces

Strengthening Local Authorities' Capabilities and Capacities Regarding the Protection of Public Space: A Co-Productive Approach

Vivian V. Gravenberch, Paul van Soomeren,
Sara Houweling, Carla Napolano, and Pilar de la Torre

INTRODUCTION

Have you ever heard the anecdote of the crime expert who went on a field visit to the European Quarter of Brussels? Together with other specialists and practitioners in the field, they went on a tour to a Public Space of Interest and the Regional Crisis Centre to observe the city's efforts and approach to the protection of public space. On their way to the airport

V. V. Gravenberch (✉) • S. Houweling
Dutch Institute for Safe and Secure Spaces (Stichting DISSS),
s-Hertogenbosch, The Netherlands
e-mail: viviangravenberch@disss.eu

P. van Soomeren
DSP-groep, Amsterdam, The Netherlands

C. Napolano • P. de la Torre
European Forum for Urban Security (Efus), Paris, France

© The Author(s) 2025   343
I. Gkotsis et al. (eds.), *Paradigms on Technology Development*
for Security Practitioners, Security Informatics and Law
Enforcement, https://doi.org/10.1007/978-3-031-62083-6_28

after a fruitful day, the crime expert was cornered at a nearby train station by loitering locals and had to lock themselves in a public restroom to avoid being robbed. The field visit took place as part of an EU-funded project named Secu4All, for which a consortium travelled through Europe to empower local and regional authorities with theoretical knowledge and practical tools to ensure the security of public spaces against potential threats. The example above underlines that the security of public space is a complex challenge, even for a city that is considered the de facto capital of Europe, and that often major contrasts occur between the security measures in public spaces, regardless of physical distance between them. Adding to this complexity is the open character of public spaces, keeping these spaces secure whilst adhering to their open and accessible nature is a precarious matter. Not to mention ethical or financial considerations, and the interests of citizens frequenting these spaces. Despite ample knowledge, experience and goodwill being present, there appears to be a gap when it comes to combining these valuable insights and different stakeholders working together in the security of public space. This chapter aims to answer the question *to which extent a partnership approach can strengthen local authorities' capabilities and capacities in protecting public space.* To begin, we will delve into the factors that define the protection of public space. Next, common approaches in defining the risks to public space will be discussed, followed by an overview of the different measures that can be implemented to treat the risks to public space. Subsequently, a practical implementation of this theoretical framework will be presented by outlining the stakeholder training as developed through the Secu4All project. Finally, the results of the stakeholder training will be discussed, as well as the conclusions drawn from this research and the discussion it provokes.

Literature Review

In the last few years, the protection and security of public spaces have continued to be a priority for European institutions and national governments, as well as local authorities who play a fundamental role in the implementation of actions. As recently stated by the Council of the European Union in the Conclusions on the Protection of Public Spaces document, local authorities are relevant actors for cooperation and the promotion of synergies in the prevention of terrorist attacks, as well as in the security and protection of public spaces. The Security, Democracy and Cities Declaration adopted in Nice on 22 October 2021 emphasises the need for cities and regions to make significant investments to protect public spaces, evaluate security needs as early as in the design stage, anticipate and respond to crises, and offer citizens the best possible protection [3]. This requires mobilising the best available expertise and technologies, depending on the context. The chapter poses three sub-questions to the abovementioned central question: (1) What defines the protection of public space? (2) What are the common approaches to identifying the risks to public space? (3) Which measures can be implemented to treat the risks to public space? By exploring and answering these questions, the chapter hopes to provide insight into how different stakeholders can work together to enhance the protection of public spaces.

Protecting Public Space

According to the European Commission, public space refers to any area that is accessible to and used by the general public, including streets, parks, squares, and other open spaces [4]. Public spaces are not only important for civic life and social interaction, but they also play a critical role in shaping the character and quality of cities and communities. However, public spaces are vulnerable to a range of safety and security threats, including crime, terrorism and petty crimes. It is crucial to take steps to protect them against these risks. Ensuring security in public space is a fundamental human right, and it is important to make cities and human settlements safer for everyone. Goal 11 of the Urban Agenda for Sustainable Development reflects this aspiration, and the EU aims to enable all citizens to live, work and participate in urban life without fear of violence or intimidation [5]. The EU Forum[1] on the protection of public spaces identifies vulnerability assessments as a good practice in protecting public spaces against terrorist threats. It is recommended that good practices in protecting public spaces include conducting vulnerability assessments, developing and implementing facility or event security plans, and identifying appropriate security measures for the specific function of the facility or event.

---

[1] Webpage of the EU Forum on the protection of public spaces. https://ec.europa.eu/newsroom/pps/items/665688/en

Defining the Risks to Public Space

Defining the risks to public space is a complex process that involves considering various factors that can contribute to safety and security threats. A first commonly used approach to identifying risks to public space has been elaborated on in the Risk Management Guidelines 31000:2018, published by the International Organization for Standardization (ISO) [8]. These guidelines provide a detailed overview of how to set up a *framework*, build upon *principles* and create a *process* for risk management. Organisations are exposed to a variety of internal and external factors that can impede their goals and objectives. Different types of risks (such as sexual harassment, feelings of insecurity, terrorist threats and different forms of petty crimes) can have an impact on their success. Managing these risks is an ongoing process that involves defining a strategy, setting objectives and making informed decisions [12]. It is important to keep in mind that risks can change over time, and so can organisations. Therefore, a continuous assessment cycle should be adopted when assessing risks for an organisation. The same approach is followed in a recently issued European standard CEN/TS 14383-2:2020² which is, like the ISO standards, available from every national standardisation body [6].

The purpose of risk management is to create and protect value, improve performance, encourage innovation and support the achievement of objectives. The *principles* of effective and efficient risk management are essential for the creation and protection of value, improved performance, innovation and the achievement of objectives. The risk management *framework* assists in integrating risk management into significant activities and functions, and its effectiveness depends on its integration into the governance of the organisation, including decision-making. The risk management *process* involves four basic elements that can help create a capable guardian against a likely offender of a crime in public space: identification, analysis, evaluation and treatment [7, 8]. *Risk identification* involves examining specific public spaces that could be vulnerable to different types of threats. Identifying vulnerabilities can help understand the perspective of potential offenders. *Risk analysis* involves analysing the impact and

---

² The new CEN TS 14383-2:2022 supersedes the old TR. It builds on ideas from risk management (ISO 31000 series), Quality management (ISO 9000 series), CPTED (ISO 22341:2021) and new approaches, new types of crime, and UN/EU standards/documents (like the ICCS).

severity of a given act and the likelihood of a specific type of threat occurring.

The new European-focused standard, CEN/TS 14383-2:2022, a Crime Prevention Through Environmental Design (CPTED) approach, places a greater emphasis on involving stakeholders such as residents and locally involved partner groups, including democratically elected local stakeholders like a city council, in the participatory process [6]. *Risk evaluation* involves determining which risks are considered acceptable and which are not [8]. This process involves deciding which degree of risk is acceptable, the chances of it occurring and its potential impact. The responsible body should work with their team and experts from different angles to determine the acceptable level of risk. Another key element of the risk management process is *risk treatment*. Risk treatment involves selecting and implementing appropriate measures to modify risks, and there are four main options for treating risks: avoiding, reducing, sharing or accepting the risk. The choice of risk treatment option will depend on the specific context and goals of the organisation, as well as the characteristics of the risk itself [6, 8].

Treating the Risks to Public Space

The multiple-helix approach emphasises collaboration between different stakeholders to protect public space, recognising that safety and security challenges are complex and require a coordinated effort [12, 13]. It involves representatives from different sectors working together to identify and address risks in public spaces through joint risk assessments, sharing information and resources, and developing partnership strategies. This approach can build trust and collaboration between stakeholders, foster a sense of shared responsibility and create safer and more secure public spaces for everyone. A very effective example of a multiple-helix approach towards public space protection is Crime Prevention Through Environmental Design (CPTED). CPTED is a multidisciplinary approach that aims to reduce and prevent crime, safety and security issues, including terrorism, and address people's feelings of insecurity in public space and soft targets. It is a feasible and effective approach, both in new and pre-existing environments. This has been shown earlier in the EU COST-action TU 1203 (Cooperation in Science and Technology) and the EU Horizon 2020 project Cutting Crime Impact [2, 3]. This approach has also been standardised both by the European Committee for

Standardization (CEN/TS 14383-2:2022) and the International Organization for Standardization (ISO 22341:2021) [6, 9]. The CPTED approach, as promoted by the International CPTED Association (ICA)[3] and also called the Security by Design approach, is a 'multidisciplinary crime prevention approach that employs urban design to diminish victimisation, discourage offender decisions that lead to criminal acts, and foster a sense of community among residents'.

METHOD AND APPROACH

Based on its previous works, and aware of the need to enhance the role of local authorities and frontline actors in charge of the protection and security of public spaces, the Secu4All project has developed and conducted a multi-stakeholder training. This training involves local authorities, law enforcement, civil protection, medical emergency services, private businesses, private security firms, residents and their organisations and other stakeholders to improve their preparedness, response and resilience to any man-made threat and risk present in public spaces. Originally the multi-stakeholder consisted of an interactive and in-person training, followed by an in situ (on-site) training. Finally, multiple field visits are carried out to exchange knowledge and expertise between project partners on the security of public spaces. Below, the design of the training will be outlined, as well as the adaptations necessary to meet local and European health and safety guidelines due to the COVID-19 pandemic.

Online Training

Due to the COVID-19 pandemic, it was necessary to organise a training programme online. This online training programme requires a virtual platform to facilitate remote learning and collaboration. The following methods and approaches have been used to ensure that the training programme remains effective and interactive during a period of working in isolation and engaging all trainees. Module 1 introduces trainees to risk and vulnerability assessment, focusing on the challenges and opportunities associated with protecting public spaces. Trainees participate in online lectures, group discussions and interactive exercises. They are given

---

[3] For further reference, see the International CPTED Association (ICA): https://www.cpted.net

assignments to complete, including a simplified version of a risk assessment, and work in small groups to encourage collaboration and discussion among trainees representing different types of stakeholders. Module 2 provides knowledge on CPTED, including the environmental context of crime and security risks, definitions, history, terminology and basic principles for process and partnerships. Module 3 aims to improve the knowledge of local authorities on technologies used to secure and protect public spaces, allowing participants to identify suitable technologies and understand ethical and legal use and data management. Module 4 aims at training local authorities to communicate effectively with different stakeholders and the population in crisis situations or terrorist attacks occurring in a public space. The module covers identifying the key definitions relevant to crisis communication, understanding the objectives of each phase of the crisis communication, identifying the relevant stakeholders and implementing the crisis communication techniques.

The online training programme was evaluated through a questionnaire consisting of four statements rated on a 1–5 point scale assessing the informativeness of the session, level of understanding of the contents, presenter's ability to explain the content in an understandable manner and relevance of the session to daily work. Participants also provided feedback and suggestions for improvement to the hosts of the online training session. The results of this evaluation are discussed in the results section of this chapter.

In Situ (On-Site) Training

In order to ensure the safety and security of public spaces, trainees are required to apply the principles of the multiple-helix approach. This approach involves collaborative efforts from various sectors such as the government, private sector and civil society. After completing online training modules, each trainee is given a workbook to record their observations in the Public Space of Interest (PSOI). The PSOI that trainees and trainers visit at the start of the day forms the basis of the rest of the training day. Trainees are expected to identify vulnerabilities and possible opportunities for security by design, as well as technological measures and crisis management tools that are already present or are lacking at the PSOI. Following this, trainees are tasked with marking three zones around the PSOI, including access roads and car parks. They must then place stickers with 'risk icons' on a map in each zone to represent different types of risks such as theft, burglary, vehicle attack, sex crimes, terrorist threats and other forms of risks to public space.

This activity serves as a warm-up exercise to prepare trainees for the in situ training modules. The four training modules cover topics such as identifying vulnerabilities, discussing together what they each define as high, medium or low risk, implementing UPDM-US, using technological solutions to enhance security and communicating during a crisis. These modules are designed to provide trainees with hands-on experience in identifying and mitigating risks in public spaces and to learn to speak each other's language. Scenarios are presented in each module that simulate real-life situations where one of the earlier identified risks actually occurs. By utilising the multiple-helix approach and applying the principles learned in the online and in situ trainings, trainees are better equipped to promote security and safety in the real world. This not only benefits the trainees themselves, but also contributes to the overall safety and security of public spaces.

Field Visits

To exchange knowledge and expertise and further describe common approaches in defining risks in public space, four field visits to major European cities were organised. In Riga, Latvia, the consortium visited the headquarters of the Riga Municipality Police (RPP). In Rotterdam, the Netherlands, the main objective was the city's central train station, *Rotterdam Central Station* and the public space surrounding the station. In Brussels, Belgium, the Robert Schuman Square, located in the area which houses several EU buildings, and the Regional Crisis Centre were observed and discussed. Finally, the consortium travelled to France to visit the Urban Supervision Centre (USC) in Nice. The goal of these visits was to answer the following question: *which measures are currently implemented to treat the risks to public space?*

One of the measures to treat risks to public space is video surveillance. In Riga, this led to the challenge of how to spend funding for video surveillance in the most effective way. The RPP now uses IT tools to determine the most effective positioning of cameras. These tools also support police work by providing an easier way to combine data, allowing for patrols to be planned more efficiently and thus supporting operators to be at the right place, at the right time. In addition, as for innovative and out-of-the-box options, the city has found that placing empty camera boxes also forms an effective deterrent. In Brussels, the Regional Crisis Centre forms the city's hub for technological security solutions. Besides video surveillance, the centre provides communication technology allowing for different services to cooperate. Also available are a temporary camera service, drones and various tools translating footage to practical intelligence. Combining technological tools with CPTED principles, the city of Rotterdam has opted for a hidden video surveillance system in its design for a renewed *Rotterdam Central Station*. For example, security cameras are placed in the small spotlights in the main hall of the train station, unnoticeable for commuters rushing past. The underlying idea is that this method of surveillance avoids fear-mongering among citizens.

CPTED and UPDM-US were topics which mostly came up during the field trips to Rotterdam and Brussels. Prior to its redevelopment which started in 2004, *Rotterdam Central Station* and its vicinity were dimly lit and uninviting. The area had deteriorated into a hub of criminal activity, including harassment and drug dealing. Thus, the redevelopment of the station did not only provide an opportunity to reorganise the public transport system, it also allowed for the area to be overhauled into an international mobility hub combining innovative architecture, liveability and security. The design combines visually pleasing natural materials with anti-terrorism features, hosts various shops, bars and restaurants attracting visitors and gives way to pedestrians and cyclists. The CPTED and UPDM-US approach is also visible in the design of the Schuman Square in Brussels. Located in the political centre of Europe, the square was designed combining aesthetics with practicality. This was done using appropriate traffic lights and lighting, placing street furniture and making strategic use of planting. Besides applying technological innovations and UPDM-US principles, the cities visited also acknowledge the strength of forming partnerships and involving citizens in its efforts to improve security.

Results

Overall, the online training programme provides a comprehensive and engaging learning experience for trainees including field visits for inspiration during the training exercises, while also ensuring that safety measures are in place, which was necessary to protect the trainees during the COVID-19 pandemic. By using a virtual platform that supports remote learning and collaboration, trainees are able to participate in the training programme from the comfort and safety of their own homes or work-places when needed.

The training programme provides trainees with the necessary knowl-edge and skills to effectively protect public spaces and improve urban safety and security. In addition, the online delivery of the training pro-gramme provides some benefits that are unique to remote learning. For example, trainees have the opportunity to learn from experts in urban safety and security from around the world and share their knowledge and experiences with others from different sectors and professions. The online format also allows for more flexibility in scheduling and reduces travel costs compared to in-person training programmes.

As mentioned in the methods, participants in the training were requested to complete an evaluation questionnaire. In this questionnaire, the participants rated various statements on a 1–5 point scale, 1 meaning 'strongly disagree' and 5 meaning 'strongly agree'. The statements referred to the session being informative, the participants understanding the con-tents as well as the helicopter view of the session, the presenter explaining the content in an understandable manner and the content being relevant to the daily work of the participants. Due to HR issues and lack of engage-ment of the participants, one training group decided to terminate the training and cancel the in situ. This illustrates the challenges that came with working from home and remote learning during the COVID-19 pandemic. Within the other participant groups, the evaluation results for the online training were overall very positive, and critical feedback was mainly aimed at technical challenges in navigating the tools that were used to meet and train online. In addition, participants generally commented that they would have preferred an in-person training to an online session, and that the density of information combined with the presentation style was sometimes perceived as demanding and tiring for the participants. Other feedback was that some of the modules could have used more prac-tical examples.

The in situ training received positive feedback for its interactive and in-person approach. The feedback highlights the need for interdepartmental cooperation in assessing the risks associated with public spaces. Recurring problems include homeless individuals engaging in pickpocketing or threatening behaviour, sexual harassment and feelings of unsafety. Risk assessments should consider various settings and involve multiple experts. Examples of ineffective vehicle mitigation measures are also highlighted. The importance of inviting an expert on vehicle mitigation thus becomes apparent. As for the variety in expertise, however, some participants commented that not each module was as applicable to their daily work (e.g. the module on crisis communication may not be as relevant to planners and/or urban designers as to police officers) [10, 11]. In addition, it was expressed that the more practical an example or exercise, the better.

Conclusions and Discussion

In conclusion, a partnership approach can significantly enhance local authorities' capabilities and capacities in protecting public space. To achieve this, it is important to (1) define what constitutes the protection of public space, (2) identify the common approaches to defining risks to public space and (3) implement effective measures to mitigate these risks. By working together, different stakeholders can enhance the protection of public spaces and create a safer environment for everyone. This study highlights the effectiveness of offline training where multiple stakeholders are brought together to collectively solve safety issues. By providing each stakeholder with snippets of information, the study demonstrates the importance of collaboration and the expertise available within a network. The study also emphasises the value of experiential learning and breaking away from traditional teaching methods, ensuring that everyone is involved and has a voice. The study further highlights the importance of a multidisciplinary, multicultural approach to problem-solving. By learning to speak each other's language, stakeholders can more effectively reach a joint conclusion. Additionally, the study found that addressing petty crime and public safety for all (including women) was of greater concern to stakeholders than terrorism. Therefore, it is recommended that a gender approach be taken when addressing safety concerns in public spaces. This study also emphasises the importance of a diverse stakeholder group and consideration of different experiences when addressing safety concerns. Finally, although not tested yet in a real-life scenario, the use of AI in the form of chatbots based on GPT technology can help bridge multidisciplinary, multi-helix and cultural differences in the security industry. It facilitates communication between different parties and can lead to new insights and innovative solutions to complex security problems. It can also be valuable in crisis situations where fast and effective communication is crucial.

Acknowledgements This project received funding from the European Union's Internal Security Fund—Police under Grant Agreement no. 952789. This chapter reflects only the authors' views, and the Research Executive Agency and the European Commission are not responsible for any use that may be made of the information it contains.

References

1. Barosso, I., Cardia, C., Nicolini, U., & Wellhoff, F. (2014). *Milan: Crime prevention through urban design*. Academic research and training. http://costtu1203.eu/downloads/cost-tu1203s-results/
2. Cutting Crime Impact. (2018). *Review of State of the Art: CP-UDP*. https://www.cuttingcrimeimpact.eu/download/may-2019_d25_1014042704.pdf
3. Efus. (2017). *The Efus Manifesto: Security, democracy and cities: Co-producing Urban Security Policies*. https://efus.eu/the-manifesto/
4. European Commission. (2017). Action Plan to support the protection of public spaces. p. 2. Retrieved on August 6th, 2024. https://eur-lex.europa.eu/legal-content/EN/TXT/PDF/?uri=CELEX:52017DC0612
5. European Commission. (2015). The European Agenda on Security.
6. European Committee for Standardization (CEN). (2022). *CEN TS 14383–2:2022*. Retrieved on November 30th, 2023. https://www.nen.nl/en/cen-ts-14383-2-2022-en-304865
7. Gravenberch, V. V. (2022). Routine activity theory: how to protect public spaces. *City Security Magazine*. https://citysecuritymagazine.com/security-management/routine-activity-theory-how-to-protect-public-places/
8. International Organization for Standardization (ISO). (2018). *Risk management-guidelines*.https://www.iso.org/obp/ui/#iso:std:iso:31000:ed-2:v1:en
9. International Organization for Standardization (ISO). (2021). *Security and resilience – Protective security – Guidelines for crime prevention through environmental design*. Retrieved on November 30th, 2023. https://www.iso.org/standard/50078.html
10. Secu4All. (2023). D4.4 Report on the training cycle implemented in Germany by DEFUS. In *Training local authorities to provide citizens with a safe urban environment by reducing risks in public spaces* (p. 14).
11. Secu4All. (2023). D5.3 Field trip: Rotterdam tools and measures preventing acts of insecurity in public spaces. In *Training local authorities to provide citizens with a safe urban environment by reducing risks in public spaces* (p. 13).
12. Van Soomeren, P., & Gravenberch, V. V. (2023). *Crime prevention can be effective and fun*. https://www.disss.eu/publications

13. Van Waart, P., Mulder, I., & de Bont, C. (2015). A participatory approach for envisioning a smart city. *Social Science Computer Review*, 1–16. Retrieved on November 30th, 2023. https://www.researchgate.net/profile/Peter-Van-Waart/publication/283256188_A_Participatory_Approach_for_Envisioning_a_Smart_City/links/5757273d08ae04a1b6b68f16/A-Participatory-Approach-for-Envisioning-a-Smart-City.pdf

Open Access This chapter is licensed under the terms of the Creative Commons Attribution 4.0 International License (http://creativecommons.org/licenses/by/4.0/), which permits use, sharing, adaptation, distribution and reproduction in any medium or format, as long as you give appropriate credit to the original author(s) and the source, provide a link to the Creative Commons license and indicate if changes were made.

The images or other third party material in this chapter are included in the chapter's Creative Commons license, unless indicated otherwise in a credit line to the material. If material is not included in the chapter's Creative Commons license and your intended use is not permitted by statutory regulation or exceeds the permitted use, you will need to obtain permission directly from the copyright holder.

CHAPTER 29

Vulnerability Assessment for Places of Worship in the EU

Konstantinos Apostolou, Joseph Levis, Christina Karafylli,
Maria Karafylli, Vivian Gravenberch, Anna van der Stok,
Rafał Batkowski, Marcin Podogrocki, Timo Hellenberg,
and Hannu Rantanen

K. Apostolou (✉) • J. Levis • C. Karafylli • M. Karafylli
Center for Security Studies (KEMEA), Ministry of Citizen Protection,
Athens, Greece
e-mail: k.apostolou@kemea-research.gr

V. Gravenberch • A. van der Stok
Dutch Institute for Safe and Secure Spaces (DISSS),
's-Hertogenbosch, the Netherlands

R. Batkowski • M. Podogrocki
University of Lodz, Lodz, Poland

T. Hellenberg • H. Rantanen
Hellenberg International, Helsinki, Finland

© The Author(s) 2025   357
I. Gkotsis et al. (eds.), *Paradigms on Technology Development*
for Security Practitioners, Security Informatics and Law
Enforcement, https://doi.org/10.1007/978-3-031-62083-6_29

INTRODUCTION

EU places of worship (PW) such as churches, mosques, and synagogues constitute cases of public spaces in need of particular attention, in order to be efficiently protected. Such places, have been frequent targets of terrorist attacks in the previous years due to their highly symbolic value [1]. As a way to increase the level of Member States' public space protection in anticipation of terrorist threats, the exchange of good practices, risk and vulnerability assessment guidance, along with protective measure guidance for local authorities, is strongly emphasized by the EU [2]. In this context, the EU-funded research project ProSPeReS (ISFP-GA 101034230) applied the vulnerability assessment checklist (VAC) developed by the European Commission's Directorates-General for Migration and Home Affairs (DG HOME) and provided to ProSPeReS for research purposes. The vulnerability assessment (VA) activities that were performed during the project, among others, consisted of in situ VA workshops with site surveys at various PWs and VA trainings targeted to the involved stakeholders. These activities allowed the identification of challenges in protecting PWs, common and distinct security gaps and needs among PWs of different types and religions, as well as the formulation of recommendations for increasing their protection level.

SCOPE

This chapter presents the applied vulnerability assessment (VA) methodology and the VA activities that were undertaken in the context of the project's Work Package 2 (WP2)—Vulnerability Assessment & Needs Analysis of Religious Sites, accompanied by relevant findings.

APPLIED VULNERABILITY ASSESSMENT METHODOLOGY

The EU vulnerability assessment checklist (VAC)[1] (Fig. 29.1) is a tool meant to enable site operators (public space or other) or managing entities, to perform a VA for their site. It consists of MS Excel spread sheets, put together in a logical series, assisting the users to consider threats as well as attack scenarios against their facilities/area of interest, while

---

[1] VAC is not publicly available and was provided to the project consortium members for research purposes.

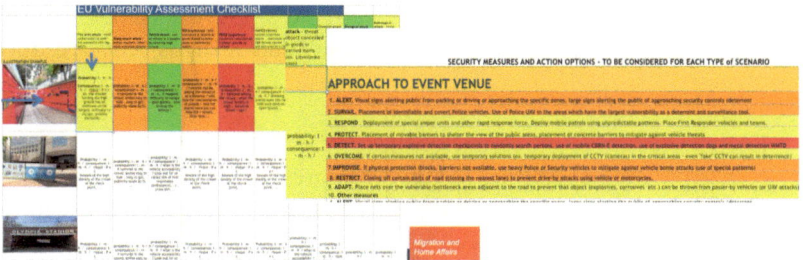

Fig. 29.1 Vulnerability assessment checklist screenshot [3]

accounting for parameters such as crowd concentration, existing security measures—and lack of— and other, concluding in vulnerabilities against these threats.

Prior to proceeding with the collection of information and analysis, the area of interest should be divided into six phases (similar to zones) according to the approach and access of pedestrians and vehicles to the main site of interest (or main facility or main area). This allows for a more thorough and effective examination of an area. The six phases are the following [4]:

- Phase 1: Access to the site—public road systems that provide access to the site
- Phase 2: Parking and transport—parking areas, garages, or public transport facilities with a high concentration of people
- Phase 3: Approach to site—public access paths for pedestrians that can lead to a high concentration of people
- Phase 4: Arrival at main site—entry or exit points
- Phase 5: Main site without access control
- Phase 6: Main site with access control[2]

The tool is mostly applicable for places with a high people concentration or large events. Essentially, the users can assess an area and its facilities based on a structured and creative step-by-step process. For a more efficient process, it can be performed by multidisciplinary working groups including experienced individuals in security issues, knowledgeable about the activities and operations of the area of interest.

---

[2] Phases 5 and 6 are alternatives to each other.

In summary, the process of the ProSPeReS VA workshops, which included information gathering, discussion, and analysis, covered the following:

1. Examination of site characteristics
2. Examination of existing security measures
3. Identification of potential threats and incident scenarios per identified threat
4. Estimation of consequences and probability (likelihood) per identified threat
5. Evaluation of results and consideration of solutions

Vulnerability Assessment Workshops

The VA workshops that were implemented in 2022, focused on the identification of security weaknesses against selected PWs and proposal of measures to mitigate them, while heavily relying on the active engagement and collaboration of the stakeholders involved in the protection of the PWs.

Three VA workshops with subsequent site surveys were held at (1) the Orthodox Church of Saint Paisios in Ioannina, Greece, (2) Archcathedral Basilica of Stanislaus Kostka in Lodz, Poland, and (3) Nozyk Synagogue in Warsaw, Poland.

The participants of the workshops consisted of

- Local law enforcement agencies (LEAs), namely police officers
- Local first responders, including representatives of the fire brigade, emergency health services, and civil protection personnel
- Religious and administrative staff of the PWs for which the VA was carried out

The total threats and attack scenarios against the sites that were selected by the participants for discussion and analysis included the following:

- *Firearms attack*
  Attack against the crowd with a concealed automatic firearm.
- *Sharp object attack*
  Attack with concealed weapons against the crowd.

- *Vehicle ramming attack*
  Attack with a vehicle against the crowd near the main street of the religious site.
- *Improvised explosives device (IED) attack*
  Attack with a bag discarded (unattended), containing explosives
- *Person-borne improvised explosive device (PBIED) attack*
  Attack by a suicide bomber against the crowd
- *Unmanned aerial vehicle improvised explosive device (UAVIED) attack*
  Attack with drone against the crowd outside the PW, carrying explosive material
- *Vehicle ramming attack*
  Attack with explosive material placed into a car parked near high crowd concentration areas
- *Chemical, biological, radiological (CBR) attack*
  Attack with chemical agents outside the PWs

The workshops were executed and moderated by the scientific members, security experts, and religious organization members of the project's consortium partners. They took place during the course of three days, and the activities were structured as follows:

1. A *theoretical training* to familiarize the participants with the concept of risk and vulnerability assessment, and the VA process as applied in the context of ProSPeReS.
2. *AVA tabletop exercise* where the participants performed a VA for a hypothetical site, based on a hypothetical attack scenario in order to gain hands-on experience.
3. Discussions with the PW operators were undertaken for *gathering information* about the sites' activities, operations, and applied protection measures during their daily activities and high-profile religious events.
4. A *site survey* where an inspection of the site was performed, focusing on applied security measures and gaps, covering the PW facilities and immediate surrounding area.
5. A *tabletop VA* for the participants' religious site, based on the vulnerability assessment checklist methodology (VAC).

ADDITIONAL CASE STUDIES

The findings of the VA workshops were complimented by additional quick VAs, performed at various PWs across Europe through additional site surveys, interviews with the site operators, and questionnaires developed for the purposes of the quick VAs. The additional sites were 10 in total, consisting of Lutheran churches, synagogues, Catholic churches, a Catholic monastery, a mosque, and an Orthodox monastery located in Finland, Poland, Denmark, the Netherlands, and Greece, respectively.

VAT LITE

Upon the completion of the VA workshop and according to the end user needs derived from the participants' feedback, ProSPeReS developed the tool VAT Lite, allowing site operators and staff to carry out quick vulnerability assessments on their own in a simplified way. The tool provides an overview of the site vulnerabilities to the operator during the PW's daily operations. It combines the main elements of the VAC and the "EU Quick Guide to support the protection of places of worship" [5] in a concise yet elaborate template. The EU Quick Guide is a checklist-based tool for carrying out a brief and simplified security and safety assessment at PWs.

The principles of VAT Lite[3] are identical to VAC's, including the steps of the VA and the PW's area segmentation. However, instead of six phases, the area is now divided into three zones (see indicative example in Fig. 29.2), a popular approach among contemporary security risk assessment practitioners.

By filling out the template once, the users can examine all three zones of their site per their vulnerabilities (lack of security measures), potential threats, and security solutions. The tool consists of four record templates that users must fill out, covering the aspects of site segmentation, vulnerability identification, threat identification, risk analysis, and risk evaluation. Examples of the VAT Lite record templates can be seen in Fig. 29.3.

[3]VAT Lite is still under development, testing, and validation process, not available outside the ProSPeReS consortium at the time of this chapter's publication.

Fig. 29.2 VAT Lite area segmentation: zone 1: building interior; zone 2: immediate exterior; zone 3: surroundings

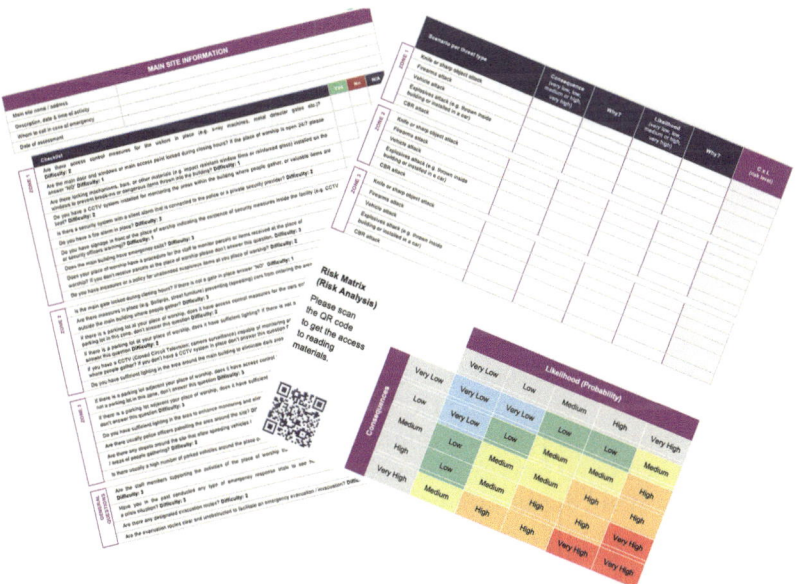

Fig. 29.3 Vat Lite record templates

Vulnerabilities of Examined Places of Worship

Naturally, the security gaps and overall vulnerabilities identified during VAs varied among the examined PWs. Some sites were security oriented and more effectively protected than others, a result of numerous factors such as local crime rate, past incidents, significance of the PW, threats experienced by the religious communities of the PW, and, finally, the security culture and experience of the PW operators/managing bodies. Security and protection in a PW concern the people present during religious events (both stakeholders and visitors) and the infrastructure, including all relevant facilities.

The list below provides an overview of the reoccurring gaps identified at most of the surveyed sites [6].

- *Security training for personnel*
  Insufficient training related to emergency incident response, evacuation drills, and risk awareness initiatives for the religious staff.
- *Incident response*
  Lack of incident response and crisis communication means and procedures.
- *Interagency cooperation*
  Unestablished/inefficient cooperation and information sharing between LEAs, first responders, and the sites' operators on a local level, in light of religious events, or daily activities.
- *Detection of dangerous items*
  Lack of measures or procedures for the detection of dangerous or suspicious items, substances, and material that may be carried into the PW.
- *Other security gaps*

  - Crowd flow and control measures
  - Limited monitoring and CCTV capabilities
  - Insufficient lighting in secluded areas
  - Limited to no security personnel
  - No access control measures or access restriction to the site for vehicles and visitors

RECOMMENDATIONS

In order to effectively enhance the protection of PW, the operators or managing bodies must adopt a holistic approach toward security, beyond the limitations of conventional security measures that in many cases are difficult to apply due to factors such as the open nature of the sites, architectural elements and requirement to preserve their aesthetics, potential obtrusive results of security measures during the activities, and feeling of discomfort that excessive measures may generate for the congregation.

Considering the above, a summary of recommendations for conventional and contemporary measures and approaches proposed for enhancing the protection of PWs is presented below, broken down into respective categories.

Planning

Identification of Stakeholders
Prior to any activities related to the security of a religious site, the PW operator should identify the actors that are involved in the protection of their site during its daily activities or religious events.

Vulnerability Assessment
Following the identification of security stakeholders and in line with good practices for the security of public spaces in general [7], the operators of PWs should initiate a vulnerability assessment (VA) for their site. This will allow them to gain an overview of their sites' actual security gaps and consider targeted solutions. A VA can take place via multidisciplinary workshops with the participation and guidance of local LEAs, first responders, or security experts. However, the operator of a PW is responsible for making a request for or initiating the VA.

Development of an Emergency Response Plan (ERP)
An ERP should be in place and updated, including evacuation procedures for the safety of both the religious staff and the congregation. Such plans can be developed through the cooperation of a site's operator, local LEAs, security experts, and first responders.

Actions

Security Awareness and Training

Through basic safety, security, and threat awareness training, the religious staff of a PW can actively contribute to the protection of the people and the site. Such training may be targeted toward fire safety, evacuation drills, emergency incident response actions and crowd guidance, identification of suspicious items and behavior, crisis communication, and site surveillance. The training can be provided by local LEAs, first responders, or security experts. Emergency response joint exercises among a site's operator local LEAs and first responders could also be planned in light of high-profile events.

Communication and Cooperation Between Stakeholders

The operators of PW are recommended to establish regular communication channels with local authorities and first responders in case of security incidents or in case of religious events where the adoption of ad hoc security measures is required.

Security Measures and Procedures

ProSPeReS has produced documents that provide an extensive list of technical measures and procedural recommendations for the PW protection. With respect to the sites examined during the VA activities, the recommendations generally focused on:

Surveillance and Monitoring

For example, using drones for surveillance of a site or during the presence of large crowds, sufficient lighting to eliminate dark areas, using modern CCTV systems and dedicated security/control rooms.

Alerts

For example, installation of alarm systems and panic buttons in critical areas.

Access Control

For example, restriction of vehicle access for crowd protection and restriction of visitor access to critical areas.

Protection

For example, protection of windows against vandalism/throwing objects by installing security window films and ballistic panels, and removal of litter bins from areas of high people concentration or evacuation routes.

CONCLUSION

This chapter presents the vulnerability assessment (VA) activities, implemented during the Work Package 2 (WP2) "Vulnerability Assessment & Needs Analysis of Religious Sites" of the EU*funded research project "ProSPeReS." A VA prior to the selection of security measures for any site of interest provides an overall picture of the site's actual needs against realistic and prioritized threats. Consequently, informed and targeted decisions can be made by a site's management toward its effective protection. Aside from the protection of the site itself, the process of the assessment can be highly beneficial for the end users. The personnel of the PW, the local LEAs and first responders who participated in the VAs at the selected sites, had the opportunity to exchange knowledge, information, and good practices, while working together in the frame of the interdisciplinary executed workshops, which can potentially improve their future cooperation. This may translate into effective preparation of joint security measures before future events, protection of the PW during events, joint response to incidents, joint trainings, and exercises. Finally, all the stakeholders who participated in the relevant activities, including the ones who were present at the additional case studies, became aware regarding the security issues surrounding their religious sites locally and on a European level, and were engaged in discussions about the novel measures and methods that can be proportionally adopted to enhance the protection of their sites while preserving their open nature and the continuity of their activities.

Acknowledgments The ProSPeReS project received funding from the European Union's Internal Security Fund—Police under grant agreement no. 101034230. This chapter reflects only the authors' views, and the Research Executive Agency and the European Commission are not responsible for any use that may be made of the information it contains.

References

1. European Commission. (2020). *Communication from the Commission to the European Parliament, the European Council, the Council, the European Economic and Social Committee and the Committee of the Regions. A Counter-Terrorism Agenda for the EU: Anticipate, Prevent, Protect, Respond.* Retrieved from https://eur-lex.europa.eu/legal-content/EN/TXT/PDF/?uri=CELEX:52020DC0795

2. Council of the European Union. (2021, June 7). *Council conclusions on the protection of public spaces.* Retrieved from https://data.consilium.europa.eu/doc/document/ST-9545-2021-INIT/en/pdf

3. DG HOME. (2019). *PRoTECT Events, Web Conference. EU Vulnerability Assessment Checklist – Practical Use.* Retrieved from https://protect-cities.eu/protect-project/public-deliverables/

4. ProSPeReS. (2021). *Deliverable 2.1 - Manual for vulnerability assessment.* ProSPeReS (GA 101034230).

5. DG HOME. (2021). *EU quick guide to support the protection of places of worship.* Retrieved from https://home-affairs.ec.europa.eu/whats-new/publications/eu-quick-guide-support-protection-places-worship_en

6. ProSPeReS. (2022). *Deliverable 2.5 Vulnerability Assessments Aggregate Report. PP. 15-16.* ProSPeReS. (GA 101034230).

7. European Commision. (2019). *Commission Staff Working Document Good practices to support the protection of public spaces Accompanying the document Communication from the Commission to the European Parliament, the European Council and the Council Eighteenth Progress Report towards an effective and genuine Security Union.* Retrieved from https://op.europa.eu/en/publication-detail/-/publication/998aeb09-4be6-11e9-a8ed-01aa75ed71a1/language-en

Open Access This chapter is licensed under the terms of the Creative Commons Attribution 4.0 International License (http://creativecommons.org/licenses/by/4.0/), which permits use, sharing, adaptation, distribution and reproduction in any medium or format, as long as you give appropriate credit to the original author(s) and the source, provide a link to the Creative Commons license and indicate if changes were made.

The images or other third party material in this chapter are included in the chapter's Creative Commons license, unless indicated otherwise in a credit line to the material. If material is not included in the chapter's Creative Commons license and your intended use is not permitted by statutory regulation or exceeds the permitted use, you will need to obtain permission directly from the copyright holder.

PART V

Effective Management of EU External Borders

PROMENADE—ImPROved Maritime awarENess by Means of Artificial Intelligence (AI) and Big Data (BD) mEthods: Detection of Abnormal Behavior of Vessels Used for Smuggling and Drug Trafficking in the Ionian Sea

Alkis Astyakopoulos, Christos Bolakis, Panagiotis Douris, Marios Moutzouris, Giovanni Laneve, Leonardo M. Millefiori, Antonio Bosisio, and Vasiliki Efstathiou

A. Astyakopoulos (✉) • C. Bolakis • P. Douris
Center for Security Studies (KEMEA), Ministry of Citizen Protection, Athens, Greece
e-mail: a.astyakopoulos@kemea-research.gr

M. Moutzouris
Satways Ltd., Irakleio, Greece

G. Laneve
Leonardo S.p.A. - Electronics Division, Taranto, Italy

© The Author(s) 2025
I. Gkotsis et al. (eds.), *Paradigms on Technology Development for Security Practitioners*, Security Informatics and Law Enforcement, https://doi.org/10.1007/978-3-031-62083-6_30

Introduction

Maritime situational awareness (MSA) is the combination of activities, events, and threats in the maritime environment that could have an impact on marine activities and affect the European Union (EU) territory. European waters are navigated daily by some 12,000 vessels, which share their positions to avoid collisions, generating a huge number of positional messages every minute. It is important that this overabundance of information does not overwhelm the marine operator in charge of decision-making. Thus, the challenge is twofold: (a) on one hand, the large-scale exploitation of heterogeneous data sources, enabling new artificial intelligence (AI)-based services for enhancing MSA; and (b) on the other hand, the seamless integration and exchange of information among maritime authorities valorizing the Common Information Sharing Environment (CISE) network.

The current landscape of AI and big data (BD) is evolving rapidly. Today, AI is heavily driven by data, and machine learning (ML) techniques play a vital role in its success. ML techniques enable machines to learn from data, make predictions, and decisions without explicit programming. These techniques have become instrumental in AI, allowing machines to progressively enhance their performance on specific tasks through experience and exposure to data. Some common ML techniques in AI include supervised and unsupervised learning, deep learning, and natural language processing (NLP).

Supervised learning involves training an algorithm on labeled data to establish a mapping between input features and corresponding output labels. Conversely, unsupervised learning trains algorithms on unlabeled data to discover patterns, relationships, or structures within the data without predefined labels [1]. Deep learning focuses on training deep neural networks (with multiple layers) to recognize complex patterns in data.

L. M. Millefiori
NATO Science and Technology Organisation (STO) Centre for Maritime Research and Experimentation (CMRE), La Spezia, Italy

A. Bosisio
Transcrime, Università Cattolica del Sacro Cuore, Milan, Italy

V. Efstathiou
Kpler, Athens, Greece

Deep learning has achieved remarkable success in areas such as image and speech recognition, natural language processing, and generative modeling. NLP aims to enable computers to understand, interpret, and generate human language. These examples highlight a few machine learning techniques, but the field of machine learning is vast and constantly evolving, with new algorithms and techniques being developed to address problems in different domains.

The performance and accuracy of ML models often improve as the available data for training increases. By training on a larger dataset, ML algorithms can capture a more robust and accurate understanding of the problem. The quality and relevance of the data are equally important. It is crucial to ensure that the data used for training is representative of the real-world scenarios and covers a wide range of possible inputs to improve the model's ability to handle different situations.

With the exponential growth of data, BD technologies have become crucial to handle and process large and complex datasets, commonly known as big data. These technologies are specifically developed to overcome the challenges associated with storing, managing, analyzing, and extracting insights from massive volumes of data that exceed the capabilities of traditional data processing systems. Nowadays, open-source distributed computing frameworks and technologies, such as Apache Hadoop and Apache Spark, enable the processing and analysis of massive datasets across clusters of computers and facilitate real-time data streaming.

Despite significant advancements in AI and big data in various domains, several challenges and gaps remain. Issues such as privacy, safety, transparency, explainability, bias, quality of data, accuracy, robustness, and security have a direct impact on people's trust in AI technology and their overall interaction experience with AI tools. It is crucial for the ongoing and positive advancement of AI to establish policies, laws, and standardized environments that ensure that AI technologies bring advantages to society and protect the public interest.

To address these concerns, the European Union (EU) is adopting a regulatory framework for AI known as the Artificial Intelligence Act (AIA) [2]. The AIA aims to establish harmonized horizontal rules for artificial intelligence based on a risk-based approach, providing a legal framework for trustworthy AI. It defines a set of objective-based requirements that AI systems should adhere to. Additionally, it establishes transparency rules for AI systems intended to interact with natural persons. According to the

AIA objectives, standardization should also play a key role in providing guidelines, requirements, and reference technical solutions to providers to ensure compliance with the regulation.

PROMENADE System Architecture and Deployment

Within the PROMENADE project, various AI-based MSA services have been developed and demonstrated in operational trials; some of them are described in the third section. These services adopt machine learning and deep learning models that have been trained on historic maritime data sets, including Radar, automatic identification system (AIS), images, videos, maritime databases, and OSINT data sources. Large dataset representative of the real-world scenarios has been used in most of the cases to ensure improved performances and accuracy of the models.

High-power computing (HPC) can play a critical role in supporting the development and training of AI models by providing the necessary computing power to process large amounts of data for training and complex algorithms. Accelerating computation, managing big data, optimizing algorithms, and scaling to new architectures are some ways in which HPC can support AI model development and training.

In PROMENADE, Leonardo S.p.A. has made available its own HPC facility DaVinci-1 during the data and computing intensive phases of AI model development and training. Moreover, PROMENADE developed a data lake infrastructure leveraging on open-source big data technologies and acting as the backbone of the system for data ingestion, data storage, data analytics, data, and streaming to integrate diverse data formats processed in the toolkit such as images, videos, JSON, CSV, and CISE data model.

Specifically, PROMENADE has used two different architectures to improve the obtained results and increase the performance of the toolkit: (a) the training architecture deployed as cloud computing infrastructure on top of Leonardo HPC platform DaVinci-1, providing virtualized services for developing the PROMENADE data lake, pipelines, and training of AI services; and (b) the operational architecture deployed at the end users' premises for trials execution with the already trained services consuming real data and enabling communication with the C2 Legacy Systems and CISE environment.

The architecture of PROMENADE exploits the full potential of the CISE data model to exchange data internally and externally. The architecture is based on microservices that provide a great degree of modularity and flexibility, thus maritime authorities can select which tools they require in their operational domain.

Data is ingested from various sources: for example, (i) video feeds from cameras offering thermal and/or optical streams to detect and classify vessels; (ii) vessel tracks from the VHF Data Exchange System (VDES) or legacy AIS provided by end users or directly from an AIS antenna connection; (iii) satellite imagery obtained to detect and classify dark vessels and correlate with vessel tracks from other sources; (iv) external maritime databases incorporated to enhance the metadata available about a vessel, thereby allowing for risky vessels to be identified early and providing decision support for operations; and (v) additional sensors such as Radar and Radar Direction Finders can be incorporated, thereby providing a fused situational picture and allowing for maximum interpretation by the AI services (see Fig. 30.1).

The AI services are categorized according to function and Joint Directors of Laboratories (JDL) categories. A number of services provide input to other services in the toolkit, thereby creating a holistic maritime awareness toolkit. The core component of the architecture is the data lake,

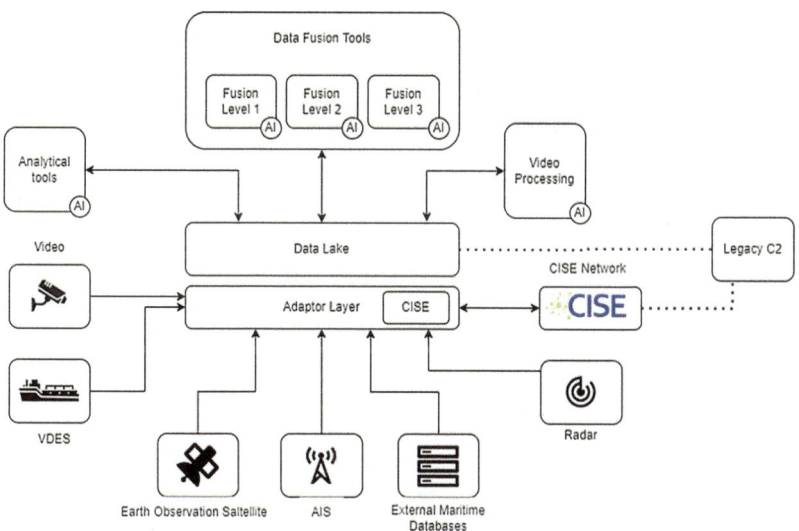

Fig. 30.1 PROMENADE high-level architecture

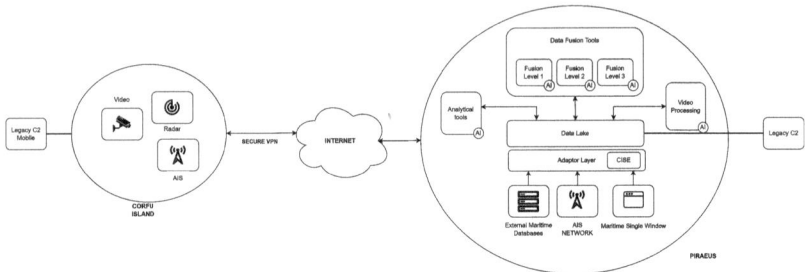

Fig. 30.2   PROMENADE Hellenic trial deployment architecture

which enables the storage of homogeneous and heterogeneous data and publishes this data for PROMENADE services to process and detect situations of interest. The data lake is based on big data technologies enhanced with data transformers and routing mechanisms, providing a transparent area of extract, transform, and load (ETL) processing.

The reference PROMENADE architecture was used as a basis to implement the operational architecture for the Hellenic trial. The deployment took place in Corfu Port Authority and Piraeus, where the national command headquarters of the Hellenic Coast Guard (HCG) is located. The two sites were interconnected via secure VPN ensuring access to the Corfu's sensors and the legacy C2 mobile (see Fig. 30.2).

PROMENADE Technological Services

PROMENADE developed a set of innovative AI/BD-based services for im-proved MSA grouped int five main categories: (a) classification services: they provide automated data processing and classification from various sources (e.g., cameras, satellite imagery, and vessel-tracking stations), related to vessel detection, route classification, vessel activity classification, and oil spill detections; (b) pattern detection services: they provide automatic pattern detection applied to multisource data fusion, extraction of patterns of life, behavior analysis, and anomaly detection, etc.; (c) risk assessment services: they provide automated data processing and risk assessment of vessels' behavior and characteristics through the use of innovative data sources and algorithms; (d) future state prediction services: they allow learning the motion of ships in particular areas of interest from historical data with the goal of predicting their future trajectories and anticipating their future behavior; (e) data infrastructure, which includes the data lake that provides the ingestion, storage, processing, and distribution of data in a BD environment and the data exchange with the CISE network.

Overall, the PROMENADE toolkit includes 22 AI-based and CISE-compliant services. Specifically, the services that have been selected to be demonstrated during the Hellenic trial are described in Table 30.1 per category.

Hellenic Trial Execution: Results and Lessons Learnt

The Hellenic trial took place in February 2023 in Greece in two locations, Corfu Island and Piraeus. It was attended by around 80 people in HCG headquarters, including policymakers, EU agencies (e.g., DG-HOME, REA, EMSA, FRONTEX, EFCA), and maritime authorities, while more than 30 operational and technical experts participated on the field from Corfu Island. The area of interest was the Ionian Sea around Corfu Island, which is known for its large amount of drug trafficking and in some cases irregular migration. Real operational assets were used by Corfu and Igoumenitsa Coast Guard Authorities to play the role of the facilitators and law enforcement. The scenarios were designed by the HCG to mimic past incidents, thereby validating the use of the PROMENADE toolkit in an operational environment. The main objective of the trial was to make use of AI and BD technologies to detect vessel abnormal behavior and identify a "drop-off" drug delivery at sea.

In order to ensure good sensor coverage of the area, a patrol van donated by the HCG has been entirely renovated and equipped with the following surveillance equipment: (i) an AIS antenna, (ii) an X-Band surface IP radar system with 48 nautical miles (NM) range, (iii) a satellite compass, and (iv) a long-range IR—Thermal Internet Protocol (IP) Pan-Tilt-Zoom (PTZ) camera. Furthermore, a display-and-control unit has been installed on the co-driver's side to allow the co-driver to control the camera, focus on the regions/objects of interest, etc. (see Fig. 30.3). It is noteworthy that the driver is not distracted by the light and brightness of the screen and the actions of the co-driver as the installation has been carried out in a way that prevents any such issues. These capabilities are leveraged using the ENGAGE platform, which enhances the end users'

Table 30.1 PROMENADE services tested in the Hellenic trial

Category	Service	Service description
(a) Classification services	Vessel detection in image data (Leonardo)	The service automatically detects, classifies, and tracks vessels within a video stream acquired by coastal surveillance electro-optical (EO) or infrared (IR) cameras. The position of each detected vessel in the video frame is highlighted by a bounding box and its classification is associated. The service generates as output the entities in CISE vessel format and the enriched stream video with bounding boxes surrounding the vessels and the corresponding track ID. Deep learning techniques are used for real-time vessel detection, classification, and tracking. Convolutional neural network layers extract and fuse features, to capture edges, complex texture, and objects shape information.
	Vessel tracking pattern (Leonardo)	The service performs the fusion of messages coming from Radar and AIS sensors in the CISE vessel format. At each timestamp, the ML algorithm predicts if two or more messages belong to the same vessel comparing multiple features, not position-based only (e.g., ship type, speed, heading, latitude, and longitude). After the association phase, all the fused messages are tracked over time to create and maintain updated the trajectory of each vessel. The service uses data-driven learning approaches for the fusion of AIS and Radar data.
(b) Pattern detection services	TRITON pattern detection service (Satways)	The service detects abnormal patterns of vessels by using input from the operator regarding the nature of the abnormality. The service can detect over 40 patterns and can be further enhanced with external databases (e.g., maritime single window) to improve its detection ability. The service is fully CISE compliant, allowing for its use in standalone or in an integrated system.
	Multisensor data fusion engine (Kpler)	The fusion engine is able to integrate positional data coming from different sources, identifying common trajectories and unifying them. It is data source agnostic, that is, it takes as input streams of tuples that can contain only the coordinates of a moving object at a given timestamp, regardless of any other contextual information. Then, the positions that are given as input from a data source A are associated to known tracks from the data source B, performing position-to-track association in real time. To improve performance and meet the requirements of real-time multisensor multiobject data fusion, the fusion engine relies on an efficient distributed architecture.

(continued)

Table 30.1 (continued)

Category	Service	Service description
(c) Risk assessment services	Risk investigator (UCSC Transcrime)	The service identifies vessels potentially involved in suspicious activities, producing a CISE-compliant risk message for every vessel queried in real time. It complements traditional approaches (e.g., check of sanction lists) with innovative ML algorithms to identify hidden patterns and red flags related to shipowner companies. The service does not only look at who controls a vessel, but also how control takes place: whether complex share-holding structures are used, high-risk jurisdictions are crossed, or opaque legal arrangements are employed to conceal the identity of a vessel's owner [3].
(d) Future state prediction services	Advanced ship prediction (NATO-CMRE)	The service addresses the task of vessel trajectory prediction using a recent deep learning algorithm [4]. The approach leverages large volumes of historical AIS data to train a pair of recurrent neural networks (the encoder and the decoder) to predict sequentially a vessel's trajectory, given a sequence of past AIS observations from the same vessel. The service advantages advances in deep learning and the combined availability of massive volumes of AIS data are paving the way to enhance maritime surveillance and predictive capabilities. The main motivation is that neural architectures have the ability to learn complex patterns directly from the data, making it possible to predict future vessel kinematic states even in complex traffic scenarios, where conventional statistical techniques would struggle to achieve satisfactory performance. Finally, the design of the algorithm allows for a probabilistic interpretation of the underlying neural network architecture, thus enabling the seamless integration of the predicted trajectories with other Bayesian techniques, including but not limited to Bayesian tracking and data fusion [5].

(continued)

Table 30.1 (continued)

Category	Service	Service description
(e) Data infrastructure	Data lake infrastructure (Leonardo)	It provides a fully containerized data lake infrastructure leveraging on big data technologies and acting as the backbone of the maritime surveillance system allowing the data ingestion, real-time data streaming, support to big data analysis support and rapid prototyping, and historical data retention. It provides the ability to store structured data in CISE formats and unstructured data such as images and videos. An efficient and reliable open-source solution has been integrated to ensure fault tolerance and meet the demands of increased data loads and scalability. The implemented data lake infrastructure is compatible with the two different architectures adopted for training in the cloud environment and for operational deployment at the end users' premises.

Fig. 30.3 Photographs of the surveillance equipment installed in the HCG patrol van

experience by offering them the opportunity to access and process the data of interest via mobile devices, such as tablets and laptops apart from the multifunctional plotter. The ENGAGE Border Management Edition (BME) developed by Satways Ltd. is a fully fledged C2 operated by the Hellenic Coast Guard for various R&D activities ensuring the latest cutting-edge technologies are tried and tested by their personnel.

In addition, there is equipment installed in the rear cabin of the van (see Fig. 30.4). An alternating current (AC) power as well as a 12 V,

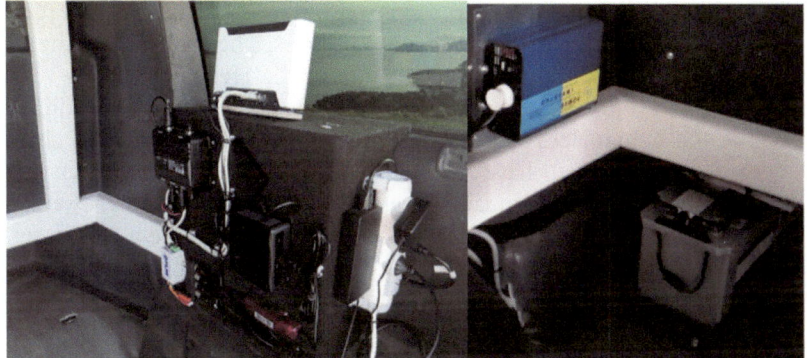

Fig. 30.4 Photographs of the telecommunication equipment and the power system battery installed in the rear cabin of the HCG patrol van

120 Ah power system battery (PSB) are also installed in the rear cabin. The additional battery ensures the smooth functioning of the patrol van and guarantees it works independently from a power autonomy standpoint as the battery is charged when the engine of the van is on. Of course, the system is designed in such a way that the engine does not need to be on as long as the battery is adequately charged. For example, the system can work for almost 15 hours with the engine of the van constantly off.

The appropriate equipment has been installed on the van and customized, leading to its transformation into a self-contained, highly autonomous and portable command-and-control (C2) system. This system aims at equipping the operational staff with all the capabilities, which are considered necessary for carrying out their operational duties, especially in cases where the required infrastructure for the support and deployment of a C2 system is unavailable (e.g., in remote and/or isolated areas). Therefore, the van constitutes a modular, easy-to-deploy, but secure design, capable of satisfying the end users' needs as a portable C2 center integrated with the HCG legacy systems. This goal is achieved by offering state-of-the-art patrolling, surveillance, detection, risk management, and future state prediction capabilities, enhanced by advanced AI as well as data fusion techniques, incorporated into the PROMENADE toolkit.

Ultimately, the operational goals through the execution of the Hellenic trial were related either to apprehend activities with drugs loaded and unloaded in remote areas with vessels that are fast enough to avoid detection or capture, or dropped off drugs that can be transported by a range of commercial vessels such as cargo ships, fishing vessels, tankers, and tug-boats, so as to be collected by the drug traffickers. The overall outcome of the AI services tested with real vessels during the Hellenic trial was very positive while the main lesson learnt was related to the dynamic adjustment of the system alert parameters to depict the 100% operational challenges that the end users have to face in the field. End users' feedback was collected through a questionnaire developed using the EU survey. Overall, the operators noted that the use of AI and big data will be a significant advantage in upgrading deprecated systems and achieving early detection of targets through valid patterns of abnormal behavior.

CONCLUSION

In the domain of maritime surveillance, PROMENADE project has made significant advancements in the field of MSA, enhancing border and external security capabilities by developing and integrating in a unique open and CISE-compliant service-based toolkit a set of innovative AI/BD-based services.

By leveraging advanced technologies, integrating diverse data sources, and conducting extensive trials, including the Hellenic one, PROMENADE has paved the way for enhanced maritime security, decision-making, and information exchange among maritime surveillance authorities. The project's achievements in providing valuable insights, predicting vessel behavior, and enabling efficient data processing will contribute to the overall goal of ensuring the safety and security of European waters and territories.

Acknowledgments This project received funding from the European Union's Horizon 2020 research and innovation program under grant agreement no. 101021673. This chapter reflects only the authors' views, and the Research Executive Agency and the European Commission are not responsible for any use that may be made of the information it contains.

REFERENCES

1. Sharma, R. (2020). Study of supervised learning and unsupervised learning. *International Journal for Research in Applied Science and Engineering Technology, 8*(6), 588.
2. European Commission. (2021, April 21). *Proposal for a Regulation laying down harmonised rules on artificial intelligence.* Retrieved from European Commission: https://ec.europa.eu/newsroom/dae/redirection/document/75788

3. Bosisio, A., et al. (2021). *Developing a tool to assess corruption risk factors in firms' ownership structure*. Final report of DATACROS Project. Milan, Italy: Transcrime - Università Cattolica Sacro Cuore: https://www.transcrime.it/wp-content/uploads/2021/09/Datacros_report.pdf
4. Capobianco, S., Millefiori, L. M., Forti, N., Braca, P., & Willett, P. (2021). Deep learning methods for vessel trajectory prediction based on recurrent neural networks. *IEEE Transactions on Aerospace and Electronic Systems, 57*(6), 4329–4346. https://doi.org/10.1109/TAES.2021.3096873
5. Forti, N., Millefiori, L. M., Braca, P., & Willett, P. (2023). Model-based deep learning prediction for maneuvering target tracking. In *Proc. 26th international conference on information Fusion (FUSION)*. Charleston, SC, USA.

Open Access This chapter is licensed under the terms of the Creative Commons Attribution 4.0 International License (http://creativecommons.org/licenses/by/4.0/), which permits use, sharing, adaptation, distribution and reproduction in any medium or format, as long as you give appropriate credit to the original author(s) and the source, provide a link to the Creative Commons license and indicate if changes were made.

The images or other third party material in this chapter are included in the chapter's Creative Commons license, unless indicated otherwise in a credit line to the material. If material is not included in the chapter's Creative Commons license and your intended use is not permitted by statutory regulation or exceeds the permitted use, you will need to obtain permission directly from the copyright holder.

CHAPTER 31

EURMARS: An Advanced Surveillance Platform to Improve the European Multiauthority Border Security Efficiency and Cooperation

Georgios Mourkousis, Matthaios Protonotarios, Chrysostomos Antoniou, Andreas Kriechbaum-Zabini, Stephan Veigl, Jonathan Boyle, Lulu Chen, George Voskopoulos, Romaios Bratskas, Laura Salmela, Jari Laarni, Antti Väätänen, Sirra Toivonen, Mikio Akagi, and Claudia-Iohana Voicu

G. Mourkousis (✉) • M. Protonotarios
Hardware and Software Engineering Epe, Marousi, Greece
e-mail: georgemr@hse.gr

C. Antoniou
European Dynamics Luxembourg Sa, Luxembourg City, Luxembourg

A. Kriechbaum-Zabini • S. Veigl
Austrian Institute of Technology Gmbh, Vienna, Austria

© The Author(s) 2025   387
I. Gkotsis et al. (eds.), *Paradigms on Technology Development for Security Practitioners*, Security Informatics and Law Enforcement, https://doi.org/10.1007/978-3-031-62083-6_31

Introduction

The EURMARS project enhances EU efforts in addressing increasingly complex security risks and threats regarding border management in the maritime domain by designing and implementing a multiauthority border surveillance platform that integrates AI, risk assessment, and visualisation innovations supported by advanced sensing technologies, such as high-altitude platform systems, satellite imagery, unmanned vehicles (UAVs), and ground-based sensors. The main end users to benefit from the project results are border authorities (BA) and agencies at the national and EU levels, such as coast guard, border guard, customs, police, fisheries control, environmental protection, and maritime safety entities. The overall concept is depicted in Fig. 31.1.

In this chapter, we describe the innovative methodological approach and the transformative integration of AI, risk assessment, visualisation novelties, and advanced sensing technologies, including high-altitude platform systems, satellite imagery, unmanned vehicles, and ground-based sensors as a blueprint for the development of next-generation 24/7 border surveillance platforms. Key innovation enablers are founded on solid and standardised best practices for user requirements extraction and extend to the development of enhanced data collection and analysis mechanisms (data fusion), improved risk assessment, real-time visualisation and monitoring, with integrated decision support system offering tailored insights and improved efficiency in situation awareness and operational capacity for a wide range of maritime security risks and threats in remote environments. The proposed blueprint additionally includes the operational validation in real-life scenarios, an AI Foresight Report and

J. Boyle • L. Chen
The University of Reading, Reading, UK

G. Voskopoulos
Geosystems Hellas IT Kai Efarmoges Geopliroforiakon Systimaton Anonimi Etaireia, Athens, Greece

R. Bratskas
SKYLD Security and Defence Ltd, Nicosia, Cyprus

L. Salmela • J. Laarni • A. Väätänen • S. Toivonen
Teknologian Tutkimuskeskus Vtt Oy, Espoo, Finland

M. Akagi • C.-I. Voicu
Trilateral Research Limited, Waterford, Ireland

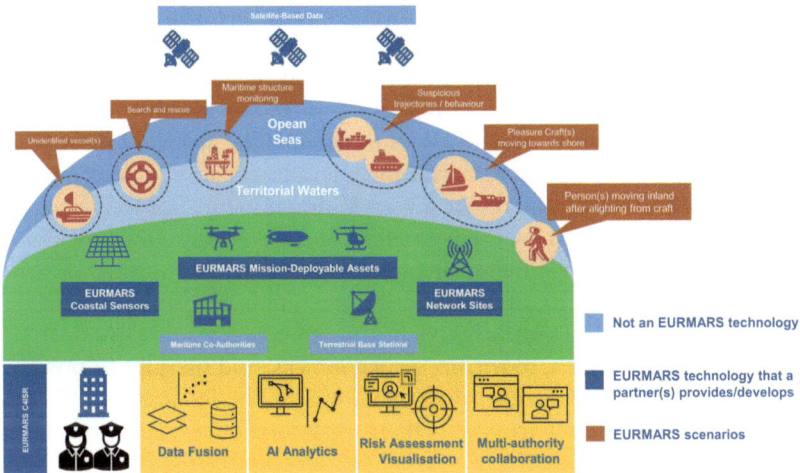

Fig. 31.1 The EURMARS concept

Blueprint containing PIA, EIA, and SIA assessments, and a best-effort attempt to create open-source domain-specific benchmark datasets and contribution to relevant international standards. By committing to implement and validate the presented approach, EURMARS aims to achieve significant advancements in data-driven decision-making, standardisation initiatives, and relevant research, with the potential to transform the border management domain and significantly improve border security.

The viability and effectiveness of the EURMARS concept will be evaluated in the following pilot use cases: (1) PUC1—Maritime Border Control: Detection of Trafficking and Other Illegal Activities; (2a) PUC2a—Search & Rescue; (2b) PUC2b—Maritime Structures and Oil Spills Surveillance and Monitoring; and (3) PUC3—Land Border Control; Illegal Crossing Outside of Business Contingency Plans.

Motivation of PUC1 is to improve the detection of abnormal behaviour of vessels involved in illegal activities, with a focus on small boats near shipping routes and shorelines. Existing controls are insufficient for detecting small boats approaching the shoreline, leading to difficulty in identifying illegal activities amidst legitimate traffic and irregular coastlines. A demonstration will be conducted to address illegal activities, including the transfer of trafficked people from larger to smaller vessels. The system's provision of timely alerts regarding high-risk vessels to

Border Force Operational Units allowing appropriate actions to be taken will be evaluated.

In PUC2a, EURMARS enhances search and rescue (SAR) operations, aligning with EU definitions of rendering assistance to distressed vessels and individuals at sea. EURMARS utilises satellite images, video, SAR data, and AIS information to detect the relevant scenarios, enhancing the effectiveness of search and rescue forces. Various satellite missions (including Copernicus, NEMO-HD μ-Satellite) enable a long endurance and wider surveillance capability to situational awareness.

The objective of PUC2b is to consistently monitor offshore structures, including gas and oil platforms, detecting emergency situations such as fires, structural failures, terrorism, armed attacks, and oil spills to prevent human loss and mitigate environmental impact. The integration of high-altitude sensors offers a groundbreaking capability for regular or real-time monitoring of structures far from land, overcoming the limitations of current systems that rely on expensive and timely helicopter deployment. The system's rapid response capability allows SAR operations to mobilise forces, including boats, helicopters, and UAVs, for timely assistance to offshore platforms in emergencies, such as evacuations or medical support. The platform's ability to detect oil spills promptly contributes to mitigating their effects on natural habitats, preventing potential environmental crises.

The challenge in PUC3 is countering illegal migration and human trafficking at state borders, specifically involving individuals arriving by boat and then moving inland to predetermined locations for illegal border crossing. EURMARS enables border guards to follow illegal activities in a joint multiauthority/country cross-border deployment. As maritime and land border authorities currently lack the necessary resources to disrupt criminal networks, the deployment of EURMARS allows the realisation of seamlessly tracking handovers between sea/land and land/sea, a capability not possible with existing systems.

METHODOLOGY FOR ELICITING REQUIREMENTS FOR BORDER SURVEILLANCE

The requirements engineering methodology has been an integral part of the project's innovation by combining leading requirements engineering practises and standards [1, 2], with insights and lessons learned gained from similar prior activities that align with the project's objectives, the available time frame (6 months), and the iterative development cycle adopted in the project including the prototype development and living labs establishment.

The adopted methodology included four phases: requirements elicitation, requirements analysis, requirements specification, and requirements validation. The requirements elicitation phase involved collecting requirements from a variety of stakeholders, including end users, domain experts, and technical experts. The requirements analysis phase included interpreting the collected requirements and identifying any inconsistencies or ambiguities. The requirements specification phase focused on creating a formal document that describes the requirements in detail. The requirements validation phase involved verifying that the requirements are complete, consistent, and feasible.

The methodology (see Fig. 31.2) was implemented using a variety of techniques, including surveys, interviews, workshops, document analysis, and prototyping. The requirements were iteratively refined and prioritised throughout the process of collecting, analysing, and converting the inputs from the stakeholders into exact requirements specifications, ensuring that they comprise the common needs of end users. High emphasis was placed on co-creative practices that intensified towards the end of the overall requirements development process. Especially for requirements prioritisation, partners were asked to provide feedback individually, and the overall material was discussed and processed together in workshops to achieve a shared view on the importance of each requirement. The process implemented to develop the requirements is illustrated in Fig. 31.2 and resulted in the specification of over 140 individual requirements structured into to nine high-level categories.

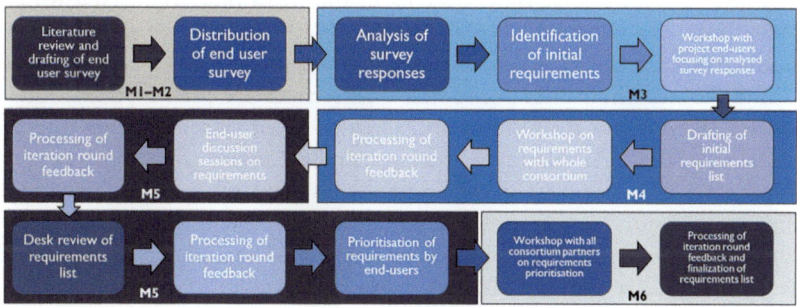

Fig. 31.2 Research process for EURMARS requirements development

MODERN SENSORS FOR BORDER SURVEILLANCE

In its core, the EURMARS platform incorporates a combination of novel AI-powered coastal ground sensor platforms, low- and high-altitude sensing systems, and satellite-based systems to feed the advanced fusion and reasoning functionality. In complement to classical optical sensors that work on different wavelengths and at different altitudes, existing relevant data sources (e.g. vessel transponders) are additionally being analysed and integrated to improve situational awareness.

Coastal Ground and Low-Altitude Sensing Systems

EURMARS is developing a mast-mounted edge-based sensor platform enabling automated vessel/boat, person, and ground vehicle detection and tracking for all different weather conditions with a focus on challenging foggy maritime situations. This is achieved by filling the classical spectral gap between visual (RGB) cameras and thermal sensors with a new sensor platform integrating SWIR and UV sensors. An AI-based algorithm is used for the detection of small vessels up to 1 km from the coast and people and vehicles near shore (YOLOX for vessel detection and other YOLO [3] derivatives for automated detection). Detections are fed to a tracking algorithm to generate short-term tracklets for vessels and ground objects. The same sensor platform will be used for classifying the vessels— a set of residual neural network models have been trained on various representative maritime vessel/ground vehicle training sets, with data augmentation where appropriate to increase the robustness of the models.

To enhance the platform's detection capabilities of targeted objects (including small vessels) in the distance of a few kilometres, EURMARS employs low-altitude sensing systems in the form of UAVs (fixed-wing, multirotor, and helicopter types) equipped with surveillance sensors such as electro-optic, thermal, and hyper-spectral cameras. UAVs are activated and deployed when unconventional occurrences are identified by other sensors like camera subsystems and satellite-based systems. In its operational role, the UAV platform, during its patrol flights, corroborates and validates events (such as vessels, individuals, and oil spills) monitoring their progression (as long as the UAV flies within the targeted area) while supporting the streaming of live video feed from the designated area. The available infrastructure integrates a robust AI-enabled computing system that allows real-time identification of events and objects of interest like ships, small watercrafts, and individuals within the maritime environment. The generated data include types and dimensions of the identified objects, precise location, trajectories, and count of individuals present in the water. The platform transforms this data in identified and categorised incidents, along with their associated metadata encompassing geographical coordinates, images, classifications, and other details.

High-Altitude Sensing Systems and Satellite-Based Sensing

The sensing assets providing the highest value in the EURMARS innovation proposition are the satellite-based systems which significantly contribute to the expansion of the platform's capabilities. Main functionalities include vessel and oil spill detection through the use of both synthetic aperture radar (SAR) and optical data from the Copernicus hub (Sentinel 1 and 2) [4], Landsat 8 [5], Landsat 9 [6], the ICEYE commercial satellite [7], and NEMO-HD (cubesat) [8]. AI/ML models are used in combination with the YOLO [3] algorithm to provide event detection in multiple output formats. Additionally, EURMARS will develop earth-LIVE, a particular simulator for a telescope equipped with COTS detectors, baffles, pointing mechanisms, and video telemetry transmitting equipment. The simulator will take as input videos from NEMO-HD of up to 3 min in duration, captured in low-resolution (40 m) and high-resolution (2.8 m) mode, and generate vessel detection events.

Relevant Data Sources Intelligence

To complement the spectrum of sensing systems from the perspective of external data acquisition, EURMARS will harness open-access data sources to enhance actionable intelligence for its purposes. The types of open data that can be processed include (i) VDES/AIS data, (ii) weather data, and (iii) CISE data. Regarding VDES/AIS data, a common data hub infrastructure exploits data acquisition from three types of sources: EURMARS' own receivers installed at the pilot sites, Cypriot and Bulgarian BAs' infrastructure and prominent commercial services (marinetraffic.com, vesselfinder.com, etc.). Present and forecasted weather data are obtained through the Copernicus Climate Change Service (C3S) [9] which provides information about the past, present, and future climate conditions in Europe. EURMARS also plans to develop CISE adaptors to be able to connect to the CISE network [10], an EU initiative promoting interoperable data exchange between concerned authorities for missions at sea. Envisioned information retrieval will regard information on vessel position, incidents, and alerts on vessel trajectories.

ADVANCED FUSION, REASONING, AND RISK ASSESSMENT FOR BORDER SURVEILLANCE

The data acquired from the sensing systems and the external data sources (candidate vessel detection positions, tracks, movement directions, object sizes and speeds, vessel/vehicle classification, velocity signatures, etc.) is fed into the multimodal data fusion platform (MDFP) for processing of the information into a single authoritative source for further analysis. A diagram of the MDFP and its connectivity with other modules within the EURMARS system is provided in Fig. 31.3.

The MDFP comprises three separate subcomponents: primary fusion module, fusion tracking module, and feedback loop.

- The *primary fusion module* performs fusion by using a Bayesian inference system coupled with computer vision techniques [11]. It uses probabilistic occupancy maps, similar to occupancy grid maps used in robotics, to aggregate information from the sensors. Subsequently, areas of high confidence are detected to generate fused detection events.
- The *fusion tracking module* parses the fused detection events to provide tracking of targets over time. The module provides multiple

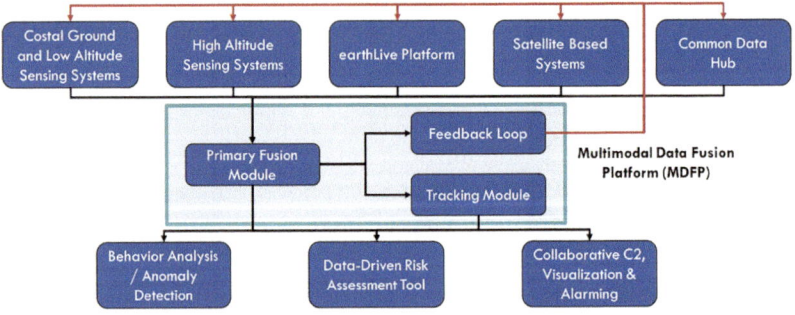

Fig. 31.3 The multimodal data fusion platform and its interfaces to other components

levels of tracking algorithms with custom rules for target recovery specific to the current scenario, for example, associating multiple tracks together if they contain the same identification via AIS. The tracked target information is then forwarded for behaviour analysis and risk assessment.

- The *feedback loop* takes the events from the previous modules and determines based on confidence scores whether additional information is required. In cases of low confidence, the relevant sensing systems are prompted to reposition or reorient to capture higher quality information; for example, reorient a pan-tilt-zoom camera towards a specific geo-location.

The MDFP can operate in two modes depending on the scenario: a live pipeline that directly processes events from available sources, and an offline mode that can use recorded information. This offline mode enables the inclusion of detected events from sensors with a large delay—such as high-quality satellite-based sensors that have a multi-hour lead time.

The final step in data processing entails the risk management, an essential procedure for identification assessment and mitigation of the potential threats that may affect the security of borders and the safety, and well-being of individuals. Existing risk assessment involving vessels for situational awareness is typically probabilistic or fuzzy-based and lacks qualitative risk assessment. EURMARS' risk assessment framework is based on FRONTEX's Common Integrated Risk Analysis Model (CIRAM) [12] that provides a standardised methodology for conducting risk analysis at the EU level and supports the operational planning and decision-making of BAs. EURMARS's risk assessment framework is a flexible data-driven vessel risk profiling tool adaptable to different scenarios and contexts. It can analyse the current situation and detect threats based on the described data fusion and assign ranks to the threats based on their potential impact for the present time and probable states in which they may evolve. Core functionality includes the automatic recognition of threats, the analysis of potential impact of threats focusing on cross-border threats and threats that may have a larger impact or an important humanitarian impact, requiring coordination and collaboration of multiple authorities.

ETHICAL ASPECTS

The ethical perspective is of utmost importance in EURMARS as the project is committed to safeguarding the ethical dimension of all procedures and results. These dimensions comprise the review of all relevant legislation, EC recommendations and best practices, coverage of all the related ethical requirements, identification and recruitment of research participants, ethical approvals for research with humans, personal data processing, and Human/Fundamental Rights Impact Assessment (HRIA/FRIA) related to AI. Procedures and criteria have been identified and utilised to identify activities requiring an ethics approval prior to commencement, to manage delivery of activities' ethics approval and devise project-tailored informed consent procedures and informed consent forms. In addition, a detailed review and analysis has been performed on all procedures entailing information on processing of personal and sensitive data. The data minimisation principle has been enforced by the detailed design and description of technical organisational and security measures implemented by partners to safeguard the data and documentation of anonymisation and pseudonymisation techniques. Finally, due to the development of AI technologies, based on the HRESIA methodology for AI [13] and the EC guidance on trustworthy AI [14], a framework was designed for HRIA/FRIA and project-specific AI-related Risk Assessment (AIRA) covering the development, deployment, and post-deployment phases of EURMARS, including detailed information on how respect for fundamental human rights and freedoms will be ensured. The framework ensures that the artificial intelligence (AI) components will be integrated through an 'ethics by design' (EbD) [15] approach, conforming to all current applicable regulations and including recommendations regarding measures for risk prevention, minimisation, as well as mitigation. The alignment with EU's AI-Act [16, 17] is evaluated in an AI ACT Foresight Compliance, Social and Ethical Impact Assessment report.

CONCLUSIONS

The EURMARS platform emphasises giving BAs the tools to track, identify, and classify vessels and, on occasion, people and ground vehicles in real time over a large geographical area in a robust and accurate way by engaging a set of sensors that range from low to high altitude and space technologies, leveraging private (earthLIVE, Stratobus), as well as national initiatives (Nemo-HD) for maritime surveillance. EURMARS covers the interface between land and sea operations and in doing so enhances the collaboration of the relevant actors via the technical infrastructure that integrates the information and makes it available to them. Besides the technical and situation awareness innovations, the project boasts specific activities to deliver on the ethics capacity, through impact assessments, recommendations, and training materials and contributing to ensuring the protection of human rights with respect to emerging technologies across and beyond Europe. In the long term, EURMARS envisions to contribute and support Actions of the EU Maritime Security Action Plan [18] by 'investigating synergies with the civilian sector, also harmonising the system requirements' and by contributing to the development of a cross-sectoral agenda for maritime security research and the dual-use of technologies contributing to maritime security'.

Acknowledgements This project received funding from the European Union's Horizon Europe research and innovation programme under grant agreement no. 101073985. This article reflects only the authors' views, and both the Research Executive Agency and the European Commission are not responsible for any use that may be made of the information it contains.

REFERENCES

1. ISO/IEC/IEEE 29148:2018, Requirements engineering. Available at: https://www.iso.org/standard/72089.html. Last accessed: 28 August 2023.
2. Wiegers, K. E., & Beatty, J. (2013). *Software requirements*. Microsoft Press., ISBN:978-0-7356-7966-5..
3. Redmon, J., Divvala, S., Girshick, R., & Farhadi, A. (2016). You only look once: Unified, real-time object detection. In *2016 IEEE Conference on Computer Vision and Pattern Recognition (CVPR), USA* (pp. 779–788). IEEE. https://doi.org/10.1109/CVPR.2016.91
4. Sentinel Online. Available at: https://sentinel.esa.int/web/sentinel/home. Last accessed: 28 August 2023.
5. Roy, D. P., Wulder, M. A., Loveland, T. R., Woodcock, C. E., Allen, R. G., Anderson, M. C., Helder, D., Irons, J. R., Johnson, D. M., Kennedy, R., et al. (2014). Landsat-8: Science and product vision for terrestrial global change research. *Remote Sensing of Environment, 145*, 154–172. https://doi.org/10.1016/j.rse.2014.02.001
6. Wu, Z., Snyder, G., Vadnais, C., Arora, R., Babcock, M., Stensaas, G., Doucette, P., & Newman, T. (2019). User needs for future landsat missions. *Remote Sensing of Environment, 231*, 111214. https://doi.org/10.1016/j.rse.2019.111214

7. Paek, S. W., Balasubramanian, S., Kim, S., & de Weck, O. (2020). Small-satellite synthetic aperture radar for continuous global biospheric monitoring: A review. *Remote Sensing, 12*, 2546. https://doi.org/10.3390/rs12162546

8. Pranajaya, F., Zee, R., Grocott, S., Rodič, T., Matko, D., Oštir, K., Peljhan, M., Urbas, A., Fröhlich, H., Blažič, S., et al. (2012). NEMO-HD: High-resolution microsatellite for earth monitoring and observation. In *Small Satellite Conference 2012*.

9. European Union, Copernicus Climate Change Service (C3S). Available at: https://climate.copernicus.eu/. Last accessed: 28 August 2023.

10. Common Information Sharing Environment (CISE). Available at: https://www.emsa.europa.eu/cise.html. Last accessed: 28 August 2023.

11. Patino, L., Hubner, M., Litzenberger, M., & Ferryman, J. (2022). Tracking of objects in a multi-sensor fusion system for border surveillance. *Journal of Defence & Security Technologies, 5*(2), 29–43.

12. Frontex. *Common integrated risk analysis model*. Available at: https://frontex.europa.eu/what-we-do/monitoring-and-risk-analysis/ciram/. Last accessed: 28 August 2023.

13. Mantelero, A. (2022). *Beyond data: Human rights, ethical and social impact assessment in AI*. Asser Press. https://doi.org/10.1007/s43681-022-00218-9

14. European Commission. (2020). *High-level expert group on artificial intelligence*. European Union.

15. European Commission. (2020). *The assessment list for trustworthy artificial intelligence (ALTAI) for self-assessment*. Available at: https://digital-strategy.ec.europa.eu/en/library/assessment-list-trustworthy-artificial-intelligence-altai-self-assessment. Last accessed: 28 August 2023.

16. European Commission. (2021). *Proposal for a regulation of the European Parliament and of the Council laying down harmonised rules on artificial intelligence (Artificial Intelligence Act) and amending certain Union legislative acts*. Available at: https://eur-lex.europa.eu/legal-content/EN/TXT/?uri=celex%3A52021PC0206. Last accessed: 28 August 2023.

17. European Commission. (2023). *Amendments adopted by the European Parliament on 14 June 2023 on the proposal for a regulation of the European Parliament and of the Council on laying down harmonised rules on artificial intelligence (Artificial Intelligence Act) and amending certain Union legislative acts*. Available at: https://www.europarl.europa.eu/doceo/document/TA-9-2023-0236_EN.html. Last accessed: 28 August 2023.

18. European Commission. (2023). *Joint Communication to the European Parliament and The Council on the update of the EU Maritime Security Strategy and its Action Plan "An enhanced EU Maritime Security Strategy for evolving maritime threats"*. Available at: https://eur-lex.europa.eu/legal-content/EN/TXT/?uri=CELEX:52023JC0008https://climate.copernicus.eu/. Last accessed: 28 August 2023.

Open Access This chapter is licensed under the terms of the Creative Commons Attribution 4.0 International License (http://creativecommons.org/licenses/by/4.0/), which permits use, sharing, adaptation, distribution and reproduction in any medium or format, as long as you give appropriate credit to the original author(s) and the source, provide a link to the Creative Commons license and indicate if changes were made.

The images or other third party material in this chapter are included in the chapter's Creative Commons license, unless indicated otherwise in a credit line to the material. If material is not included in the chapter's Creative Commons license and your intended use is not permitted by statutory regulation or exceeds the permitted use, you will need to obtain permission directly from the copyright holder.

CHAPTER 32

BorderUAS Project: Semiautonomous Border Surveillance Platform Combining a Lighter-Than-Air (LTA) Unmanned Aerial Vehicle (UAV) with Ultra-High-Resolution Multisensor Surveillance Payload: A Comprehensive Overview

Ioannis Athanasakis, Dimitrios Myttas,
Theodore D. Katsilieris, Elisavet Bellou, Michalis Zervakis,
Marios Antonakakis, Nikolaos Koutras,
George Boulougaris, Marios Georgiou, Iva Salom,
Dejan Todorovic, Ivan Salajster, Giovanni Nico,
Olimpia Masci, Ioannis Kontopodis, Francisco Iriarte,
and Peter Leskovsky ⓘ

I. Athanasakis • D. Myttas (✉) • T. D. Katsilieris • E. Bellou
Center for Security Studies (KEMEA), Athens, Greece
e-mail: d.myttas@kemea-research.gr

M. Zervakis • M. Antonakakis
Digital Image and Signal and Processing Lab, School of Electrical and Computer
Engineering, Technical University of Crete, Akrotiri Campus, Chania, Greece

© The Author(s) 2025   401
I. Gkotsis et al. (eds.), *Paradigms on Technology Development*
for Security Practitioners, Security Informatics and Law
Enforcement, https://doi.org/10.1007/978-3-031-62083-6_32

INTRODUCTION

At the onset of the 2020s, EU Member States and their non-EU (Third) Country Neighbors have come to share 14.647 km of land borders and 67.571 km of maritime borders (coastline) [1]. The width, landscape bio-diversity, and seasonal terrain's adversities of EU's borderlands have, before long, come to impact the frontier surveillance duties ascribed to various agencies operating within their premises, that is, National Police Units, Gendarmerie, and Frontex. Faced with such territorial hardships and an ever-increasing number of illegal activities, such as drug and human trafficking, border crossings, and state-backed hybrid threats, Border and Coast Guard Authorities have become increasingly reliant on state-of-the-art equipment (i.e., UAVs, mobile equipment, sensors, machine learning, etc.) to counter the constantly shape-shifting threats [2]. The recent tech-nological advancements in the field of unmanned aerial vehicles (UAVs), often referred to as drones, have emerged as a promising tool in border surveillance given their versatility and mobility, coupled with AI-backed applications and a multiplicity of integrated sensors. In this regard, UAVs can provide EU Border Security Practitioners with significantly improved, timelier, situational awareness, and decision-making capacity. To this end, our work explores an ultra-high-resolution multisensor surveillance pay-load supporting border surveillance, search and rescue applications, and

N. Koutras • G. Boulougaris • M. Georgiou
ADDITESS Advanced Integrated Technology Solutions & Services Ltd,
Nicosia, Cyprus

I. Salom
Institute Mihajlo Pupin, University of Belgrade, Belgrade, Serbia

D. Todorovic
Dirigent Acoustics, Belgrade, Serbia

I. Salajster
AVT, Zagreb, Croatia

G. Nico • O. Masci
DIAN_S.r.l., Matera, Italy

I. Kontopodis • F. Iriarte • P. Leskovsky
Vicomtech Foundation, Basque Research and Technology Alliance (BRTA),
Donostia-San Sebastian, Spain

rough terrain detection(s) using a lighter-than-air (LTA) UAV to complement the efforts of Border Police Agencies during and after the project
[3]. By employing a synthetic aperture radar (SAR), shortwave/longwave
infrared (SWIR/LWIR) and acoustic cameras for direct target detection,
as well as optical and hyperspectral cameras for indirect detection (via vegetation disturbance), this project will use innovative data models (AI) to
identify illegal crossing patterns and preferred routes and advanced audio/
video analytics and storage in the C2 center to make borderlands safer and
more secure. The remainder of this chapter consists of the literature review
section and the BorderUAS' methodology analysis, expanding on the
analysis and performance of the payload components. The proofs-of-
concept (SAR, acoustic sensor) are also included along with preliminary
results. The chapter's final segment touches upon a series of concluding
points and future work until the project's completion.

RELATED WORK

Border surveillance and control (land, air, and maritime) is conducted by
the European or national authorities as depicted in Fig. 32.1 [4].

In particular, the case of airborne border surveillance has garnered significant interest from EU agencies. This interest is fueled by remarkable

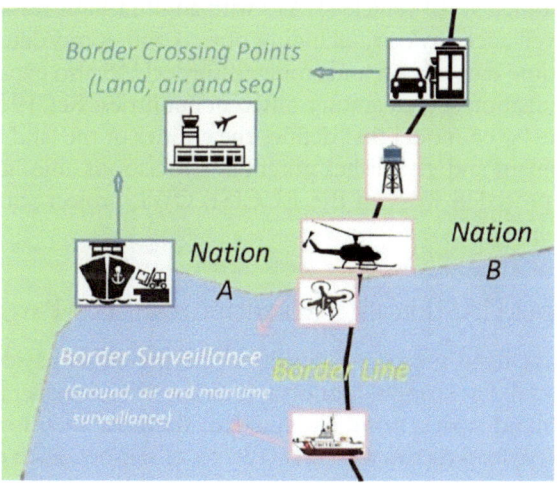

Fig. 32.1 Border management concept

progress witnessed in the drone industry, coupled with advancements in computer vision technology. A comprehensive overview of this tendency is analyzed in [5], where it is stated, among others, that Frontex and the European Maritime Safety Agency began using drones for border surveillance in 2018 after many years of research and pilot projects. The EU-funded program FOLDOUT [6] presented an under-foliage detection method of illegal activity at borders and traces the movement and routes prior to arrival in border areas. To achieve this, a holistic system employing fused advanced ground, aerial, and space-based sensors mounted on Stratobus™ and satellite platforms was deployed in different countries for testing the performance under different weather conditions. In the same context, [7] examines the use of UAVs to support maritime safety and search and rescue operations by providing live video streams and images from the area of operations, within the EFFECTOR project. They developed an embedded system that employs machine learning algorithms, allowing a UAV to autonomously detect objects in the water and keep track of their changing position through time. Another innovative project in the field of sea border surveillance is [8], developing unmanned vehicles (UxV)—aerial, sea surface, and submarine ones, to enhance current maritime border surveillance operations regarding the detection of irregular migrants and narcotics smugglers. In addition to the state-of-the-art, BorderUAS combines for the first time a multirole lighter-than-air (LTA) unmanned aerial vehicle (UAV) with an ultra-high-resolution multisensor surveillance payload, including objects and events detection, providing real-time data stream and storage. BorderUAS also expands on the data harmonization shared among different countries; Ref. [9] developed a knowledge–based model that defines the entities of the land border surveillance domain and establishes potential associations among them, following the preceded work of the EUCISE-OWL project (http://www.eucise2020.eu).

BorderUAS Project Methodology and Payload

The project at hand presents a novelty of having all its payload components mounted on a single platform to allow for interoperability, thus making it stand apart from like-minded ventures mentioned above. BorderUAS payload carries with it two arrays of sensors. Each component

Table 32.1 Overview of BorderUAS' payload components

Payload component	Component's colocation	Functionality
Acoustic sensor	NADIR (immotile)	Mapping, direct detection
CCTV optical RGB sensor	PTZ Board (motile)	Direct detection, tracking
Hyperspectral sensor	PTZ Board (motile)	Mapping, detection
LWIR, SWIR cameras	PTZ Board (motile)	Direct detection, indirect detection, tracking
Optical high-resolution RGB sensor(s)	NADIR (immotile)	Direct detection, tracking
SAR/RAR sensor	Slant positioned (immotile)	Mapping, detection

is intended to support one (or more) of the border surveillance functions that BorderUAS seeks to empower. Table 32.1 provides a list of sensor functionality and their positioning thereof.

The footage secured by the payload's components is then processed by means of an algorithmic design. The feed processing is there to enable end users to determine the degree of a threat each detected event may represent. In a nutshell, the techniques that will be applied to acoustic and visual feedback based on component functionality are laid down as follows:

- Data acquisition, storage, and streaming
- Object detection, tracking, characterization, and reidentification
- Data and metadata fusion for the reduction of false positives or negatives
- Geo-localization of detections
- Photogrammetry and multimodal terrain mapping

The payload components, along with the algorithmic design applied in the feed received from them, will both be used in a set of field-trial scenarios designed to test them in real-life conditions. The scenario concepts were the outcome of end user-directed questionnaires [10] and are designated to feature detection/tracking of individuals/objects in open areas, hardly accessible areas or under canopy, as well as indirect (non-real time) detection of individuals and events through observations of the land cover.

BORDERUAS PAYLOAD-RELATED FUNCTIONALITIES

The payload components of BorderUAS are briefly described below.

Acoustic Sensor (Proof of Concept) For the purpose of mapping hardly accessible border areas, especially so during S&R operations, BorderUAS features an acoustic sensor (AS) designed and developed to capture acoustic signal from far distances that can be related to specific events and observations on the ground. Considered are working machinery, passing vehicles, or yelling people. The AS comprises a microphone array and uses a beamforming algorithm to generate a visual presentation of the acoustic field, that is, acoustic mapping of the UAV's borderland surroundings, in real time.

Besides the acoustic map, which is georeferenced using coordinates from UAV's GPS, the sensor also allows for real-time augmented listening to specific locations on the ground. The overall concept of the acoustic map acquisition is depicted in Fig. 32.2, where detected ground noise has been geo-localized based on the GPS position of the sensor and the individual results of the beam-forming algorithm.

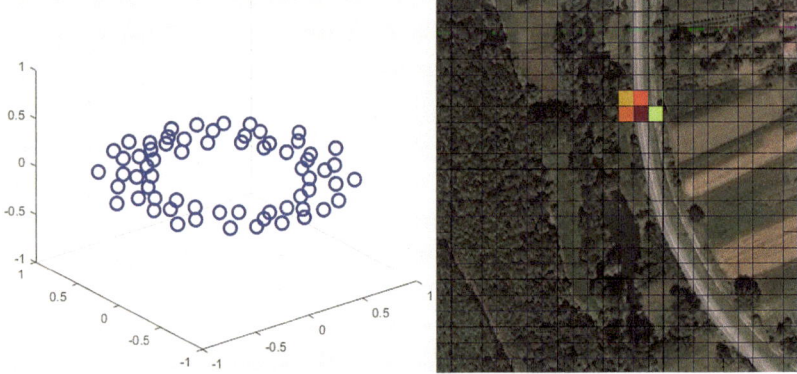

Fig. 32.2 Concept of the acoustic mapping: the spatial configuration of the AS microphone array adopted to the UAV haul is shown on the left and the resulting acoustic map, marking sound sources, is shown on the right, overlayed on the top of the satellite imagery

LWIR/SWIR Novelty BorderUAS will also feature two IR sensors (LWIR, SWIR) for enhancing border surveillance under nightly conditions. LWIR comprises two cameras, one with fixed and another with a zoom lens, operating at a 7–14 µm wavelength range and will be used for direct detection of individuals/objects via its thermal vision property. SWIR comprises a camera with a fixed lens that operates at a 0.9–1.7 µm wavelength range and uses night vision to detect individuals/objects in ambient lighted terrains and through a glassy surface (i.e., a car's windshield). The three cameras' output will be fused (LWIR/SWIR) to allow for freely switching between thermal and night vision whilst tracking the same target.

Hyperspectral Cameras Hyperspectral (HS) cameras are used for terrain mapping and the case of indirect detection via vegetation disturbance (i.e., stepping on a stuck of tree vines, etc.). Their output is a mosaic type of snapshot that pans out 16 bands of a 4 × 4 grid. Visual as well as near-infrared spectra are captured (see Fig. 32.3). The information captured by the different wavelengths helps in enhancing vegetation imaging by discerning between fresh, dry, or different types of greenery or foliage. An example is the normalized difference vegetation index (NDVI) that is used to quantify vegetation greenness, density, and greenery health. Ultimately, the HS sensor will be used in conjuncture with other payload components in producing multilayered map(s) of a given surveyed area that can be visually assessed by the LEA.

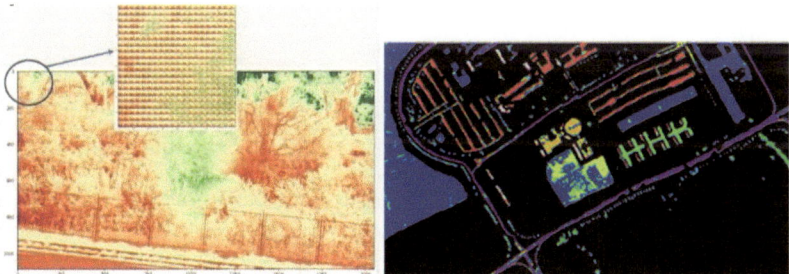

Fig. 32.3 The result of HS camera's preliminary testing, presenting NDVI (left) and segmentation (right) analysis of the input data represented as a multi-layered map

SAR (Proof of Concept) The land observations through optical imaging have received a lot of attention due to their high spatial resolution and low cost. However, cloudy weather and night observation are critical aspects of optical images. A solution is provided by active remote sensing with synthetic aperture radar (SAR) that works with signal wavelengths that penetrate clouds and smoke.

The most important factor of SAR instruments is the backscattering coefficient that depends on vegetation, humidity, ground terrain, transmit frequency, and the polarization of the frequency. Some of the most common SAR active systems operate in L (1–2 GHz), C (4–8 GHz), and X (8–12 GHz) frequency bands. A collection of SAR ground-based received images is given below. Figure 32.4 shows how the soil and vegetation response varies in all three frequencies. Figure 32.5 shows the contrast between the backscattering coefficients of a water surface and an urban area. The larger the wavelength, that is, lower frequency, the better is the penetration through foliage or ground. To this end, L-band SAR is

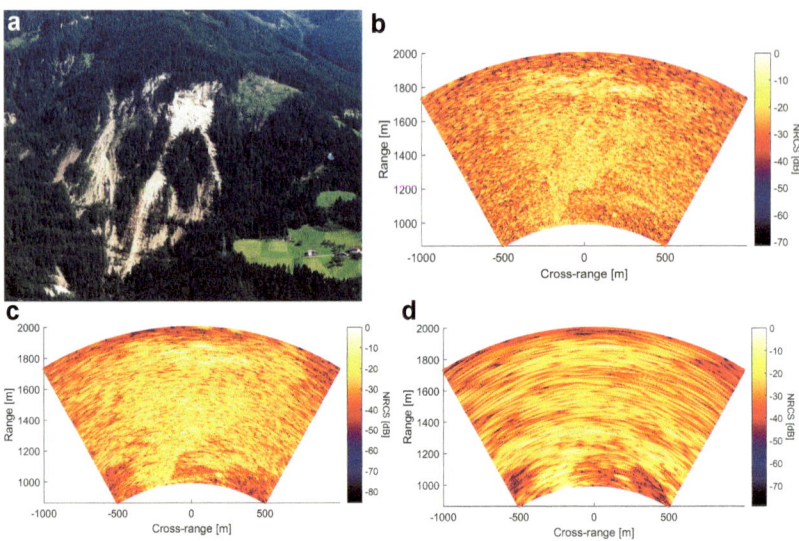

Fig. 32.4 Examples of SAR images collected in three different frequencies: (**a**) scene; (**b**) Ku-band; (**c**) C band; (**d**) L band. The captured area extends in range from 1 to 2 km

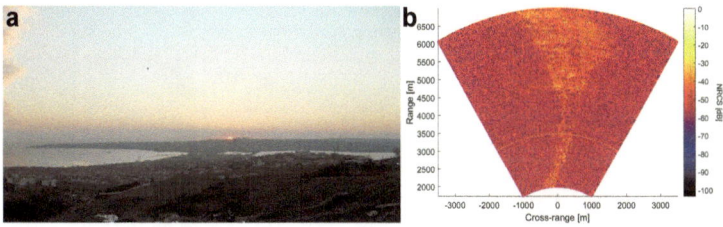

Fig. 32.5 Ku-band SAR image of a coastal area: (**a**) scene; (**b**) amplitude of SAR image. Radar observation from an altitude of 65 m above the scene with a range extension from 2 to 7 km

deployed on the UAV platform that allows for stationary and continuous monitoring of area of interest at the time when a specific event occurs.

Raw data is processed to generate both full-resolution SAR images and stacks of real aperture radar (RAR) profiles with only range resolution. Both products are georeferenced and are updated about every 5 min. Detection techniques involve target detection via position change made visible in a pair of SAR images through anomaly detection. Additionally, RAR profile stacks generated from a raw dataset allow for target movement detection within the time window of raw data file collection.

Optical High-Resolution RGB Sensor To best acquire video stream from very large distances, a set of optical sensors that supports a resolution beyond the standards was deployed. Two high-resolution cameras are used, along with a set of interchangeable lenses and one processing unit that serves as the control and processing unit (see configuration in Fig. 32.6). Camera functionalities include, but are not limited to, data storage, preprocessing, live streaming, and perception used for object tracking and characterization of detections.

After the preprocessing of the image data, the data must be stacked in a way for automatic feature extraction, simultaneous detection, and localization of an object. Due to the ultra-high resolution of each input image and the use of a resource-limited embedded system, real-time processing is very laborious. In fact, any modern DNN-based real-time detection algorithm has been designed to handle inputs of much lower resolution. We implemented a three-step optimized scheme for handling ultra-high-resolution images for real-time object detection on our embedded system. We split the input image into K subframes, applying YOLOv5 on each one

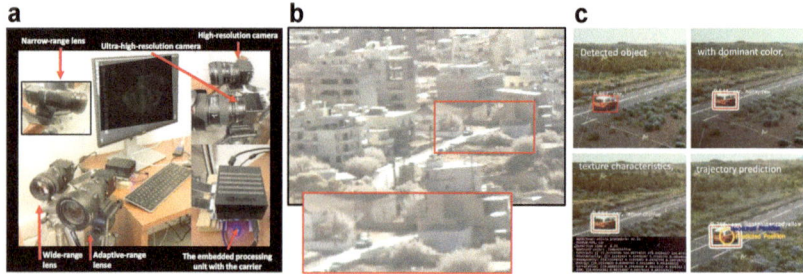

Fig. 32.6   HW setup of the high-resolution optical sensor (**a**). Examples of target detection in the horizon (1 km distance) direction (**b**) and tracking and characterization of target (**c**)

of the K subframes, and finally merging all the intermediate detections for determining a final detection on the initial input image (for more details, see Antonakakis et al. [11]).

PRELIMINARY RESULTS

BorderUAS platform is an event-driven system of systems. Sensors' output creates events upon data processing and fusion enhancing them with information from the analytics subsystems. All events and detections are displayed within the Command and Control Centre (C2) for assessment by a LEA operator.

The C2 is a web-based, platform-independent, application designed to enable Border Guard Authorities involved in the future field trials to interact with the UAV platform's payload and effectively handle events unfolding during scenario execution. The major components featured by the C2 are side-menu, real-time video stream visualization, event list and management, and GIS interface where events, targets, and points of interest are visualized at their global world location (see Fig. 32.7). Additional functionalities supported through particular views are dedicated to sensor and video management, business intelligence, and exporting or reporting on the found incidents.

The components constituting the payload of the airborne vessel have already entered into their individual testing phase(s), with some initial results being briefly outlined below.

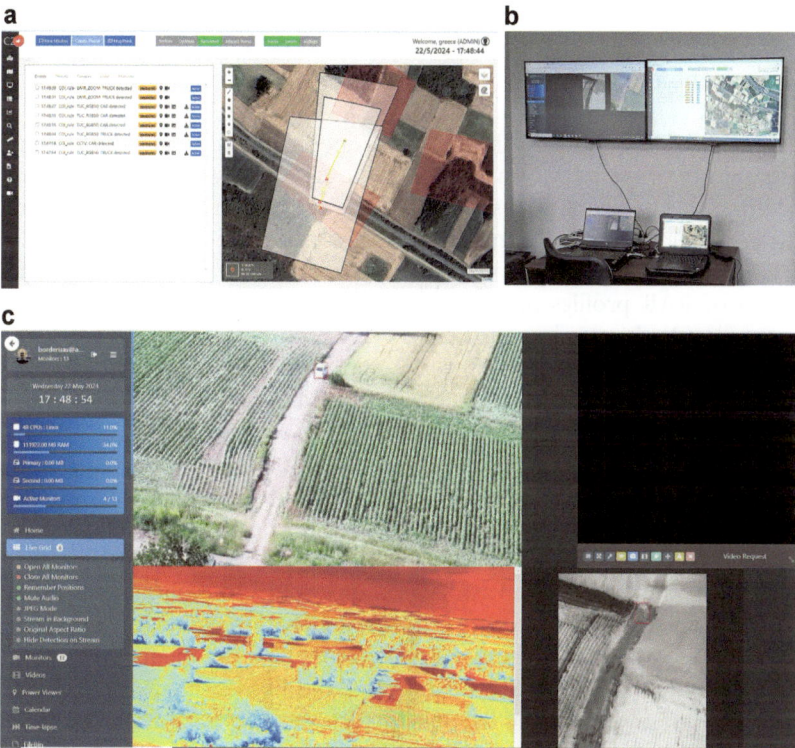

Fig. 32.7 C2 GIS/event management front-end, introducing live video feed, event notifications, and related map visualization

Acoustic Sensor Work with the AS so far has involved the use of a beam-forming algorithm coupled with array of microphones to improve the signal-to-noise ratio output along with the proper Microphone EMI and wind shielding. Moreover, the use of LMS algorithm can help with noise cancellation coming from the engines of the airborne vessel. Concerning the analysis of received feed will entail K-NN, K-means, & GSOM algorithms to perform noise classification to spot the events of interest.

Hyperspectral Camera Testing involved the application of principal component analysis (PCA) and support vector machine (SVM) algorithms to perform image segmentation using the HS camera and providing information about the land cover (see Fig. 32.3). Along with the false-color visualizations, these will be used to evaluate the accuracy of image segmentation captured by the HS camera.

MIMO SAR The added value of the proposed MIMO radar is twofold. First, it can provide SAR images when the LTA UAV is still over the area of interest. Usually, SAR systems need the movement of the platform to generate the synthetic aperture as in the case of space- and airborne SAR systems. In addition, the possibility to generate both an SAR image and stacks of RAR profiles provides a means of detecting anomalies in the scene via a backscattering coefficient every 5 min and tracking the movement of moving targets within the 5-min time window needed to complete the raw data acquisition (see Figs. 32.4 and 32.5).

EOS Sensors (IR and High-Resolution RGB) Initial test sessions have used the video feeds, along with optimized YOLO4 and YOLOv5 algorithms' real-time airborne detections of targets. Detection is then followed by object tracking by assigning it a unique ID; this is done using the 2D Kalman filter, along with an association function (OverUnion algorithm), to form a relation between the former's predictions and the YOLO's detections. So far, the main objective is to improve the training on detections through the public Stanford dataset and by creating a separate one tailored to the needs of the BUAS scenarios to follow (see Fig. 32.6).

Conclusions

The BorderUAS Project seeks to provide Border Surveillance Authorities with a lighter-than-air UAV platform of 24/7 Operational Autonomy featuring a multiplicity of payload components. The UAV's payload output is further analyzed and refined by means of algorithmic processes revolving around machine learning and AI support. End users (LEAs) designated to participate in field trials for testing the concept's operational added value in borderland surveillance will be able to utilize a state-of-the-art Command and Control (C2) System to both have access to payload component feed and help automatize, in an interoperable fashion, the content of alerts, events, and border mission guidelines for in-field units to follow.

Future research expanding on the given EU project's area of expertise may unfold along the lines of the following topics:

- Organizing and carrying out field trials to validate the solution at hand
- Building an ecosystem including industry partners that may wish to employ the payload's capabilities in the private sector
- Posthumously testing the bundled payload components on other unmanned or manned airborne/ground vessels to check as to the interoperability, versatility, and viability of the solution at hand
- Training the models (i.e., for detection, tracking, reidentification, etc.) using additional (public, etc.) datasets for improved sensor/ camera output analysis.

Acknowledgments This research received funding from the European Union's Horizon 2020 Research and Innovation Framework Programme under grant agreement no. 883272, project BorderUAS.

REFERENCES

1. Buzmaniuk, S. (2021). The Union's external borders: A European debate revisited. *Fondation Robert Schuman, 585*(European issues), 1–7.
2. Barnes, S., Georgadze, A., et al. (2023). Cosmic-ray tomography for border security. *Instruments, 7*, 13.
3. https://borderuas.eu/project/
4. Chatzis, P., & Stavrou, E. (2022). Cyber-threat landscape of border control infrastructures. *International Journal of Critical Infrastructure Protection, 36*, 100503.
5. Koslowski, R. (2021). Drones and border control: An examination of state and non-state actor use of UAVs along borders. In *Research handbook on international migration and digital technology* (pp. 152–165). Edward Elgar Publishing Limited.
6. Bolakis, C., Mantzana, V., et al. (2022). Correction to: FOLDOUT: A through foliage surveillance system for border security. In *Technology development for security practitioners* (pp. C1–C2). Springer.
7. Vasilopoulos, E., Vosinakis, G., et al. (2022). Autonomous object detection using a UAV platform in the maritime environment. In *International conference on research challenges in information science* (pp. 567–579). Springer.
8. Bauk, S., Kapidani, N., et al. (2019). Autonomous marine vehicles in sea surveillance as one of the COMPASS2020 project concerns. *Journal of Physics: Conference Series, 1357*(1), 012045.
9. Kontopodis, I., Leskovsky, P., et al. (2022). BorderUAS: A knowledge-based representation of the border surveillance domain. In *IEEE international conference on imaging systems and techniques (IST 2022)* (pp. 1–6). IEEE.

10. Kampas, G., Vasileiou, A., et al. (2022). Design of sensors' technical specifications for airborne surveillance at borders. *Journal of Defence & Security Technologies, 5,* 58–83, 65.
11. Antonakakis, M., Zervakis, V., Petrakis, M., et al. (2022, June). *Real-time object detection using an ultra-high-resolution camera on embedded systems.* Paper presented at the 2022 IEEE International Conference on Imaging Systems and Techniques (IST 2022), pp. 1–6.

Open Access This chapter is licensed under the terms of the Creative Commons Attribution 4.0 International License (http://creativecommons.org/licenses/by/4.0/), which permits use, sharing, adaptation, distribution and reproduction in any medium or format, as long as you give appropriate credit to the original author(s) and the source, provide a link to the Creative Commons license and indicate if changes were made.

The images or other third party material in this chapter are included in the chapter's Creative Commons license, unless indicated otherwise in a credit line to the material. If material is not included in the chapter's Creative Commons license and your intended use is not permitted by statutory regulation or exceeds the permitted use, you will need to obtain permission directly from the copyright holder.

CHAPTER 33

Best Practices for Creating and Maintaining Clusters in EU-Funded Projects: Insights from the H2020 BES Cluster

Chatzimallis Charalampos, Kyrkou Danai,
Kampouridou Anneta, Themistocleous Christianna,
Astyakopoulos Alkiviadis, Nikolaidou Artemisia,
Rosgova Mirela, Yayilgan Sule Yildirim,
and Velanas Pantelis

C. Charalampos (✉) • K. Danai • K. Anneta
ViLabs OE, Thessaloniki, Greece
e-mail: chatzimallis@vilabs.eu; danaikyrkou@vilabs.eu; kampouridou@vilabs.eu

T. Christianna • V. Pantelis
European University Cyprus, Nicosia, Cyprus
e-mail: C.Themistocleous@research.euc.ac.cy; P.Velanas@research.euc.ac.cy

A. Alkiviadis • N. Artemisia • R. Mirela
Center for Security Studies, Athens, Greece
e-mail: a.astyakopoulos@kemea-research.gr; a.nikolaidou@kemea-research.gr;
m.rosgova@kemea-research.gr

Y. S. Yildirim
Norges teknisk-naturvitenskapelige universitet, Trondheim, Norway
e-mail: sule.yildirim@ntnu.no

© The Author(s) 2025   415
I. Gkotsis et al. (eds.), *Paradigms on Technology Development*
for Security Practitioners, Security Informatics and Law
Enforcement, https://doi.org/10.1007/978-3-031-62083-6_33

INTRODUCTION

In EU projects, clustering performs a crucial role in enabling collaboration and innovation among different stakeholders in a particular industry or field. Clusters are groups of companies, research institutions, and other organisations that are geographically and sectorally concentrated and work together to achieve common goals. Clusters and sister project communities are a significant part of the dissemination and communication strategies in EU-funded projects to promote outputs and exchange knowledge.

However, maintaining such communities and ensuring a meaningful impact during or after the funding period can be challenging. This chapter provides best practices for creating, managing, and sustaining clusters, engaging projects, supporting joint activities, and supporting sustainability. The chapter draws on insights from the H2020 BES Cluster, which consists of security-related H2020 and Horizon Europe projects collaborating to support communication and dissemination activities, exchange good practices and methodologies, and explore possibilities to combine pilot activities. The chapter offers a comprehensive review of the outcomes and impact of the H2020 BES Cluster projects, highlighting their significance and potential for further exploitation. It also discusses strategies for early exploitation and effective dissemination of ongoing projects.

Although the H2020 BES Cluster focuses on border security, its methodology and practices can be customised and applied to different domains. This chapter aims to promote collaborative methodologies and draw attention to the tangible results and benefits of EU-funded projects, introducing and strengthening innovations and improvements to the security and border control system within the EU.

Clustering activities can bring various benefits to the EU and its member states. First, they can enhance competitiveness by creating economies of scale and scope, increasing productivity, and promoting innovation through knowledge transfer and collaboration among cluster members. In addition, clustering can foster regional development by creating jobs and driving economic growth. Clusters can also contribute to the development of regional innovation systems, which are essential for supporting innovation and entrepreneurship [1].

By bringing together companies and research institutions from different sectors, clusters can facilitate the exchange of knowledge and expertise that can lead to the development of more sustainable products and services. By incorporating clustering activities into EU projects, the EU can support the growth and development of different industries and sectors, while also promoting sustainable and inclusive economic growth.

METHODOLOGY

Building a cluster consists of the same characteristics as building a community. In this case, the aforementioned community consists of EU-funded projects, in the sector of Border External Security (BES). Community building is the process of creating and fostering a sense of belonging among individuals or institutions with shared interests or goals. The BES Cluster reviewed a number of community-building methodologies and best practices, drawing inspiration from the European Commission's "Communities of Practice Playbook" [2] and Wenger-Trayner's "Introduction to Communities of Practice" [3] to finally structure its own customised approach.

The BES Cluster set seven steps to follow toward building and maintaining its community of projects:

1. *Identify the target audience*: The first step was to identify the target audience and understand their needs, interests, and goals. This involves conducting research on the relevant stakeholders that participate in the consortia of the projects that belong in the security sector. The target audience includes research performing organisations (RPOs) specialising in security, law enforcement agencies (LEAs), and public authorities.

2. *Develop a mission and values statement*: The next step was to develop the mission and values of the cluster and define its purpose and goals. This served as a guiding principle for community members. Mission and values should be clear, concise, and inclusive, reflecting the diverse perspectives and needs of the community.

3. *Create a communication plan*: A vital step was to develop a communication plan. This plan outlined the methods and channels of communication, including social media, a mailing list, and events, as well as the frequency and tone of communication. This plan also provisioned the recruitment of new members to the cluster in order to ensure its continuing growth.

4. *Establish a leadership team*: The BES Cluster was an initiative of the EU-funded project METICOS. The dissemination and communication managers of the project took over the role of the leadership

team. The team included individuals with diverse skills and backgrounds who can work together to achieve the community's goals, always taking into consideration the input and feedback of the cluster members.

5. *Encourage participation*: The cluster made sure to be visible and vibrant in order to encourage the existing members to participate in cluster activities, such as online workshops, physical events, or panels.

6. *Foster a culture of inclusivity*: The BES Cluster made sure to create a culture of inclusivity for building a welcoming and supportive community. This involves actively promoting diversity and equity. The projects support each other's work while ensuring that all cluster members feel valued and respected.

7. *Evaluate and adapt*: Evaluating the effectiveness of the community-building process is essential for identifying areas of improvement and adapting the methodology as needed. This involves collecting feedback from members, monitoring engagement levels, and monitoring community's growth and impact.

Since this is an ongoing process, the steps are being re-evaluated and readjusted according to the needs of the cluster.

CASE STUDY: MEETING THE BES CLUSTER

The BES Cluster was created in 2020, and it was an initiative started by the METICOS project, as it was foreseen in its Grant Agreement. The project seeks collaboration and synergies with other projects and organisations in order to further enhance existing endeavours and jointly shape the future vision of the security sector. In this vein, the METICOS dissemination/communication team promoted interactions with similar projects (FP7, H2020 and Horizon Europe) and with other stakeholders like government organisations or departments (e.g. of Interior or Public Safety), policymakers, and RPOs, to exchange non-confidential information with them in order to combine their research and outputs.

The H2020 BES Cluster consists of projects collaborating to support each other, identify solutions to upcoming challenges, secure effective dissemination and valuable exploitation potentials, and generate knowledge that, along with the developed solutions, will change the current state in the areas and fields that the projects are working upon. But what do the BES Cluster members win? First, they have the opportunity to exchange information relevant to communication and dissemination and engage in other projects' activities. As a result, the cluster boosts the exchange of ideas on good practices and methodologies and the exploration of possibilities and opportunities to combine pilot activities, as well as the potential to produce common policy suggestions. Finally, the projects cooperate for a wider impact of the sustainability/exploitation plans of the projects to find and propose effective measures for evidence-informed policy suggestions.

The BES Cluster Achievements

In the last 3 years of the existence of the BES Cluster, several joint dissemination and piloting activities have taken place. The projects are constantly looking for new opportunities to cooperate and evolve together. These efforts have resulted in a range of successful initiatives that demonstrate the cluster's dedication to collective progress. In this vein, this chapter provides some examples of such activities.

One notable accomplishment involves the demonstration activities at Piraeus Cruise Port on 22 and 23 July 2021. This activity was organised by the TRESSPASS Project, while METICOS representatives participated as observers, gaining valuable insights into real seaport border crossing points and the interactions and perceptions of travellers in risk-based screening situations. The goal was to enhance the utilisation of existing infrastructure and facilities by port operators and border authorities, increase their capacities and throughput, and improve border and customs control processes while minimising resource requirements and without causing delays for cruise ship passengers. As the insights generated by TRESSPASS were directly shared with METICOS and the rest of the cluster members, this success story highlights the power of knowledge exchange within the H2020 BES Cluster. By observing such an operation, the METICOS representatives benefited from the lessons learned and the best practices. This input was taken into consideration when METICOS was designing its pilot methodology.

Additionally, the first PROMENADE Workshop organised in December 2021 has been supported by the BES Cluster and broadly disseminated through its network and communication channels. The workshop was held remotely due to COVID-19 and represented an excellent opportunity to present the project to the public and engage potential actors interested in the project. Its aim was to gather experts on border surveillance,

AI & big data technologies, policymakers, and practitioners, to get their valuable insights on the findings regarding the requirements of the PROMENADE system. The user community is an essential driver to ensure the effectiveness of the collected requirements enabling the project to shape an innovative toolkit for improving vessel tracking, behaviour analysis, and automatic anomaly detection, meeting expectations from a broader list of stakeholders and potential future adopters of the PROMENADE system. To this end, the workshop objectives were to (i) present the initial analysis of the requirements collected through the Requirement Gathering Process, (ii) discuss, validate, and fine-tune the analysis of the requirements with the user community, and (iii) foster user community interactions using live and interactive polls. The event was attended by 88 participants covering all target audiences from 12 European countries, including FRONTEX, that is, end users (23%), EU organisations (13%), large industries (10%), SMEs (19%), RTOs (11%), academia (4%), and others (20%). After the workshop, a database accessible to project partners was created to keep the traceability of engaged stakeholders and establish communication channels for future project activities and trials.

Furthermore, the H2020 BES Cluster actively participated in two NESTOR workshops. The first one, which was related to creating the Roadmap for Border Management Standardisation, took place at the CCMC in Brussels in cooperation with the CEN-CENELEC Sector Forum on Security Brussels, on 17 February 2023. The full-day workshop had a total of 85 participants, of which 25 were on-site and 60 joined online. The event was attended by policymakers (10%), BM practitioners (28%), research organisations (21%), industry (22%), NGOs (3%), and others.

The second NESTOR workshop, namely Demo Day & Final Workshop, was held on 24 April 2023 in Athens. It was a full-day hybrid event demonstrating the operational capabilities of the NESTOR platform as they were developed and tested throughout the land and maritime trials in Lithuania, Cyprus, and Greece. More than 190 attendees, 140 on-site in Athens and 50 online, participated having the opportunity to see the NESTOR project achievements, the trials' demonstration through six dedicated videos and the live presentation of all NESTOR project assets, including the BC3i platform, surveillance cameras, mixed reality HoloLens glasses, RF sensors and unmanned vehicles (tethered drones, autonomous ground vehicle and underwater vehicle). The event was attended by universities (11%), EU agencies (7%), research organisations (20%), industries (34%), public entities/ministries (26%), policymakers (1%), and NGOs (1%).

As a result of such activities, the cluster managed to have a great impact in terms of reach and engagement. Also, it attracted the attention of the policy officers, as well as the EU agencies, such as Frontex and DG Home, as they saw the potential of further exploiting the cluster to create a hub that will be a reference point for current and future security projects. At this moment, the cluster counts 20 EU-funded project members, creating a network of more than 287 partners, including research organisations, public entities, private companies, and law enforcement agencies. These partners cover 37 countries, carrying out more than 45 piloting activities. The projects are creating different tools that are being tested in the security sector, focusing on border management. However, many of these tools are versatile and can be adapted and customised to be employed in other industries, such as IT, education, health, energy, and logistics.

The Future of the BES Cluster

Many of the projects that initially supported the BES Cluster, including its leader, the METICOS project, are approaching their completion. A plan on how to preserve the BES Cluster has been drafted.

The BES Repository

The members of the BES Cluster aspire to establish a website or a portal that all projects will be able to store their results, knowledge, and outputs, so that future security projects can use it as a library.

The BES Cluster's library may be set up to facilitate effective information organising and retrieval. The library's content will include project results that can be shared publicly. They will be categorised and arranged into sections to make sure viewers can explore and discover the particular resources they are looking for easily. By creating sections dedicated to research papers, technical reports, case studies, best practices, and policy documents related to border external security, the library becomes a well-structured repository.

Additionally, associating each item in the library with metadata such as title, author(s), date of publication, keywords, and abstract can significantly enhance search and discoverability. The metadata summarises basic information about data, which can make it easier to find, use, and reuse particular instances of data. Metadata use on web pages has a lot of potential, while using appropriate keywords and tags further enhances accessibility by enabling users to search for specific topics or keywords within the library [4].

Dissemination, Communication, and Exploitation Activities
Another way to continue boosting the sustainability of the cluster is to promote joint communication, dissemination, and exploitation activities [5].

As the METICOS project nears its completion, a workshop focused on knowledge and recommendations for stakeholders is planned. This workshop aims to showcase the successful testing of the METICOS solution and the METICOS Social Sensing toolkit across all pilot initiatives. The organisers extended a warm invitation to end users of BES Cluster projects to participate in this event. Furthermore, the possibility of collaborating with other BES projects is also open. This collaborative approach promotes a comprehensive and inclusive platform for sharing insights and fostering cooperation within the BES Cluster community, as well as promoting discussions and initiating new project ideas and proposals for the open BES-related calls.

Additionally, the BES Cluster is exploring collaboration opportunities with partners from BES projects at the Ninth Italian Conference on Computational Linguistics (CLiC-it 2023) that will be held in Venice from 30 November to 2 December 2023. The conference is a reference forum for researchers working in the fields of computational linguistics (CL) and natural language processing (NLP). It promotes and disseminates high-level research on automatic language processing, targeting state-of-the-art theoretical results, experimental methodologies, technologies, and application perspectives. A paper is planned to be submitted, related to how to do social media analysis based on perception extraction architecture for acceptance of border control technologies. The potential for collaboration with partners from BES Clusters' projects is being explored.

In conclusion, by promoting joint communication, dissemination, and exploitation activities, and through workshops and collaborations, the BES Cluster aims to enhance sustainability and foster cooperation within its community. The upcoming events present valuable opportunities to showcase successful initiatives and explore further collaborations.

CONCLUSIONS

This chapter has provided insights into the importance of clustering or creating sister project communities as a dissemination and communication strategy in EU-funded projects. It has focused on the H2020 BES Cluster, which consists of security-related projects collaborating to support communication, exchange good practices, and explore joint exploitation activities. The chapter has reviewed the outcomes and impact of the BES Cluster, showcasing successful initiatives such as demonstration activities and workshops. Through these activities, the cluster has facilitated knowledge exchange and collaboration among its members, leading to tangible results and benefits in the security and border control system.

While this chapter has provided an overview of the BES Cluster and highlighted its achievements, there are additional topics and information that could not be included due to the word limit. Some of these topics could include the followed methodology in creating such clusters, a deeper analysis of specific joint exploitation activities, the impact of the cluster on policy development, case studies of successful collaboration between cluster members, and the long-term sustainability of the cluster beyond the completion of individual projects.

In conclusion, the H2020 BES Cluster serves as a successful example of how clustering and sister project communities can effectively promote collaboration, knowledge exchange, and the dissemination of results in EU-funded projects. By following the "Best practices" identified in this chapter, future projects can enhance their clustering efforts and maximise the impact and sustainability of their outcomes. The BES Cluster has demonstrated the potential for creating a collaborative environment, driving innovation, and fostering improvements in the security and border control system within the EU.

Acknowledgements These projects received funding from the European Union's Horizon 2020 research and innovation programme under grant agreements nos. 883075, 101021851, and 101021673.

REFERENCES

1. Delgado, M., Porter, M. E., & Stern, S. (2014). Clusters, convergence, and economic performance. *Research Policy, 43*(10), 1785–1799.
2. European Commission. (n.d.). *Communities of practice playbook: Methodology.* Retrieved from https://op.europa.eu/webpub/jrc/communities-of-practice-playbook/en/methodology.html
3. Wenger-Trayner, E., & Wenger-Trayner, B. (2015). *An introduction to communities of practice: A brief overview of the concept and its uses.* Retrieved from https://www.wenger-trayner.com/introduction-to-communities-of-practice
4. Duval, E. (2001). Metadata standards: What, who & why. *Journal of Universal Computer Science, 7*(7), 591–601. Retrieved from https://www.researchgate.net/profile/Erik-Du-val/publication/220348691_Metadata_Standards_What_Who_Why/links/09e4150dc24e459dc9000000/Metadata-Standards-What-Who-Why.pdf
5. European Commission, Executive Agency for Small and Medium-Sized Enterprises, Haardt, J., Weiler, N., Scherer, J., et al. (2019). *Making the most of your H2020 project – Boosting the impact of your project through effective communication, dissemination and exploitation.* Retrieved from https://data.europa.eu/doi/10.2826/045684

Open Access This chapter is licensed under the terms of the Creative Commons Attribution 4.0 International License (http://creativecommons.org/licenses/by/4.0/), which permits use, sharing, adaptation, distribution and reproduction in any medium or format, as long as you give appropriate credit to the original author(s) and the source, provide a link to the Creative Commons license and indicate if changes were made.

The images or other third party material in this chapter are included in the chapter's Creative Commons license, unless indicated otherwise in a credit line to the material. If material is not included in the chapter's Creative Commons license and your intended use is not permitted by statutory regulation or exceeds the permitted use, you will need to obtain permission directly from the copyright holder.

PART VI

Enhancing the Defense Capabilities
of the EU

CHAPTER 34

Frugal and Robust AI for Defence Advanced Intelligence

Andromachi Papagianni, Konstantinos Ioannidis,
Theodora Tsikrika, Stefanos Vrochidis,
and Ioannis Kompatsiaris

INTRODUCTION

Artificial intelligence (AI) technology has greatly influenced various aspects of modern society [1] and economy [2]. Recent technological advancements have significantly improved decision-making and support systems, as well as autonomous processes, by utilising different types of data, such as text, sound, visual content, and video footage. To provide accurate outcomes, AI models require collection and preparation of a number of representative data that leads to a costly and timely process in terms of hardware and human resources. Alternatively, frugal AI approaches [3] have emerged to address the data necessity issue, having as a main

A. Papagianni (✉) • K. Ioannidis • T. Tsikrika • S. Vrochidis • I. Kompatsiaris
Information Technologies Institute, Centre for Research and Technology Hellas,
Thermi, Thessaloniki, Greece
e-mail: andromap@iti.gr; kioannid@iti.gr; theodora.tsikrika@iti.gr; stefanos@iti.gr;
ikom@iti.gr

© The Author(s) 2025   427
I. Gkotsis et al. (eds.), *Paradigms on Technology Development*
for Security Practitioners, Security Informatics and Law
Enforcement, https://doi.org/10.1007/978-3-031-62083-6_34

advantage the use of fewer data for training as learning procedures entail very few samples, with the ultimate challenge being the development of a powerful model, capable of continuously learn with minimal human interactions and no intervention of experts.

Hence, the usage level of an AI model to be exploited under certain conditions is application-oriented as most models comprise supervised approaches and are data-driven. Within the defence domain [4], data may be characterised as limited or incomplete, as well as sensitive in nature, requiring security clearance for proper labelling by dedicated personnel leading to an insufficient developing process hindering the insertion of such technologies in this domain. Additionally, data may be specific to certain types of military sensors, such as infrared or multispectral sensors. When employing AI algorithms in military applications, it is essential that any recommendations and decisions made are compliant with safety and security regulations, considering the complex and continuously changing environment. Furthermore, an AI framework must be interpretable from the perspective of operators such as commanders.

Overall, data restrictions and singularity of the application affect severely the use of such models in a defence-oriented application. To this end, recent advancements in AI development, which specifically address these challenges, may be beneficial to overcome the aforementioned restrictions. The main principles of frugality, robustness, and explainability in AI have already proven to be auspicious solutions for these tasks. In this chapter, relevant topics and basic principles of such technologies are thoroughly analysed and presented acting as a proof of concept to strengthen their applicability in the defence domain. More specifically, an implementation approach is presented that involves all the aforementioned principles. The analysis is performed considering the specificity of the domain. A feasibility study is also presented to apply combinations of services in order to introduce higher semantically components and extract additional knowledge.

The rest of the chapter is organised as follows. The second provides insights into the unified approach to address the core challenge of few data availability for the defence domain while the third section presents the expected impact derived from its implementation. Finally, the fourth section concludes the chapter by describing the main results of this study.

METHODOLOGICAL DESCRIPTION

In general, artificial intelligence (AI) corresponds to the engineering and scientific methodologies that result in a system, which poses intelligent reasoning and behaviour. Recent technological advances have established AI as a technology with an impressive impact on almost every aspect of the modern socio-economic environment, thus significantly enhancing the capabilities of, among others, decision-making and support systems and autonomous processes, based on various types of data (text, visual content, video footage, etc.). State-of-the-art machine learning (ML) and deep learning (DL) techniques rely on mathematical methods that aim at extracting and exploiting application-relevant information from a large dataset. In this manner, AI systems are able to learn patterns and structures in an automated way by extracting correlations between complex data through the training/learning process. Hence, the availability of an amount of data comprises a major requirement to ensure a sufficient performance.

In a military context, data can be characterised as scarce or incomplete, sensitive (i.e. labelling can be performed only by experts with security clearance), and specific (i.e. data depend on specific types of military sensors, e.g. IR, multispectral, etc.) [5]. The recommendations and decisions to be proposed by AI algorithms in military applications must be acceptable, with respect to the safety rules and security regulations related to their uses and mission in a complex and rapidly changing environment. Moreover, from the operator's point of view such as a commander, an AI framework must be acceptable with respect to its interpretability.

The described framework (Fig. 34.1) addresses the above challenges by incorporating novel approaches, algorithms, and tools as products of research to deliver some of the most prominent issues relevant to AI and increase their applicability and impact on defence systems. The framework relies on the following five pillars identified considering AI advancements and the peculiarity of the defence domain:

- *Data acquisition and improvements:* Low availability of data can be addressed by applying data augmentation schemes while the timely process of annotation can be delivered with automated processes and innovative techniques.
- *Frugality in AI:* To achieve high levels of accuracy when deploying an AI framework and exploiting it as an asset in modern warfare,

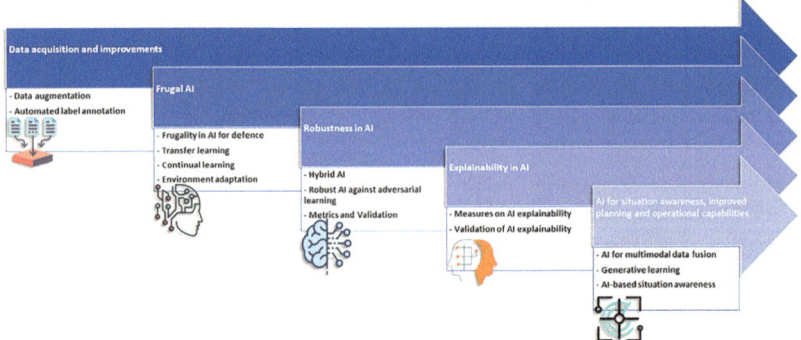

Fig. 34.1 Schematic depiction of AI-based framework in defence

operation-specific data must be available. Thus, the framework involves the application of dedicated models such as few-shot detections, domain adaptation, and transfer learning techniques.

- *Robustness in AI*: Towards the same objective of reducing the size of the training dataset, principles such as discriminative, generative, and evolutionary learning will contribute to produce *robust intelligence, surveillance, target acquisition, and reconnaissance (ISTAR)* operations.

- *Explainability in AI*: A comprehensive representation of the models' outcomes is more than essential for an operator of a defence system. Hence, the appropriate measures and validation approaches are mandatory for quantifying their performance.

- *Improved situational awareness and mission planning*: Additional knowledge can be extracted by ensuring a robust data exchange between the developed models. Hence, multimodal fusion schemes, threat assessment, and optimised mission planners can benefit from the application of AI models and deliver improved and comprehensive situational awareness.

Since the research is focused on defence applications, the first step will be to determine the user requirements of the defence domain which will drive the data needed to be processed. So, the defined end user requirements will then guide all the conducted research, designs, developments, and validations as this information is reflected in the required datasets.

Data augmentation techniques incorporate generative learning principles [7] to produce additional data not only to increase the size of the available training set but also to reduce overfitting and improve the generalisation capabilities of the model. Random oversampling (ROS) comprises a naive approach which duplicates images randomly from the minority class until a desired class ratio is achieved. Intelligent oversampling techniques date back to synthetic minority over-sampling techniques (SMOTE) proposed by Chawla et al. [8]. SMOTE and its updated model, Borderline-SMOTE [9], develop new instances by interpolating new points from existing instances with the use of K-nearest neighbours. On the contrary, by their introduction from the research community, generative artificial nets (GANs) [5] displayed a clear superiority in producing synthetic data, mostly visual-related data. Architectures such as DCGAN [6] (Fig. 34.2), Progressively Growing GAN [10], CycleGAN [11], and Conditional GANs [12] seem to have the most application potential in the field. Similar progress has been observed in automating the process of annotation as it comprises a timely and resource-demanding (e.g. dedicated personnel) phase. To this end, deep learning architectures as proposed in [13, 14] can be rather beneficial.

Frugal AI comprises the most crucial aspect of the presented framework. Frugal principles adopted in a detector model can contribute to reducing the necessity of large datasets. Considering also the lack of data for defence applications, the reuse of existing models adapted for these applications can be an efficient substitution (similar to the schema in Fig. 34.3). Hence, many AI models have been proposed which were

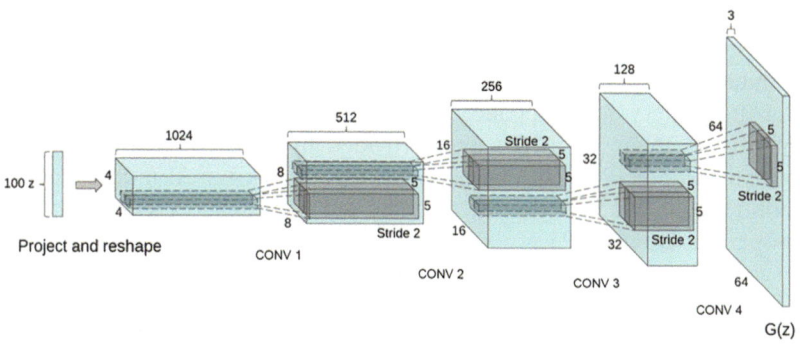

Fig. 34.2 DCGAN, generator architecture [6]

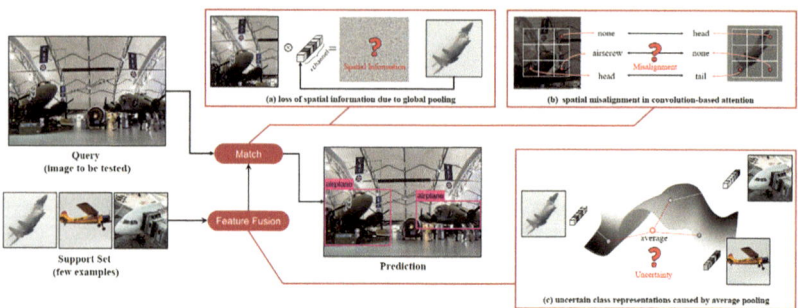

Fig. 34.3   Example of frugal AI for computer vision applications [15]

tailored to support these applications like in [16, 17]. A similar rationale is used when the models incorporate transfer learning principles where the training process is not mandatory as pre-trained models are used for different applications. Indicatively, the models proposed in [18] display a sufficient performance, and by rectifying their behaviour, they could be utilised in the defence domain. Continual learning models [18] can contribute to higher performances with less data by continuously training the current version of a model when additional data are available. In more requirement-oriented models, environment adaptation [19] is another category of models that implements frugality and can be rather useful in defence applications.

Robust AI is most commonly envisaged with the application of hybrid AI principles. As is also implied by the used terminology, different types of learning are combined as well as the combination of data-driven learning with expert or domain knowledge. Thus, hybrid AI methods target to decrease the effects of dataset bias and overfitting. Various hybrid AI architectures have been proposed in numerous civilian applications [20]. For the military domain, concepts of integrating the advantages of different ML strategies with the ones provided by structure reasoning have been developed [21]. Additionally, the robustness of an AI model is commonly quantified with the use of dedicated metrics and the adoption of proper validations [22].

Explainable AI focuses mostly on interpreting appropriately and comprehensively a model's detection outcome. For example, in computer vision, heatmap-based methods have been proposed that rely on saliency-like analysis [23]. Towards this objective, validation comprises a significant factor interpreting correctly the outcomes of a detection model whereas in defence applications considering the challenges, its significance is even larger [24].

Lastly, *AI for improved situational awareness* can lead in the upcoming years to the interest of the defence research community as its maturity level is still low compared to other domains. As in many applications, various sensors can be utilised having the same objective, and so, the collected data are fused to a higher integration layer. The most representative multimodal data fusion deep learning models from the perspectives of the model task, model framework, and evaluating data are based on deep belief networks [25], stacked autoencoders [26], etc. Another significant category of this pillar involves generative learning tools which account for enhanced threat assessment capabilities during mission definition and planning, incorporating cognitive analysis to provide quantitative and qualitative assessments of the potential threats such as in underwater mine warfare [27]. Finally, AI-based mission planning [28], a critical aspect in warfare, can complement the C2 functionalities towards delivering a complete decision support framework.

EXPECTED IMPACT

All the aforementioned works can comprise a proper research base that is expected to produce advances in the development of new technologies which will be successfully applied in the context of AI for defence applications, covering all stages of a military operation supply chain. The framework and its core pillars described in the methodology section will contribute to the establishment of a strong AI-based defence industry enhancing the autonomy in the field of AI for military applications, leading to more efficient protection of critical infrastructures and military establishments and equipment. Furthermore, the frugal AI techniques will provide updates on the collection of data, contributing to a severe issue particularly in the defence domain while data-related restrictions can be overpassed.

Apart from the defence industry, deployment in a wide range of applications can be accelerated as the main attribute of the AI models is flexibility and efficient adaptation in unknown environments and different operating conditions. New business models and opportunities could emerge that will reduce the unit cost and competition growth among industries as one of the advantages of the models to be developed is that they will be built on the top of established knowledge and recent technology advances, ensuring effective services for potential adopters.

Beyond the utilisation in the defence domain, the present study will contribute to further exploring and understanding the usage of synthetic data and generative learning techniques. The data augmentation schemes that are studied are expected to create a pipeline of processing steps that can be used as guidance during the training phase of similar models. When applied to the input data, the robustness of the system could be increased and the overfitting of the model on the input data decreased. The methods introduced in the present study will be based on state-of-the-art data augmentation techniques that will explore further the utilisation of synthetic and cross-spectral data for deep learning. In addition, as far as annotation speed is concerned, the research will focus on increasing the annotation speed of tasks involving multimodal data by extending their usage to a production-level annotation pipeline of real-world datasets and using model architectures that capture inter-modality. The study will also deal with existing models' knowledge and continuous learning in order to optimise the models that will acquire new knowledge without forgetting past knowledge when a new one is introduced. Focus will be given to the adaptation of few-shot learning techniques and selective prediction to increase the model's self-awareness and aim to benefit the current state-of-the-art models.

CONCLUSIONS

In conclusion, the work presented in the current study aims to deliver new approaches, algorithms, and tools as research products based on the aforementioned pillars to address some of the most prominent issues relevant to AI applications and increase their applicability and impact on defence systems. Frugal AI models can account for operation-specific data and achieve high levels of accuracy when an AI framework is deployed in modern warfare. On the other hand, an operational-ready AI framework must remain reliable in case of natural or artificial perturbations and the guidelines of the framework must be trusted and understood to be acceptable by humans. Therefore, the developed models can focus on providing robustness and explainability.

Finally, the ultimate goal of the research is to contribute to the digitalisation of the defence domain by inserting novel and adjustable to current infrastructure systems in order to facilitate processes in the defence domain and exploit the benefits of human and machine cooperation.

Acknowledgements This project received funding from the European Defence Fund programme under grant agreement no. 101103386. The views and opinions expressed are, however, those of the author(s) only and do not necessarily reflect those of the European Union or the European Commission. Neither the European Union nor the granting authority can be held responsible for them.

REFERENCES

1. Van Wynsberghe, A. (2021). Sustainable AI: AI for sustainability and the sustainability of AI. *AI and Ethics, 1*(3), 213–218.
2. Furman, J., & Seamans, R. (2019). AI and the economy. *Innovation Policy and the Economy, 19*(1), 161–191.
3. Smolensky, P., McCoy, R., Fernandez, R., Goldrick, M., & Gao, J. (2022). Neurocompositional computing: From the Central Paradox of Cognition to a new generation of AI systems. *AI Magazine, 43*(3), 308–322.
4. Taylor, T. (2019). Artificial intelligence in defence: When AI meets defence acquisition processes and behaviours. *The RUSI Journal, 164*(5–6), 72–81.
5. Goodfellow, I., Pouget-Abadie, J., Mirza, M., Xu, B., Warde-Farley, D., Ozair, S., et al. (2014). *Generative adversarial nets* (Advances in neural information processing systems) (Vol. 27). Curran.
6. Radford, A., Metz, L., & Chintala, S. (2015). Unsupervised representation learning with deep convolutional generative adversarial networks. *arXiv preprint*, arXiv:1511.06434.
7. Shorten, C., & Khoshgoftaar, T. M. (2019). A survey on image data augmentation for deep learning. *Journal of Big Data, 6*(1), 1–48.
8. Chawla, N. V., Bowyer, K. W., Hall, L. O., & Kegelmeyer, W. P. (2002). SMOTE: Synthetic minority over-sampling technique. *Journal of Artificial Intelligence Research, 16*, 321–357.
9. Han, H., Wang, W. Y., & Mao, B. H. (2005). Borderline-SMOTE: A new over-sampling method in imbalanced data sets learning. In *International conference on intelligent computing* (pp. 878–887). Springer Berlin Heidelberg.
10. Karras, T., Aila, T., Laine, S., & Lehtinen, J. (2017). Progressive growing of gans for improved quality, stability, and variation. *arXiv preprint*, arXiv:1710.10196.
11. Zhu, J. Y., Park, T., Isola, P., & Efros, A. A. (2017). Unpaired image-to-image translation using cycle-consistent adversarial networks. In *Proceedings of the IEEE international conference on computer vision* (pp. 2223–2232). IEEE.
12. Mirza, M., & Osindero, S. (2014). Conditional generative adversarial nets. *arXiv preprint*, arXiv:1411.1784.

13. Berger, D. R., Seung, H. S., & Lichtman, J. W. (2018). VAST (volume annotation and segmentation tool): Efficient manual and semi-automatic labeling of large 3D image stacks. *Frontiers in Neural Circuits, 12,* 88.

14. Dupont, C., Ouakrim, Y., & Pham, Q. C. (2021). UCP-net: Unstructured contour points for instance segmentation. In *2021 IEEE international conference on systems, man, and cybernetics (SMC)* (pp. 3373–3379). IEEE.

15. Chen, T. I., Liu, Y. C., Su, H. T., Chang, Y. C., Lin, Y. H., Yeh, J. F., et al. (2021). *Dual-awareness attention for few-shot object detection.* IEEE Transactions on Multimedia.

16. Arruda, V. F., Paixao, T. M., Berriel, R. F., De Souza, A. F., Badue, C., Sebe, N., & Oliveira-Santos, T. (2019). Cross-domain car detection using unsupervised image-to-image translation: From day to night. In *2019 International joint conference on neural networks (IJCNN)* (pp. 1–8). IEEE.

17. Liu, Y. C., Ma, C. Y., He, Z., Kuo, C. W., Chen, K., Zhang, P., et al. (2021). Unbiased teacher for semi-supervised object detection. *arXiv preprint,* arXiv:2102.09480.

18. Bischke, B., Helber, P., Folz, J., Borth, D., & Dengel, A. (2019). Multi-task learning for segmentation of building footprints with deep neural networks. In *2019 IEEE international conference on image processing (ICIP)* (pp. 1480–1484). IEEE.

19. Yang, T., Tang, H., Bai, C., Liu, J., Hao, J., Meng, Z., et al. (2021). Exploration in deep reinforcement learning: A comprehensive survey. *arXiv preprint,* arXiv:2109.06668.

20. Xie, Y., Gardi, A. G., & Sabatini, R. (2021). Hybrid AI-based demand-capacity balancing for UAS traffic management and urban air mobility. In *AIAA AVIATION 2021 FORUM* (p. 2325). AIAA.

21. Dijk, J., Schutte, K., & Oggero, S. (2019). A vision on hybrid AI for military applications. In *Artificial intelligence and machine learning in defense applications* (Vol. 11169, pp. 119–126). SPIE.

22. Garg, S., Balakrishnan, S., Lipton, Z. C., Neyshabur, B., & Sedghi, H. (2022). Leveraging unlabeled data to predict out-of-distribution performance. *arXiv preprint,* arXiv:2201.04234.

23. Li, J., Zhang, C., Zhou, J. T., Fu, H., Xia, S., & Hu, Q. (2021). Deep-LIFT: Deep label-specific feature learning for image annotation. *IEEE Transactions on Cybernetics, 52*(8), 7732–7741.

24. Yang, S. C. H., Folke, T., & Shafto, P. (2021). Abstraction, validation, and generalization for explainable artificial intelligence. *Applied AI Letters, 2*(4), e37.

25. Amer, M. R., Shields, T., Siddiquie, B., Tamrakar, A., Divakaran, A., & Chai, S. (2018). Deep multimodal fusion: A hybrid approach. *International Journal of Computer Vision, 126,* 440–456.

26. Khattar, D., Goud, J. S., Gupta, M., & Varma, V. (2019). MVAE: Multimodal variational autoencoder for fake news detection. In *The world wide web conference* (pp. 2915–2921). ACM.
27. Williams, D. P. (2016). Underwater target classification in synthetic aperture sonar imagery using deep convolutional neural networks. In *2016 23rd international conference on pattern recognition (ICPR)* (pp. 2497–2502). IEEE.
28. Lucas Martínez, N. (2021). *Contributions to adaptive mission planning for cooperative robotics in the internet of things* (Doctoral dissertation, ETSIS_Telecomunicacion).

Open Access   This chapter is licensed under the terms of the Creative Commons Attribution 4.0 International License (http://creativecommons.org/licenses/by/4.0/), which permits use, sharing, adaptation, distribution and reproduction in any medium or format, as long as you give appropriate credit to the original author(s) and the source, provide a link to the Creative Commons license and indicate if changes were made.

The images or other third party material in this chapter are included in the chapter's Creative Commons license, unless indicated otherwise in a credit line to the material. If material is not included in the chapter's Creative Commons license and your intended use is not permitted by statutory regulation or exceeds the permitted use, you will need to obtain permission directly from the copyright holder.

CHAPTER 35

Methodological Approach for Designing an Artificial Intelligence Repository for Defence Applications

Georgios Kampas, Marios Moutzouris, Leonidas Perlepes, Konstantinos Gyftodimos, Dimitrios Papageorgiou, Alexandros Savvopoulos, Pantelis Michalis, Georgios Eftychidis, Antonis Kostaridis, and Dimitrios Diagourtas

G. Kampas (✉) • M. Moutzouris • L. Perlepes • K. Gyftodimos •
D. Papageorgiou • A. Savvopoulos • P. Michalis • G. Eftychidis • A. Kostaridis •
D. Diagourtas
Satways Ltd., Neo Irakleio, Athens, Greece
e-mail: g.kampas@satways.net; g.kampas@hotmail.com; m.moutzouris@satways.
net; l.perlepes@satways.net; k.gyftodimos@satways.net; d.papageorgiou@satways.
net; a.savvopoulos@satways.net; p.michalis@satways.net; g.eftychidis@satways.net;
a.kostaridis@satways.net; d.diagourtas@satways.net

© The Author(s) 2025   439
I. Gkotsis et al. (eds.), *Paradigms on Technology Development
for Security Practitioners*, Security Informatics and Law
Enforcement, https://doi.org/10.1007/978-3-031-62083-6_35

INTRODUCTION

Artificial intelligence has made great leaps of progress since the time of the Logic Theorist programme (Simon, Newell, Shaw 1956) where a machine was created to think like a human. Since then, there has been a continuous evolution of AI in terms of algorithms and the associated computing power required to make it possible. These days one of the core pillars of AI are data. Without data, AI cannot be trained, and any inference is based on false information. Defence organizations are using AI in different spaces such as detection, planning, and field operations. The management of this data requires a structured storage area, and thus, designing and developing an AI repository for defence applications requires careful consideration of several factors. A well-designed AI repository needs to store and manage large volumes of data considering various parameters such as the type and volume of data to be stored, the purpose of the AI algorithm to be developed, the format in which the data is to be stored, and to provide access to this data.

In most cases, the data must be annotated with accurate labels, and the labelling process should comply with ethical and legal standards. Appropriate data management practices shall be defined regarding cataloguing, metadata, storage, and documentation. Access to the data should be restricted and secured to prevent unauthorized use and potential security risks, while specific authentication mechanisms should be applied keeping in mind the application domain. Versioning is also important to capture the evolution of the data and finding changes to datasets and manage incompleteness. Additionally, the repository should follow community-endorsed interoperability best practices to facilitate data exchange and reuse within and across relevant disciplines, such as security applications, enabling the researchers to advance their scientific work on a need-to-know basis. Finally, documentation of the data provenance and quality assurance processes should be meticulously maintained to ensure transparency and reliability of the AI models developed from the data. All the aforementioned parameters should be taken into account to select the best possible design approach for any individual repository.

METHODOLOGICAL APPROACH

The methodology for designing and developing the FDR involves several key steps to achieve its main objectives. The *initial methodological step* focuses on the *identification and collection of datasets* from various existing sources to meet the user requirements and use case needs. The datasets may include non-cooperative and cooperative tracking data, ground, marine, and airborne electro-optical data, radio signals, and other relevant data sources.

Once the datasets are gathered, a *thorough assessment* is conducted to evaluate the quality and relevance of the data as the *second step*. This assessment involves examining factors such as data accuracy, completeness, consistency, and data source reliability. It ensures that the selected datasets meet the desired criteria and are suitable for the intended AI algorithms and applications.

Data preparation is a crucial, *third step* that follows the assessment of data and involves cleaning, preprocessing, and transforming the datasets to make them suitable for AI algorithms. This process includes but is not limited to tasks such as data cleaning, normalization, and data formatting to ensure consistency and compatibility across different datasets.

Before storing the data, it is of utmost importance to clearly articulate *the purpose of the AI algorithms* that will be applied to the datasets. This involves identifying the specific objectives, tasks, or analysis that the AI algorithms will perform on the data. Defining the purpose of the algorithms contributes to the design of the repository and organization of the data in a way that aligns with the intended use cases.

This *fifth step* involves the actual *design and development of FDR*. The repository is designed according to the principles defined in the previous steps as well as to the chosen repository characteristics, including scalability, security, metadata management, versioning, data transfer, and collaboration features. The development process includes implementing the necessary software components, user interfaces, and backend infrastructure to support the repository functionalities.

The *collected and prepared datasets are stored* in the developed FaRADAI Dataset Repository. The repository should provide secure and scalable storage infrastructure capable of accommodating the volume and diversity of the collected datasets. Proper data organization, indexing, and storage practices are implemented to ensure efficient data retrieval and management.

The *seventh step* follows the development phase, where *thorough testing and validation of the FDR functionalities* are performed. This involves conducting several tests to ensure that the repository functions are as intended, including dataset uploading and downloading, metadata management, searchability, access control, versioning, and collaboration features.

Once the FDR is tested and validated, *access is granted* to the intended users. User roles and access rights are defined, allowing authorized partners to securely access and collaborate on the datasets stored in the repository. Proper access controls and authentication mechanisms are implemented to ensure data security and privacy. Once access is granted to the users, the *final step* includes the *monitoring* and tracking of users' activities within FDR. This foresees implementing logging and auditing mechanisms to record user interactions, including dataset access, downloads, and modifications. By monitoring access, any unauthorized activities can be detected, ensuring in parallel the security and integrity of the data.

By following the aforementioned methodological steps, as also illustrated in Fig. 35.1, the design and development of the FDR can be carried out effectively and ensure the availability and accessibility of the relevant datasets for analysis and collaboration.

IMPLEMENTATION PRINCIPLES AND DECISION TREE

In this chapter, the implementation principles that were considered to follow the appropriate design and development path up to the final release of FDR and the final step are described. In some of the following steps, there

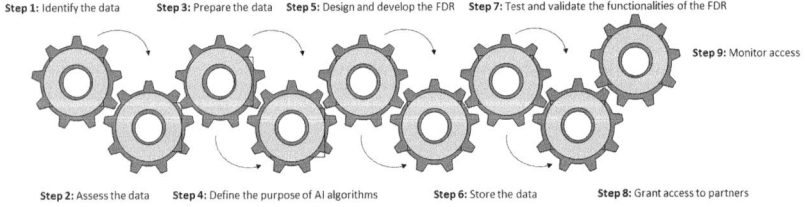

Fig. 35.1 Methodological approach in steps for designing and developing the FDR

were various approaches to follow and were evaluated according to the specific needs of the project.

Identification and Collection of Datasets The outcome of successful data identification and collection sets the foundation for FDR design. It directly influences the choice of storage solutions. For example, if data sources are distributed across various locations or systems, alternatives such as distributed databases or cloud-based storage can be explored to ensure efficient data access and retrieval. Moreover, with a vast volume of data, deduplication strategies might be implemented to optimize storage space, removing redundancy in datasets. Another alternative to consider is federated databases, allowing data to remain at its source but still be accessible through the FDR. These choices align the design with the specific needs and characteristics of the collected datasets (Fig. 35.2).

Data Assessment Thorough data assessment informs the FDR's design in several ways. High-quality, well-structured data may require less extensive preprocessing, affecting decisions regarding data cleaning and normalization. In cases where data quality is lower, the FDR's design should incorporate advanced data cleaning algorithms to handle data imperfections and outliers effectively. The alternatives may involve developing custom data cleaning routines that align with the unique characteristics of the data or using specialized data transformation techniques to enhance the data's readiness for AI algorithms.

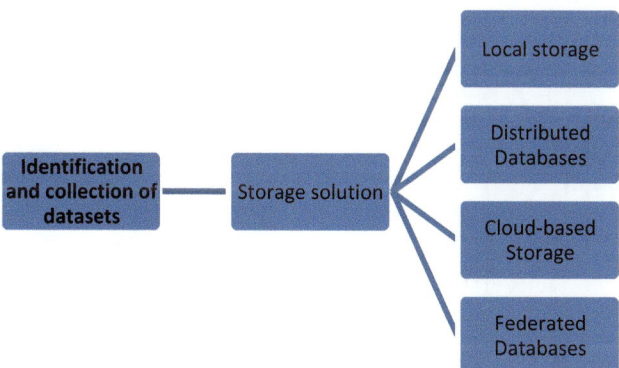

Fig. 35.2   Step 1: identification and collection of datasets

Preparation The preparation of data is critical for seamless integration into the FDR. Cleaned, normalized, and AI-ready datasets expedite processing and analysis. The design must ensure that the data is compatible with the repository's structure. For datasets arriving in various formats, the FDR might include format transformation modules to standardize data. Alternatives involve focusing on a specific data format to minimize format transformation, but this may require stricter data source requirements or additional data preparation at the source. The choice between these alternatives depends on the extent of data format variation and the trade-offs between standardization and source data flexibility (Fig. 35.3).

Definition of the Purpose of AI Algorithms Clarity in articulating the objectives of AI algorithms shapes the design of the FDR. Clear objectives facilitate data categorization and indexing, allowing for efficient data retrieval and usage. Alternatives might include advanced search functionalities if the use cases are complex or subject to frequent changes in algorithm objectives. In cases where algorithms serve multiple, distinct purposes, the FDR design may emphasize robust tagging and metadata management, enabling flexible search and retrieval based on different criteria (Fig. 35.4).

Datasets Storage The secure and scalable storage of data is crucial, especially for defence applications dealing with substantial data volumes. The FDR's design shall account for storage that can accommodate growth. For enhanced data security, alternatives include encrypted storage solutions.

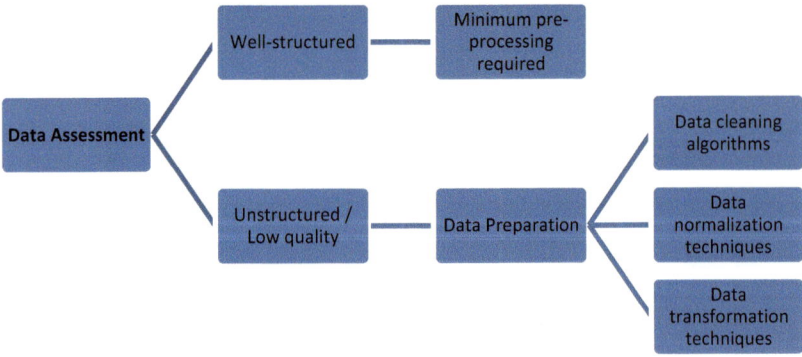

Fig. 35.3 Steps 2 and 3: data assessment and preparation

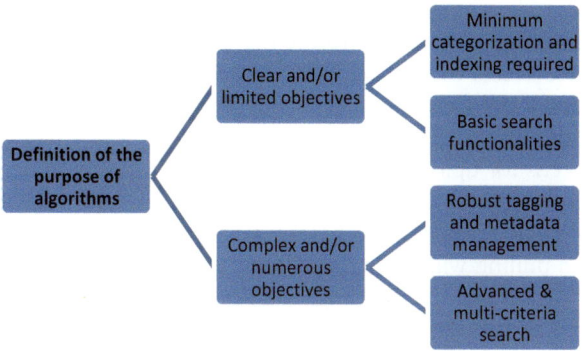

Fig. 35.4 Step 4: definition of the purpose of AI algorithms

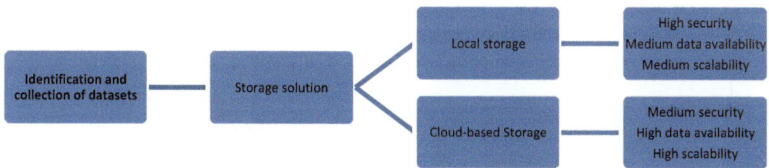

Fig. 35.5 Step 6: identification and collection of datasets

Integrating cloud storage might also be considered, and a multi-cloud approach can provide redundancy and improved data availability. The choice between on-premises and cloud-based storage is a critical decision, and it influences both the repository's architecture and its scalability (Fig. 35.5).

Design and Development of FDR The design options for FDR components are directly shaped by the outcomes of the previous steps which are summarized here. For data-intensive tasks, such as AI algorithms that require substantial computational power, the FDR might opt for an asynchronous processing unit (APU) with a queue implementation to efficiently manage processor and storage-intensive tasks. When ease of data access is a priority, REST API Services might be preferred for communication between the various components of the FDR. Access to these services is only for the components of the FDR and no external access is allowed. To support user-friendliness, a Web Interface can be considered for time-efficient dataset management and administrative tasks. The web interface should make available to the visitor which datasets they have created (or have access to) and to manage their access rights. It should also allow certain users to manage access to the FDR. Additionally, a client module with command-line utilities could be developed to cater to specific user needs for uploading, downloading, and managing datasets. The client module can simplify interaction with FDR, thereby making its adoption much quicker and easier. The paradigm used with Git repositories of pushing and pulling will also be used here as technical persons are familiar with this. In this regard, versioning is also mandatory to ensure traceability and understand the various changes that have been made to the dataset. Retrieving the dataset(s) is only available through this client application, thereby restricting access to the datasets. The client application can also be distributed through secure means and its usage could have an expiry date, thereby deactivating old versions. Security is an important factor, therefore restricted access, token-sharing, and expiry should also be implemented. For the purposes of the FDR, a cloud-based solution, instead of an on-premises deployment, was selected for storing the datasets. The FDR could also support other storage options depending on the application domain. Keeping these in mind, the preliminary design review of the foreseen FDR is illustrated in Fig. 35.6.

Testing and Validation of the FDR Functionalities The outcomes of testing and validation directly guide design refinements. If performance issues are identified during testing, alternatives can be explored. For example, optimizing database indexing can enhance data retrieval speed or implementing caching mechanisms for frequently accessed data can improve response times. Thorough testing can and will uncover specific requirements, driving the selection of design alternatives that align FDR's capabilities with the performance and functionality demands of the users (Fig. 35.7).

Granting Access Access control requirements, driven by the specific needs of project partners, significantly influence the design decisions related to user roles, permissions, and access to the repository. Alternatives include the implementation of role-based access control (RBAC) for fine-grained permissions, enabling granular control over who can access and manipulate specific datasets. Additionally, employing single sign-on (SSO) solutions can streamline partner access by allowing users to access the FDR

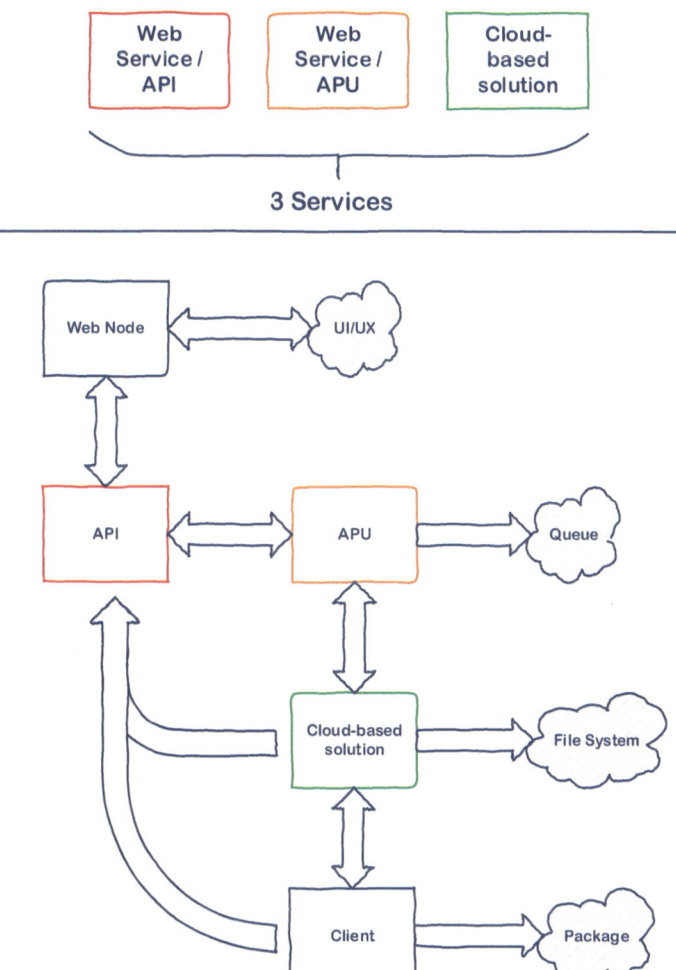

Fig. 35.6 Step 5: design and development of FDR

using their existing credentials. The choice of access control mechanisms directly impacts FDR design for user management, ensuring the protection and controlled access of sensitive data.

Access Monitoring The implementation of real-time alerts and notifications for access monitoring directly enhances data security and

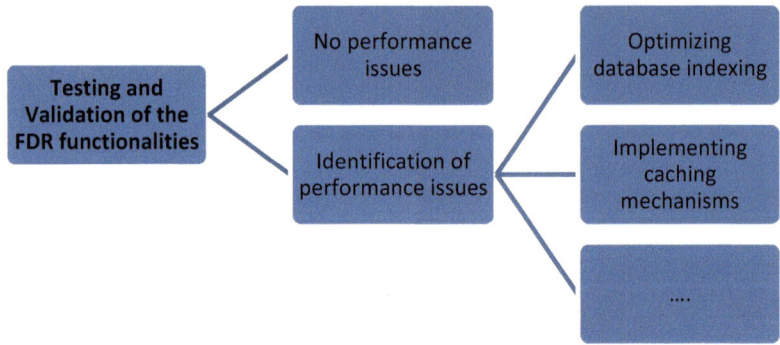

Fig. 35.7   Step 7: testing and validation of the FDR functionalities

accountability. Alternative approaches may include the utilization of machine learning-based anomaly detection algorithms to identify suspicious access patterns and potential security breaches. Another alternative is the implementation of blockchain-based access auditing, which provides an immutable ledger of data access events, offering enhanced data security and integrity. The choice of the most appropriate alternative should align with the level of access control and monitoring required for the project's security and compliance needs.

Conclusions

This chapter details a systematic process that encompasses several key methodological steps. It begins with the identification and collection of diverse datasets from varied sources, spanning non-cooperative and cooperative tracking data, electro-optical data, radio signals, and more. These datasets are meticulously assessed to ensure their quality and relevance, aligning them with specific user requirements and use cases. Subsequently, data preparation measures are implemented, including data cleaning, normalization, and format standardization, rendering the datasets compatible with AI algorithms. To achieve the highest level of efficacy, the repository is designed with several critical characteristics. Security remains paramount, with robust access controls, user authentication, and role-based permissions to protect sensitive defence-related data. Metadata management and searchability ensure efficient dataset organization and retrieval. The FDR also supports versioning, format transformation, and long-term preservation to accommodate evolving AI needs and guarantee data integrity. Moreover, the chapter underscores the significance of monitoring data access within the repository. This monitoring mechanism enables the timely detection of unauthorized or unusual access patterns, reinforcing data security and user accountability.

In summary, this chapter provides insights into the methodological approach underpinning the creation of an AI dataset repository tailored for defence applications. The FDR's design principles prioritize security, accessibility, and scalability, making it an invaluable asset for frugal and robust AI in the defence sector. Furthermore, the aforementioned methodology outlined in this study offers a versatile blueprint that extends beyond its immediate application, holding significant promise for similar use cases within the broader AI domain. Its systematic approach to data handling, encompassing identification, assessment, preparation, storage, access control, and monitoring, can be readily adapted to a multitude of AI applications. As the field of AI continues to evolve and confront new challenges, the adaptability of this methodology positions it as a foundation for building AI repository solutions that can flexibly respond to the ever-changing landscape of data and technology. It stands as a testament to the potential for sustainable and effective data management in the AI domain, enabling not only the optimization of current applications but also the preparation for future AI endeavours and their evolving data needs.

Acknowledgements This project received funding from the European Defence Fund programme under grant agreement no. 101103386. The views and opinions expressed are, however, those of the author(s) only and do not necessarily reflect those of the European Union or the European Commission. Neither the European Union nor the granting authority can be held responsible for them.

BIBLIOGRAPHY

1. Deloitte. (2021). The Age of With – The AI advantage in defence and security, (4), 4–20. *Deloitte Analytics*.
2. Van Wynsberghe, A. (2021). Sustainable AI: AI for sustainability and the sustainability of AI. *AI and Ethics, 1*(3), 213–218.
3. Shorten, C., & Khoshgoftaar, T. M. (2019). A survey on image data augmentation for deep learning. *Journal of Big Data, 6*(1), 1–48.

4. Tsay, J., Braz, A., Hirzel, M., Shinnar, A., & Mummert, T. (2022). Extracting enhanced artificial intelligence model metadata from software repositories. *Empirical Software Engineering, 27*(7), 176.
5. Shin, P. W., Lee, J., Kim, J., Shin, D., Lee, Y., & Hwang, S. H. (2020). A research in applying big data and artificial intelligence on defence metadata using Multi Repository Meta-Data Management (MRMM). *Journal of Internet Computing and Services, 21*(1), 169–178.

Open Access   This chapter is licensed under the terms of the Creative Commons Attribution 4.0 International License (http://creativecommons.org/licenses/by/4.0/), which permits use, sharing, adaptation, distribution and reproduction in any medium or format, as long as you give appropriate credit to the original author(s) and the source, provide a link to the Creative Commons license and indicate if changes were made.

The images or other third party material in this chapter are included in the chapter's Creative Commons license, unless indicated otherwise in a credit line to the material. If material is not included in the chapter's Creative Commons license and your intended use is not permitted by statutory regulation or exceeds the permitted use, you will need to obtain permission directly from the copyright holder.

CHAPTER 36

Military-Grade, Data Exchange Platform (DXP) Enhancing Cybersecurity Automation and Information Sharing, and Its Application on Autonomous Military Systems

Christos Skoufis, Marios Sophocleous, and Frini Lazarou

INTRODUCTION

The current state of the art in cybersecurity faces several challenges. Security requirements are often underestimated or ignored, leading to vulnerabilities that can be exploited by cyber threats [1]. The design and implementation of security measures are typically underfunded, resulting in inadequate protection for digital assets [2]. Inherent vulnerabilities in system components and supply chain weaknesses further contribute to

---

Empowering Defence Systems: HERMES European Defence Industrial Development Programme (EDIDP) Project Advances Cybersecurity Automation and Information Sharing

---

C. Skoufis (✉) • M. Sophocleous • F. Lazarou
eBOS Technologies Ltd, Nicosia, Cyprus
e-mail: christoss@ebos.com.cy

© The Author(s) 2025
I. Gkotsis et al. (eds.), *Paradigms on Technology Development for Security Practitioners*, Security Informatics and Law Enforcement, https://doi.org/10.1007/978-3-031-62083-6_36

451

security risks [3]. Modern cyber threats are complex and sophisticated, capable of bypassing traditional security solutions [4].

In the military domain, the increasing use of state-of-the-art information and communication technologies (ICTs) introduces new threat vectors and amplifies the impact of cyber threats on defence capabilities. Defence systems heavily rely on interconnected Command, Control, Communications, Computers, Intelligence, Surveillance, and Reconnaissance (C4ISR) systems, which must be secured despite constrained budgets and limited resources [5]. Additionally, weapon systems and logistical supply chains face cyberattack risks due to their dependence on automation and information management [6]. The asymmetric threats of greatest concern in the military domain are offensive cyber operations targeting C4ISR, weapon systems, and logistical supply chains [7]. The development of a proficient cyber force has become increasingly common and straightforward for many nations, highlighting the growing importance of enhancing cyber defence capabilities. As warfare evolves with the integration of robotics and artificial intelligence, autonomous systems will play an increasingly significant role not only in combat but also in logistics and support functions.

The reliance on computer and electronic systems in the military introduces new pressure points and unpredictability in future wars. Critical infrastructures, supply chains, and military systems can be compromised, necessitating a high peacetime cyber readiness and resilience level. Timely and accurate information and data remain central to achieving information superiority, ensuring effective use of force, and defending military systems. The development of new cyber defence capabilities, such as cyber situational awareness technologies, defensive cyber technologies, and predictive analysis, has been identified as a priority in EU Capability Development [8]. Overall, the evolving landscape of cybersecurity in the military domain requires proactive measures to address cyber threats, enhance information superiority, and strengthen cyber defence capabilities in order to ensure the security and effectiveness of military operations. It has been observed that projects attempting to address the underlying operational requirements for these objectives continue to take the same general approach, namely the implementation of siloed data storage and exchange mechanisms, constrained by interfaces whose interoperability is controlled through proprietary solutions, ad hoc approaches, or traditional standards development processes.

The cyber threat and incident response information-sharing work through a platform, as performed in HERMES, is used to significantly enhance the work conducted on similar technologies for information sharing of active defence measures and cyber threat intelligence [9]. Furthermore, HERMES progresses the existing research for a software solution for enabling real-time cyber threat hunting and live incident response, based on shared cyber threat intelligence [10]. This is done by mixing user-facing functionality on specific subsets of cybersecurity, such as cyber situation awareness, threat hunting, continuous monitoring, end-point detection, and incident response with information sharing, by developing in-system cybersecurity information-sharing mechanisms that address information-sharing interoperability challenges.

In this chapter, the subject of 'cybersecurity solutions for the protection of future security and defence systems' is addressed. We propose the development of a foundational system for data representation, storage, and exchange across all cybersecurity solutions. Such an approach is extremely challenging; however, it should not be dismissed because of its complexity. By tackling the key issues head-on with a blank sheet approach, the underlying principles developed over several years in earlier work present an opportunity to achieve significant advantages over the approach taken by existing cyber defence products and solutions. Furthermore, the Data Exchange Platform (DXP) developed under the HERMES project focuses on the general military use case but more specifically aims to demonstrate its capabilities in the use cases using autonomous military systems (AMS).

CHALLENGES IN AUTONOMOUS MILITARY SYSTEMS

The complete design of the DXP platform will be achieved by the end of the HERMES project, based on common requirements agreed upon by the participating Member States during the studies phase. Through engagements with participating Member States and stakeholders, the DXP will validate the underlying novel approach proposed to address the need for cyber resilience in AMS. DXP is a military-grade, enterprise system comprised of different software components distributed throughout an organisation, including across security domains. It is used by various experts to collect, curate, and distribute cybersecurity information on the specific domain of AMS, and more specifically, unmanned ground vehicles (UGVs).

The problems associated with data security affect all areas of cybersecurity. DXP, as a foundation for building solutions for cybersecurity operations, can be applied to different areas such as intelligence sharing and information warfare. However, DXP mostly focuses on the specific domain of AMS. The focus on UGVs is motivated by the following two factors:

1. The urgent need for cyber defence solutions that can operate as autonomously as the systems they are designed to safeguard. In such scenarios, where high-security environments are complex and challenging, there is a significant need for data exchange between distinct entities [11].
2. The lack of sufficient solutions in this field means that DXP is a new novel design developed using the latest technologies and knowledge.

Moreover, DXP aims to enable automation and autonomy in cybersecurity operations, improve controlled information sharing of high-quality cybersecurity data, and facilitate burden-sharing collaboration and outsourcing of cybersecurity data management. Unlike traditional approaches that rely on interoperability standards, DXP takes a disruptive paradigm shift by separating data representation, storage, and exchange from the uses made of exchanged data. It recognises the complexity of exchanging cybersecurity data and offers a foundational system that can be used by all, providing common data representation, storage, and exchange capabilities. This allows applications to obtain their data not from multiple sources but from a unique system, which takes care of the common data representation, storage, and exchange issues. This significantly reduces the integration effort and resources needed, while allowing this effort to be applied to the development of better applications for specific needs.

DXP Use Cases

DXP outputs are demonstrated through two use cases (UCs). The first UC is a general one showing the use of DXP as a foundational data management service across multiple organisations, while the second UC is more specific for the dissemination of cybersecurity data to autonomous military systems.

DXP General UC

The general UC is illustrated in Fig. 36.1, which shows two organisations, A and B, and a number of cybersecurity application vendors. To illustrate the distributed nature of DXP, Organisation A is shown to be located in two sites, 1 and 2. In this diagram, DXP is illustrated at the conceptual level by the following components in dark blue:

- HERMES Data Store (HDS), for storing all data within DXP.
- HERMES Data Management (HDM), which provides automated data management functions defined in policies.
- HERMES User Interface (HUI), which provides the functionality for policy administrators, data curators, and quality assurance experts to manage the data throughout its life cycle.
- HERMES Data Exchange (HDX), responsible for exchanging data with other instances of DXP according to the defined policies.
- HERMES Application Programming Interface (HAPI), which provides access to data to cybersecurity applications ('App x').

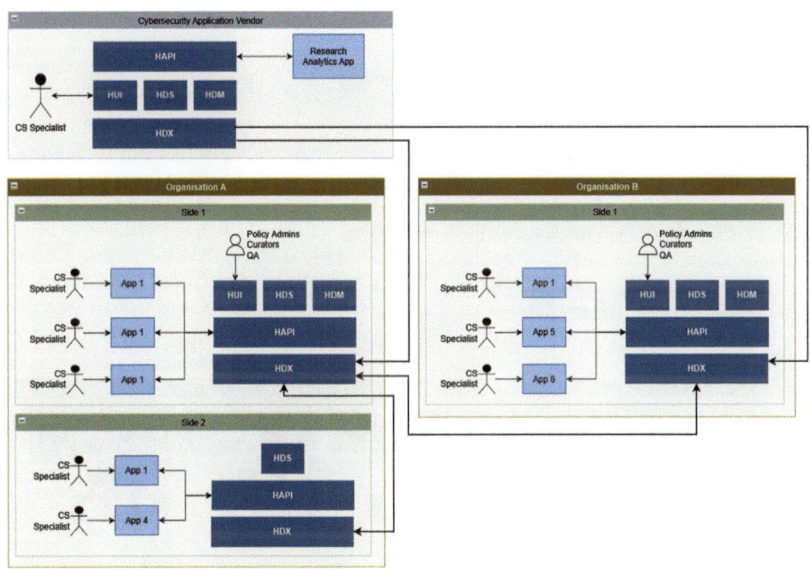

Fig. 36.1 DXP general UC

The various cybersecurity applications represent existing cybersecurity solutions operated by cybersecurity professionals ('CS Specialist'). While these represent applications such as today's antivirus software, intrusion detection systems, end-point protection software, and security incident and event monitoring software, for example, these have been modified to use DXP as their source of data through the HAPI. As vendors update the datasets used by their products, DXP brings this data to the applications installed in end user facilities while enforcing licence agreements. Because cybersecurity data is concentrated in DXP, data management activities such as correlation and quality assurance can be done via functionality offered by the HUI, including correlation with private data held in the HDS but never meant for sharing. The HDX component handles data flow across sites as instructed by policies. It also mediates exchanges between organisations where exchange agreements have been put in place for information sharing, collaboration, and outsourcing.

DXP UC for Autonomous Military Systems
In this UC, as illustrated in Fig. 36.2, DXP provides the ability to channel data from various sources and communities to UGVs as below:

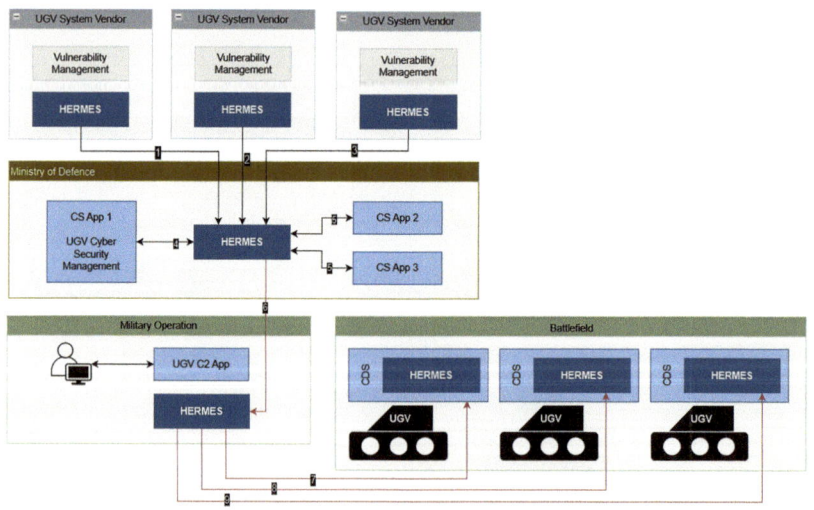

Fig. 36.2   DXP UC for autonomous military systems

- The UGV System Vendor holds information about potential vulnerabilities in its systems, which it will share with its customers through the DXP functionality. This functionality provides confidentiality and data exchanges are performed according to policies that cover licensing, copyrights, and authorised uses, amongst other things. This gives vendors the ability to prevent customers from further sharing this information, at least from a contractual and system point of view. This is data flow #1 in the diagram.
- This type of information includes generic data, such as details on vulnerabilities and security updates for operating systems and widely used software libraries, and it is shared within the global cybersecurity community through organisations like the Forum for Incident Response and Security Teams (FIRST) or national Computer Security Incident Response Teams (CSIRTs) [12]. This is data flow #2 in the diagram.
- In terms of illustrating the various sources of cybersecurity data, it also shows generic cyber threat intelligence being shared by Allies via DXP. This is data flow #3 in the diagram.

All of this data and information is received at an office within the Ministry of Defence of the country operating the UGVs, for example, responsible for managing the UGV program. This could be done via a dedicated cybersecurity application ('CS App 1') connected to DXP, shown as data flow #4. Other cybersecurity applications ('CS App 2' and 'CS App 3') could also use DXP as their source of data for other purposes, shown as in data flow #5 in the diagram. The application used to manage the cybersecurity of the UGVs could be provided by the UGV vendor, providing the overall functionality to maintain the fleet of UGVs during its life cycle, including the functionality required to manage the cybersecurity of the fleet.

Once the staff managing the UGV fleet have considered the available information and decided how to address cybersecurity issues, the data would be passed to the operators of the UGV on deployed operations. This is data flow #6. It is expected that this would happen at a higher security classification, which would be facilitated within the DXP installed at the Ministry of Defence. Data flow #6 is, therefore, shown as a red line.

The operators of the UGVs employed in military operations can then use a Command-and-Control application ('UGV C2 App') to consider the information sent by the managers of the UGV fleet and merge it with the current mission parameters and local threat information available in the theatre. Based on the assessed risks, which ultimately incorporate data from the UGV manufacturer, the cybersecurity community, Allies, UGV programme managers, and operational considerations, the operators can decide which available measure to take and use HERMES to transfer the data directly to the deployed UGVs. This is data flow #7, also in red as it is expected to be done at a higher classification level.

Moreover, Fig. 36.2 shows a smaller DXP application being part of a Cyber Defence System ('CDS' in the diagram), also most likely provided by the UGV vendor, showing how the components of DXP can be designed to be minimalistic for constrained environments and embedded into other applications.

CONCLUSION

The proposed DXP system is built upon prior North Atlantic Treaty Organization (NATO) work known as the 'Cyber Security Data Exchange and Collaboration Infrastructure (CDXI)' [13]. Technical reports related to this work were published as part of the Allied Command Transformation Programme [14] and the Multi-National Cyber Defence Capability Development Project. While these publications are available from the NATO Communications and Information Agency and NATO Nations, they are not publicly accessible due to their classification.

Within the EU context, the HERMES project can greatly benefit other initiatives such as the 'Cyber Threats and Incident Response Information Sharing Platform (CTIRISP)' and the 'Integrated Unmanned Ground System (UGS)' project under the PESCO framework. The focus of such projects is on specific aspects, like cyber threats and incident response information sharing. By incorporating insights from the DXP system, these projects can potentially overcome the challenges faced by other solutions in the same domain.

The HERMES project not only focuses on the design of the HERMES Data Exchange Platform itself but also includes a demonstration of its value in the context of cyber defence for UGVs. If implemented, HERMES would be considered an enabling capability for cyber-responsive operations, aligning with EU Capability Development Priorities [8].

Overall, the HERMES project aims to address the limitations in information sharing, collaboration, automation, and autonomy in cyber defence by providing a dedicated system for the secure exchange and management of cybersecurity data.

REFERENCES

1. Dunn Cavelty, M. (2007). *Cyber-security and threat politics: US efforts to secure the information age* (1st ed.). Routledge. https://doi.org/10.4324/9780203937419

2. Anderson, R. (2001). Why information security is hard – An economic perspective. In *Seventeenth annual computer security applications conference* (pp. 358–365). IEEE. https://doi.org/10.1109/ACSAC.2001.991552

3. do Amaral, T. M. S., & Gondim, J. J. C. (2021). Integrating Zero Trust in the cyber supply chain security. In *2021 Workshop on communication networks and power systems (WCNPS)* (pp. 1–6). IEEE. https://doi.org/10.1109/WCNPS53648.2021.9626299

4. Ponemon Institute. (2020). *The third annual study on the state of endpoint security risk.* Morphisec and Ponemon Institute. [Online]. Available: https://www.morphisec.com/hubfs/2020%20State%20of%20Endpoint%20Security%20Final.pdf

5. Adams, G., Ben-Ari, G., Logsdon, J., & Williamson, R. (2004). *Bridging the gap. European C4ISR capabilities and transatlantic interoperability*, p. 194. [Online] Available: https://apps.dtic.mil/sti/pdfs/ADA450221.pdf

6. Boulanin, V., & Verbruggen, M. (2017). *Mapping the development of autonomy in weapon systems.* https://doi.org/10.13140/RG.2.2.22719.41127

7. Schmitt, M. (2017). Tallinn manual 2.0 on the international law applicable to cyber operations. In *Tallinn manual 2.0 on the international law applicable to cyber operations* (pp. I–II). Cambridge University Press.

8. *The EU capability development priorities, 2018 CDP revision.* European Defence Agency, [Online]. Available: https://eda.europa.eu/docs/default-source/eda-publications/eda-brochure-cdp

9. PESCO. *Cyber threats and incident response information sharing platform project* [Online]. Available: https://www.pesco.europa.eu/project/cyber-threats-and-incident-response-information-sharing-platform/

10. Work Programme EDIDP-CSAMN-SSS-2019. *Software suite solution, enabling real-time cyber threat hunting and live incident response, based on shared cyber threat intelligence* [Online]. Available: https://ec.europa.eu/info/funding-tenders/opportunities/portal/screen/opportunities/topic-details/edidp-csamn-sss-2019

11. Oreyomi, M., & Jahankhani, H. (2022). Challenges and opportunities of autonomous cyber defence (ACyD) against cyber attacks. In H. Jahankhani, D. V. Kilpin, & S. Kendzierskyj (Eds.), *Blockchain and other emerging technologies for digital business strategies. Advanced sciences and technologies for security applications.* Springer. https://doi.org/10.1007/978-3-030-98225-6_9

12. Forum for Incident Response and Security Teams. [Online]. Available: https://first.org

13. Cyber Security Data Exchange and Collaboration Infrastructure (CDXI). NATO Communications and Information Agency. [Online]. Available: https://www.ncia.nato.int/cyber-security-cdx.html
14. NATO Allied Command Transformation Programme of Work. NATO Allied Command Transformation. [Online]. Available: https://act.nato.int/programme-of-work

Open Access This chapter is licensed under the terms of the Creative Commons Attribution 4.0 International License (http://creativecommons.org/licenses/by/4.0/), which permits use, sharing, adaptation, distribution and reproduction in any medium or format, as long as you give appropriate credit to the original author(s) and the source, provide a link to the Creative Commons license and indicate if changes were made.

The images or other third party material in this chapter are included in the chapter's Creative Commons license, unless indicated otherwise in a credit line to the material. If material is not included in the chapter's Creative Commons license and your intended use is not permitted by statutory regulation or exceeds the permitted use, you will need to obtain permission directly from the copyright holder.

PART VII

Increased Cybersecurity

CHAPTER 37

SECANT: Cyberthreat Intelligence in IoMT Ecosystems

Arnolnt Spyros, Dimitrios Kavallieros, Theodora Tsikrika, Stefanos Vrochidis, and Ioannis Kompatsiaris

INTRODUCTION

The landscape of cyberattacks is constantly evolving. Cyberattacks evolve in terms of sophistication while the effort to perform cyberattacks is decreasing given that attackers use different set of tools and tactics in order to attack their targets [1, 2]. Well-resourced threat groups can use a variety of Tactics, Techniques and Procedures (TTPs) to attack organisations in various sectors. Among the most critical and most targeted sectors is the healthcare sector [3–5].

The continuous adaptation of automated systems in the healthcare domain with regards to medical data processing and sharing poses critical threat [6]. Specifically, the adaptation of the Internet of Medical Things (IoMT) devices by healthcare organisations increases the threat landscape considering that IoMT devices are susceptible to various cyberattacks.

A. Spyros (✉) • D. Kavallieros • T. Tsikrika • S. Vrochidis • I. Kompatsiaris
Information Technologies Institute, Centre for Research and Technology Hellas, Central Macedonia, Thessaloniki, Greece
e-mail: aspyros@iti.gr

© The Author(s) 2025   463
I. Gkotsis et al. (eds.), *Paradigms on Technology Development for Security Practitioners*, Security Informatics and Law Enforcement, https://doi.org/10.1007/978-3-031-62083-6_37

Considering the vulnerable nature of IoT, the integration of such solutions might expose the organisation to numerous threats [7]. Specifically for AI, healthcare industry applications are considered one of the most critical fields [8]. Furthermore, there is a constantly increasing number of common medical devices, which are connected to the Internet, hospital networks, as well as other medical devices in order to provide features that facilitate the provision of healthcare [6], and, therefore, the organisations become further susceptible to more cyberthreats. A critical measure to address the ever-evolving threat landscape is the gathering and analysis of Cyber Threat Intelligence (CTI). In particular, over the past years, CTI has emerged as a critical component of an organisation's security. The content of CTI contains information that can help identify, assess, monitor and respond to cyberthreats [1] in a timely manner.

RELATED WORK

In order for organisations to prevent, respond or mitigate the cybersecurity threats that affect their assets, they need to be informed about cyberthreat trends and defend themselves against a wide range of adversaries with various levels of motivations, capabilities and access to resources [2]. Therefore, organisations must collect relevant CTI data from different sources and utilise the gathered knowledge in terms of enhancing the overall security of the organisation. Among other organisations, the Cybersecurity and Infrastructure Security Agency (CISA) maintains the ICS-CERT Alert that provides comprehensive timely notifications concerning critical infrastructure, including alerts affecting medical devices. Furthermore, most manufacturers of medical devices maintain a repository with CTI [6].

The authors in [6] propose a system for gathering CTI concerning various medical devices from sources that include medical manufacturer and ICS-CERT vulnerability alerts. The solution utilises a named entity extractor to extract CTI and other cybersecurity information from the gathered data. Subsequently, the CTI is augmented with data from sources such as Wikidata[1] and public medical databases (e.g. US FDA AccessGUDID Database[2]) by integrating this information in a cybersecurity knowledge graph (CKG) from previous research of the authors.

---

[1] https://www.wikidata.org/wiki/Wikidata:Main_Page
[2] https://accessgudid.nlm.nih.gov

A novel platform called Intelligent Mitigation Platform for Advanced Cyber Threats (IMPACT) is proposed in [9] which, according to the authors, is well-suited for the increasingly connected complex healthcare ecosystem. The proposed platform utilises a decentralised Collaborative Intrusion Detection Networks (CIDN) architecture to allow IDS nodes to gain knowledge by sharing information. Moreover, the proposed solution leverages machine learning (ML) and federated learning, allowing the detection and prevention of sophisticated multi-stage attacks. Therefore, the platform is able to meet the needs of complex healthcare ecosystems, enhancing their resilience against sophisticated cyberattacks. A CIDN comprises several IDS workers that collect and share security events, as well as analysis units that provide functionalities for correlation of events and extracting useful threat intelligence information.

The authors in [10] perform a systematic literature review to investigate the threat intelligence issues and solutions concerning vulnerabilities that affect IoMT devices. The authors propose path search algorithm that incorporates threat intelligence, solutions, stakeholders (i.e. medical practitioners, system or network administrators, and patients) and infrastructure.

MAIN AIM

Healthcare is considered to be among the most critical and targeted domains, with an annual increase of cyberattacks against this domain. Especially during the period of the COVID-19 pandemic, the cybersecurity incidents in healthcare increased significantly [11, 12]. Successful cyberattacks against healthcare organisations can lead to devastating results for the system, as well as the loss of human lives (e.g. in case of violating the integrity of lab results). Consequently, healthcare organisations must safeguard their systems by implementing both proactive and reactive measures. An important asset in this matter is the use of CTI, which allows the organisation to maintain visibility in the modern threat landscape. The TIM module of SECANT provides functionalities that enable the gathering, enrichment, sharing and utilisation of high-quality CTI.

SECANT

There has been an unprecedented number of changes in the healthcare sector during the last years such as new models of remote delivery of services and the excessive use of IoT devices for different functionalities. While the various changes can offer many advantages, at the same time they introduce novel cybersecurity threats. Furthermore, the integration of IoT devices can facilitate critical procedures such as the clinical status reporting of patients which require continuous medical supervision (e.g. in-house surveillance) [13]. Nevertheless, considering the vulnerable nature of IoT, the integration of such solutions might expose the organisation to numerous threats [7]. A large amount of sensitive data that is produced and exchanged among IoT devices is usually transmitted over insecure networks (e.g. wireless network). Consequently, this raises several security and privacy issues, resulting in more susceptible infrastructures against a vast amount of cyberattacks. Despite these critical issues, the level of security awareness is still disproportionately low compared to the criticality and potential of a security breach in critical sectors.

SECANT is an EU-H2020 project, which recognises these challenges and aims to deliver a holistic framework for cybersecurity risk assessment to enable strengthening the understanding of cybersecurity risks, at both human and technical levels, as well as for enhancing the digital security, privacy and personal data protection in complex ICT infrastructures such as the healthcare domain[3,4] SECANT comprises of four major pillars which contribute towards the enhancement of the organisations' cybersecurity capabilities providing (i) collaborative threat intelligence collection, analysis and sharing; (ii) innovative risk analysis specifically designed for interconnected nodes of an industrial ecosystem; (iii) cutting-edge trust and accountability mechanisms for data protection and (iv) security awareness training for more informed security choices.

Threat Intelligence Module (TIM)

The Threat Intelligence Module (TIM) of SECANT enables the collection, extraction, enrichment and sharing of CTI from both external (i.e. online) and internal sources. External sources include sources such as vulnerability databases, CERT feeds, databases with proof of concept (PoC) exploits, social media, forums and relevant web pages from the Surface, Deep and the Dark Web. Figure 37.1 illustrates the UI of the web crawlers of TIM. On the other hand, internal sources of TIM include different

---

[3] https://secant-project.eu/challenges-and-vision
[4] https://secant-project.eu/objectives

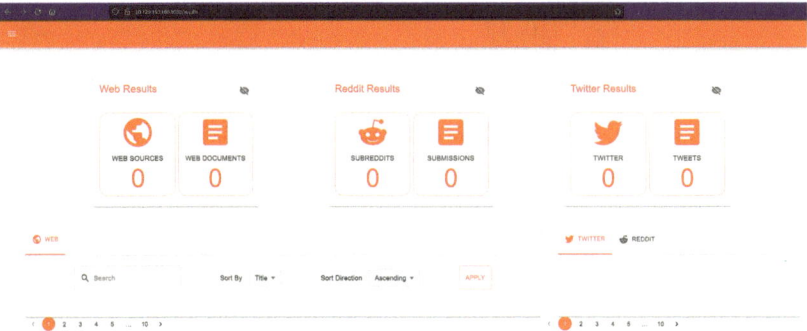

Fig. 37.1   TIM web crawlers

honeypot instances. TIM filters the collected data to avoid storing Personal Identifiable Information (PII) by leveraging rule-based techniques and extracts CTI from the collected sources using rule-based and ML-based techniques.

Subsequently, the collected data from all sources is further analysed and enriched by leveraging correlation and dynamic taxonomy allocation techniques. Possible correlations between the information are identified leveraging both simple (e.g. MISP correlation) and advanced (e.g. ML-based) techniques. The collection process is achieved both (i) manually through a user-friendly GUI as well as (ii) automatically on a daily basis, leveraging appropriate scripts and configurations.

The generated CTI is stored on MISP as MISP events and is available via the MISP platform and MISP API as depicted in Fig. 37.2. The MISP platform was selected since it offers many advantages compared to other available platforms, including the ability to include technical and non-technical information. MISP also facilitates the interoperability of TIM since it supports the export of the stored data in various data formats, including the STIX[5] standard.

The functionalities provided by TIM allow the organisation to remain updated on the current as well as emerging cyberthreats. The user is able to gather, utilise and share CTI in a secure and efficient manner. Furthermore, TIM enriches the extracted CTI, thus allowing the creation

---

[5] https://soasis-open.github.io/cti-documentation/stix/intro.html

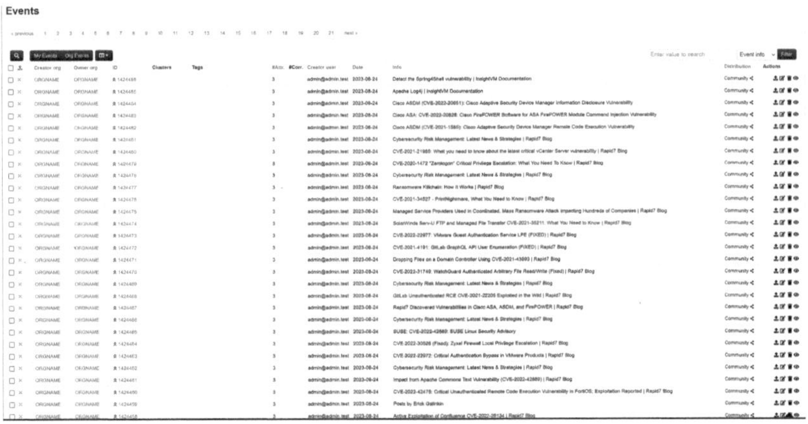

Fig. 37.2   Genereated CTI stored as MISP events

of more complex rules to prevent, identify and mitigate cyberthreats within the infrastructure of the organisation.

DEMONSTRATION CASES

SECANT will be demonstrated and validated across four realistic use cases which are discussed in this section. In particular, the performance of the relevant modules will be validated in four realistic pilot use case scenarios applied within the healthcare domain. With regards to TIM, the evaluation of the module will be performed within three out of four use cases, namely (i) Protecting the Connected Ambulance of the Future, (ii) Cybersecurity for Connected Medical Devices and Mobile Applications, (iii) Health Data Protection in the Healthcare Supply Chain and (iv) Cybersecurity Training.

The scope of the first use case is to provide monitoring of the continuous assessment of the devices that are present within a smart ambulance environment in order to protect the systems which also include the data of the patients. The second use case of SECANT focuses on the monitoring and risk assessment of an infrastructure that handles electronic health records. In both use cases TIM facilitates the identification of vulnerabilities as well as the risk assessment process by providing enriched CTI. Specifically, TIM composes enriched CTI that is sent to IPL. IPL, in turn, sends the enriched CTI, including CVEs, CPEs and other CTI data, to the TVIA in order for the latter to map the existing vulnerabilities to the discovered assets. TIM can also send the list of threats and vulnerabilities to CO-CRAE in order to allow further configuration from the security administrator.

The fourth use case aims to provide an educational framework platform, covering the needs of both experts and non-expert users. Towards this direction, two distinct scenarios have been defined. The first scenario enables cybersecurity expert users that aim to gain insight into the SECANT solution to safeguard their infrastructure through the cyber range which is a safe environment though it is prone to mistakes. In conjunction with the other modules that are deployed in the cyber range, TIM facilitates the identification of vulnerabilities and the risk assessment process by providing updated and enriched CTI. The second scenario of the fourth use case aims to increase the cybersecurity awareness of regular users in terms of identifying possible cybersecurity attacks and what actions they could take in order to elevate and safeguard the posture of their organisation.

OPEN ISSUES

Despite the vulnerable nature of IoMT devices, their use is constantly increasing. While IoMT devices pose serious security risks, they facilitate critical medical tasks including the monitoring of the patient and in some cases allow the medical staff to save human lives. Therefore, the use of these devices cannot be limited. Furthermore, there is a constantly increasing number of common medical devices, which are connected to the Internet, hospital networks, as well as other medical devices, in order to provide features that facilitate the provision of healthcare [6]. This raises more security risks since there are various identified security gaps in contemporary interconnected medical systems, thus increasing the attack surface of the organisation.

Medical organisations can gather valuable CTI from several sources. Some manufacturers of medical devices such as Philips[6] maintain a repository with CTI [6]. Infrastructure Security Agency (CISA)[7] maintains the

---

[6] https://www.philips.com/a-w/security/security-advisories.html
[7] https://www.cisa.gov/about-cisa

ICS-CERT Alert[8] that provides timely notifications concerning critical infrastructure, including alerts affecting medical devices. Nevertheless, there is a lack of sufficient IoMT CTI sources while the available sources might not include adequate technical details, resulting in the extraction of CTI with limited amount of Indicators of Compromise (IoC) which could decrease its actionability.

Apart from the technical security issues there is also a critical issue concerning the lack of security awareness from the medical staff. Considering the complex infrastructures of the healthcare domain and the increasing integration of IoMT and other IoT devices, it is crucial to train both the security expert and the non-expert (i.e. medical) staff how to identify and react appropriately to cybersecurity incidents.

CONCLUSIONS

This chapter presented the TIM module of the SECANT platform including the provided functionalities and the use cases for the evaluation of TIM's performance. SECANT introduces a holistic approach which addresses the security issues concerning technical and human factors. TIM provides a variety of functionalities to the end users facilitating the collection and utilisation of CTI including (i) manual and automatic collection of information regarding threats and vulnerabilities from different sources, (ii) incorporation of new sources in an effortless manner, (iii) automatic extraction, correlation, as well as enrichment of the composed CTI and (iv) storing and sharing CTI in a secure and efficient manner.

Furthermore, the chapter stresses the advantages of utilising CTI in order to enhance the resilience of an organisation against cybersecurity attacks. In particular, CTI enables the organisation to be informed regarding the modern threat landscape and facilitate the implementation of appropriate security measures such as threat assessment as well as mitigation strategies. The IoCs that are included in the CTI can be leveraged to implement proactive measures (e.g. firewall rules, IDS rules). In conjunction with IoCs, CTI content could include the TTPs of the attacker, thus allowing the creation or improvement of security measures.

TIM module provides functionalities for the collection, enrichment and collection/acquisition, storage, enrichment and sharing of CTI. The quality of the generated CTI is significantly improved through the

---

[8] https://www.cisa.gov/uscert/ics/alerts

enrichment, resulting in more comprehensive information regarding the cyberthreat and increased actionability.

REFERENCES

1. Chantzios, T., Koloveas, P., Skiadopoulos, S., Kolokotronis, N., Tryfonopoulos, C., Bilali, V. G., & Kavallieros, D. (2019). The quest for the appropriate cyber-threat intelligence sharing platform. In *Proceedings of the 8th international conference on data science, technology and applications (DATA)*. SciTePress.
2. Alshamrani, A., Myneni, S., Chowdhary, A., & Huang, D. (2019). A survey on advanced persistent threats: Techniques, solutions, challenges, and research opportunities. *IEEE Communications Surveys & Tutorials, 21*(2), 1851–1877.
3. Kioskli, K., Fotis, T., & Mouratidis, H. (2021). The landscape of cybersecurity vulnerabilities and challenges in healthcare: Security standards and paradigm shift recommendations. In *ARES 2021: The 16th international conference on availability, reliability and security*. Association for Computing Machinery.
4. Aijaz, M., Nazir, M., & Anwar, M. N. (2021). Classification of security attacks in healthcare and associated cyber-harms. In *2021 First international conference on advances in computing and future communication technologies (ICACFCT)*. IEEE.
5. Bhuyan, S. S., Kabir, U. Y., Escareno, J. M., Ector, K., Palakodeti, S., Wyant, D., Kumar, S., Levy, M., Kedia, S., Dasgupta, D., & Dobalian, A. (2020). Transforming healthcare cybersecurity from reactive to proactive: Current status and future recommendations. *Journal of Medical Systems, 44*(5), 98.
6. Sills, M., Ranade, P., & Mittal, S. (2020). Cybersecurity threat intelligence augmentation and embedding improvement – A healthcare usecase. In *2020 IEEE international conference on intelligence and security informatics (ISI)*. IEEE.
7. Tao, H., Bhuiyan, Z. A., Rahman, A., Wang, G., Wang, T., Ahmed, M., & Li, J. (2019). Economic perspective analysis of protecting big data security and privacy. *Future Generation Computer Systems, 98*, 660–671.
8. Muthu, B., Sivaparthipan, C. B., Manogaran, G., Sundarasekar, R., Kadry, S., Shanthini, A., & Dasel, A. (2020). IOT based wearable sensor for diseases prediction and symptom analysis in healthcare sector. *Peer-to-Peer Networking and Applications, 13*, 2123–2134.
9. Kolokotronis, N., Dareioti, M., Shiaeles, S., & Bellini, E. (2022). An intelligent platform for threat assessment and cyber-attack mitigation in IoMT ecosystems. In *2022 IEEE globecom workshops (GC Wkshps)*. IEEE.
10. Mathew, A. (2020). Threat intelligence and internet of medical things (IoMT). *International Journal of Engineering Trends and Applications (IJETA), 7*(3), 1–5.

11. Djenna, A., Harous, S., & Saidouni, D. E. (2021). Internet of things meet internet of threats: New concern cyber security issues of critical cyber infrastructure. *Applied Sciences, 11*(10), 4580.

12. Georgiadou, A., Michalitsi-Psarrou, A., Gioulekas, F., Stamatiadis, E., Tzikas, A., Gounaris, K., Doukas, G., Ntanos, C., Ribeiro, L. L., & Askounis, D. (2021). Hospitals' cybersecurity culture during the COVID-19 crisis. *Healthcare, 9*(10), 1335.

13. Gao, J., Wang, H., & Shen, H. (2020). Smartly handling renewable energy instability in supporting a cloud datacenter. In *2020 IEEE international parallel and distributed processing symposium (IPDPS)*. IEEE.

Open Access   This chapter is licensed under the terms of the Creative Commons Attribution 4.0 International License (http://creativecommons.org/licenses/by/4.0/), which permits use, sharing, adaptation, distribution and reproduction in any medium or format, as long as you give appropriate credit to the original author(s) and the source, provide a link to the Creative Commons license and indicate if changes were made.

The images or other third party material in this chapter are included in the chapter's Creative Commons license, unless indicated otherwise in a credit line to the material. If material is not included in the chapter's Creative Commons license and your intended use is not permitted by statutory regulation or exceeds the permitted use, you will need to obtain permission directly from the copyright holder.

PART VIII

Strengthened Security Research and Innovation

CHAPTER 38

MultiRATE: EU R&D&I Readiness Level Evaluation Framework

Dimitrios Kavallieros, Katerina Valouma, Ilias Gkotsis,
Eleni Darra, Theodora Tsikrika, Dimitrios Diagourtas,
Stefanos Vrochidis, Antonis Kostaridis,
and Ioannis Kompatsiaris

Challenge

A plethora of different metrics focused on the evaluation of the maturity of products, systems, frameworks and procedures exist nowadays. Even though significant efforts have been made to integrate widely used frameworks, methodologies and indicators for measuring the maturity of products and systems, there has not been any framework or methodology that integrated multiple readiness level domains under a holistic framework

D. Kavallieros (✉) • E. Darra • T. Tsikrika • S. Vrochidis • I. Kompatsiaris
Center for Research and Technology Hellas-CERTH, Information and
Technologies Institute, Thessaloniki, Greece
e-mail: dim.kavallieros@iti.gr

K. Valouma • I. Gkotsis • D. Diagourtas • A. Kostaridis
Satways Ltd, Neo Irakleio, Greece

© The Author(s) 2025   475
I. Gkotsis et al. (eds.), *Paradigms on Technology Development
for Security Practitioners*, Security Informatics and Law
Enforcement, https://doi.org/10.1007/978-3-031-62083-6_38

and, most importantly, tailored to security solutions. As a result, the development of a robust scaling framework is of utmost importance to achieve the improved cross-disciplinary assessment of the maturity of novel solutions (developed for security practitioners).

Moreover, there is a need for more efficient use of maturity assessment frameworks to convey technology readiness, synchronize parallel projects, forecast deployment and support decision-making in security investment planning.

THE MultiRATE APPROACH: FOCUSING ON TECHNOLOGICAL, MANUFACTURING AND COMMERCIALIZATION ASPECTS

MultiRATE conducts research in multiple fields to create a solid ground for the development of the holistic, homogeneous and harmonized RL evaluation methodology. The fields of research regarding the RLs will be the following: technology (TRL), societal (SocRL), security (SecRL), legal, privacy and ethics (LPERL), integration (IRL), commercialization (CRL) and manufacturing (MRL). In each field, appropriate indicators will be identified per level (e.g. RL1–9) to accurately evaluate the solutions developed for the security domain. Where appropriate, the consortium will identify the indicators for both security practitioners and non-practitioners. Moreover, the project aims to evaluate solutions in all the aforementioned fields by developing a unique methodology and specific indicators for each field based on the different aspects and objectives. MultiRATE will integrate the above RLs under a holistic RL evaluation methodology and tool, which will be used by the EU R&D community to harmonize their approach during security solutions development.

MultiRATE solution is designed in an agile and robust manner, and will be tested in multiple domains (e.g. cybersecurity, border management, fight against crime and terrorism) and through multiple rounds of testing, maximizing its accuracy. In the following sections, three very important readiness level assessors—TRL, CRL and MRL—will be discussed. These can be used as a set for assessing the maturity of a technological solution along the innovation pathway.

Technology Readiness Level

NASA introduced the TRL in the 1970s and had seven levels, which were formally defined in the late 1980s focused on space missions and the relevant technological requirements and components. In the 1990s NASA altered this to a nine-level scale (the lowest level is TRL 1 while the highest is TRL 9), and in 2013 the International Organization for Standardization produced the ISO 16290:2013 standard [Space Systems—Definition of the Technology Readiness Levels (TRLs) and their criteria of assessment] [1]. The focus of the TRL was to evaluate the maturity of developments in space technology. The TRL was introduced in EU-funded projects in 2014 [2] after the term of each level was altered to fit the Horizon 2020 programme, but the overall meaning of each level remained the same. Nevertheless, no assessment guideline or criteria per level were developed, and thus, each R&D consortium was assessing the TRL of the technologies they were developing. MultiRATE developed a concrete methodology, initially focused on assessing the maturity of software solutions and components designed and developed under the horizon Europe Cluster 3: 'Civil security for society', based on a set of indicators per level, which can be categorized into four groups as presented below (Table 38.1).

Each level is compiled by indicators falling under one of the four categories. Furthermore, to be more simplistic and easier to use, it follows the progression of each indicator from one level to the next level (where relevant) using specific keywords either individually or in combination as depicted in Table 38.2.

Table 38.1 MultiRATE TRL indicators' categories

Category	Explanation
Technology preparation and requirements	This category includes indicators regarding requirements, specifications, architectural design and data flows
Reporting documents	This category includes indicators regarding supporting documentation such as installation guides and user guides
Operability and continuity	This category includes indicators regarding deployment, operation and maintenance matters
Evaluation and usability	This category includes indicators focused on end user testing and evaluation

Table 38.2   Keywords affecting the status of MultiRATE TRL indicators

Keyword	Status	Explanation
Identify/prepare	Open	The relevant information described in the indicator is somehow identified in a readable form and prepared under specific circumstances
Updated	Open	The identified information is updated
Accepted	Open	The information described under these indicators is tested and accepted by the appropriate people/bodies
Finalized	Closed*	The information described under these indicators is marked by appropriate people/bodies, as final for the project *Under special circumstances it could be open and go back to previous stages

Manufacturing Readiness Level

In 2003, the Government Accountability Office (GAO) recommended in GAO Report 03-476 [3] to establish cost, schedule and quality targets for product manufacturing early on in technology development in order to obtain process maturity. The report suggests that design and manufacturing knowledge should be obtained early in product development for a product to be successful. In response, the Joint Defense Manufacturing Technology Panel developed MRL definitions as well as manufacturing readiness assessments (MRAs). This MRL scale helps programme managers assess manufacturing risks, which will facilitate the identification of areas that require additional management attention or investment. Manufacturing readiness is as important to the successful development of a system as the readiness and capabilities of the system. Though MRLs were created from the manufacturing perspective to evaluate—the manufacturing readiness of a product and supplement existing TRLs, they, too, have limitations. A limitation of the MRLs is that the lower MRL levels can be difficult to correlate to corresponding TRL numbers due to the technology's immaturity. MultiRATE adjusts these existing methodologies and develops indicators, specifically for EU projects' security-related technologies, seeking to harmonize the levels of the MRL with the TRL and the rest of the RLs.

The proposed assessment approach comprises five phases, strategically designed to analyse current conditions and pinpoint manufacturing risks. The primary objective is to empower end users in formulating plans to mitigate or eliminate these risks, facilitating evidence-based decision-making. The development of the MultiRATE MRL framework aligns with the five manufacturing maturity phases, namely Material Solutions Analysis, Technology Maturation and Risk Reduction, Engineering and Manufacturing Development, Production and Deployment, and Operations and Support. These phases are intricately evaluated through the 10-level MultiRATE MRL scale, serving as a tool to gauge the manufacturing maturity of projects within the EU security domain and aid the identification of associated risks.

To enhance the efficacy of the MRL Criteria Matrix, a set of indicators has been established, thoroughly aligned with the criteria. These indicators (accompanied by a total of 68 questions) provide measurable parameters, enabling a more precise evaluation of manufacturing maturity of state-of-the-art technologies in the security field. A high-level overview of the selected eight indicator categories is presented in Table 38.3, aiding project consortia, in comprehending the impact of each one.

The utilization of MRLs proves to be an effective methodology for confirming manufacturing feasibility, establishing a predictable schedule, managing and potentially reducing costs, identifying and mitigating risks, enhancing product quality and optimizing manufacturing processes and supply chains. In summary, adopting MRLs empowers end users to track progress, identify areas for improvement and make informed, data-driven decisions to enhance their manufacturing capabilities.

Table 38.3 MultiRATE MRL indicators' categories

Categories of indicators	Descriptions
Manufacturing process	Process design, optimization and validation
Supply chain readiness	Assessment of the readiness and availability of necessary components, materials and suppliers
Production capacity	Assessment of production capacity to meet anticipated demand
Quality control and assurance	Evaluation of the quality control measures and assurance processes in place
Manufacturing equipment and machinery	Availability and readiness of manufacturing equipment and machinery
Cost and efficiency	Assessment of the cost-effectiveness of the manufacturing process
Training and workforce	Workforce skills and knowledge alignment with manufacturing requirements
Regulatory and compliance	Determine if the manufacturing process complies with relevant regulations and standards

Commercialization Readiness Level

The CRL framework assesses various indicators, which influence the commercial and market conditions beyond just the technology maturity. This enables key barriers to be addressed to support the commercialization of a technology. CRL refers to how ready a product or service is, to go to market as a commercial offering for a group of customers. CRL is a more recent concept, and definitions and operationalization of commercialization readiness are therefore far less widely accepted and solid than TRL. However, the focus on market readiness has become increasingly prevalent as, for example, expressed by Horizon2020s' sharpened focus on the market aspect of product development [4]. Thus, MultiRATE addresses this by developing an assessment methodology tailored to evaluate the commercial readiness of security-related solutions in European R&D through the identification of specific indicators and harmonization of the CRL with other project's RLs.

Such a comprehensive methodology assesses market access, financial capital, manufacturing possibilities and users' profits, emphasizing the commodification of technology. MultiRATE commercialization process, compliant with that of literature and online tools [5–7], follows the phases of Ideation and Research, Development and Assessment, and Marketing and Sales. Adapted to the security research domain, it is further analysed in nine steps, including criteria assessment and guiding the user from research idea to a viable market, reducing risk, facilitating evidence-based decisions and ensuring a holistic evaluation of the element's commercial potential.

The dedicated MultiRATE CRL scale offers a robust framework for evaluating security innovations for successful market deployment, recognizing the pivotal criteria influencing commercialization readiness. These criteria address technological innovation, market potential, IP protection, regulatory compliance, business model, scalability, consortium expertise and financial viability, collectively shaping the commercialization readiness level (Table 38.4). Stakeholders can leverage these criteria for valuable insights, aiding informed decisions about commercialization and funding prospects.

Independent of the steps followed during commercialization process, the aspects that have been identified to affect both the level but also the evolution of the process include the market potential and awareness, the regulations and certifications that should be complied with or acquired,

Table 38.4 MultiRATE CRL indicators' categories

Categories of indicators	Descriptions
Solution definition/ design/development	Assessment of the level that the product/service is well-defined, adapted to the end user needs
Market potential and competitive landscape	Evaluation of the solution's market size, demand and growth potential
Team/consortium expertise	Assessment of the expertise and background of the project consortium members
Intellectual property (IP)	Assessment of the extent of IP protection and the potential for commercial exploitation
Go-to-market	Evaluation of the business model and the scalability potential of the solution
Manufacturing/supply chain	Evaluation of the supply chain readiness, suppliers' capacity and end user engagement, but also of the quality of the processes

the business models and commercialization strategies developed, and the relationships established both with suppliers and customers.

A Technology, Manufacturing and Commercialization Assessment Case

Several stakeholders and use cases that could benefit from the MultiRATE approach have been identified. For example, when writing a proposal, consortium members should identify the RLs of the proposed solutions and the efforts needed to achieve the proposed RL. Another case would be that of project reviews, during which the EU agencies and reviewers (with the support of consortium members) could identify the RL of the developed solutions and compare them with the envisioned or previously assessed RLs.

In order to conduct the assessment of an element's readiness level, two main aspects should be considered, that of the nature of the element (software, hardware, methodology, etc.) and that of the environment in which it is supposed to operate (e.g. security domain, operational conditions). Based on the above, the most appropriate assessment criteria (as further described in the section above) are identified and a tailored assessment process is defined (i.e. planning, data collection, data validation, attribute rating and reporting).

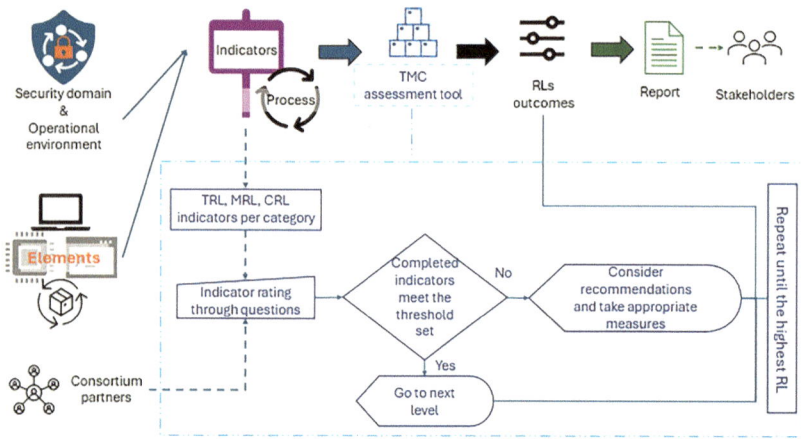

Fig. 38.1 T-M-C RL assessment methodological process

As depicted in Fig. 38.1, potential users such as consortium members may use the TRL, MRL and CRL by answering specific questions associated with each category of indicators, identifying the corresponding level of maturity for their product. As a very simplified example, if the following questions are true, then the respective indicator is considered complete, and the corresponding maturity level is achieved:

- Demo solution available for demonstration in an intended operational environment → Yes → TRL6
- Are manufacturing process and tooling mature? → Yes → MRL6
- Full and complete understanding of the competitive landscape, target applications, competitive products/services and market has been achieved → Yes → CRL5

A set of questions are provided per indicator category and level of the assessment tool. Considering the specific threshold of indicator needed to be completed, the user will be able to get a report on the levels achieved so far and plan for further activities and efforts that will leverage the maturity of the element assessed, from the technological, manufacturing and commercialization point of view. It must be underlined that achieving a maturity level in the TRL, for example, is not directly linked with a level in the MRL or CRL. Thus, a solution can be more mature in one of the three assessment aspects.

Conclusions

Research and innovation, including that for security and resilience of citizens and infrastructure, proved to be among the most powerful of European policies to boost the union's economies and competitiveness at the global scale. To support the exploitation and market uptake of security products and services, assessment of the readiness of such R&I projects' results is very important both by the projects themselves and the European Commission's services.

MultiRATE identified that EU research projects even in the same domain of security have distinct characteristics and differences in terms of market, stakeholders, financing and innovation development and uptake. These differences reflect the unique challenges and opportunities within each domain (e.g. disaster resilient societies, fight crime and terrorism) and highlight the need for tailored approaches to innovation development and uptake.

The main factors influencing the uptake of EU-funded security research outcomes include the following (Table 38.5).

As presented in this study, an evaluation framework of more than TRL, which when used alone has proven to provide limited or unclear outcomes (e.g. TRL-9 does not mean ready-for-the-market), is of utmost importance in order to have a holistic assessment of the maturity of the project's solution. Thus, when combined with the MRL, assessing the development process, and the CRL to track and support the pathway to go-to-market, shall provide to users and consortia a useful tool to enhance their solutions maturity, but also to the EC services a more detailed evaluation assessment.

Table 38.5 Hindering and enabling factors influencing the uptake of EU-funded security research outcomes

Hindering factors	Enabling factors
Market fragmentation	End user involvement
Quality of information flows	Partnerships and collaboration
Insufficient output maturity for uptake	Testing and demonstrations
Lack of foresight and evolving end user requirements	Funding and procurement mechanisms
Protection and clarity of IP rights	Communication and dissemination of information
Challenges associated with public acceptance	
Restrictions of an institutional market	

Acknowledgements This study was funded by the European Union (GA 101073929). The views and opinions expressed are, however, those of the author(s) only and do not necessarily reflect those of the European Union or REA. Neither the European Union nor the granting authority can be held responsible for them.

References

1. International Standard Organisation. (2013). *ISO 16290:2013 Space systems—Definition of the Technology Readiness Levels (TRLs) and their criteria of assessment.* International Standard Organisation.
2. Banke, J. *Technology readiness levels demystified*, 7 August 2017. [Online]. Available: https://www.nasa.gov/topics/aeronautics/features/trl_demystified.html. Accessed May 6, 2023.
3. U. S. G. A. Office. (2003). *Assessments of major weapon programs.* U.S. General Accounting Office.
4. Cyberwatching.eu Consortium. (2018). *D2.3 Methodology for the classification of projects/services and market readiness,* cyberwatching.eu
5. Paun, F. (2012). The demand readiness level scale as new proposed tool to hybridise market pull with technology push approaches in technology transfer practices. In *Technology transfer in a global economy, edition 127* (pp. 353–366). Springer.
6. U. o. York. *Commercialisation process,* [Online]. Available: https://www.york.ac.uk/staff/research/commercialising-research/commercialisation-process/. Accessed July 2023.
7. Dent, D., & Pettit, B. (2011). *Technology and market readiness levels.* Dent Associates Ltd.

Open Access This chapter is licensed under the terms of the Creative Commons Attribution 4.0 International License (http://creativecommons.org/licenses/by/4.0/), which permits use, sharing, adaptation, distribution and reproduction in any medium or format, as long as you give appropriate credit to the original author(s) and the source, provide a link to the Creative Commons license and indicate if changes were made.

The images or other third party material in this chapter are included in the chapter's Creative Commons license, unless indicated otherwise in a credit line to the material. If material is not included in the chapter's Creative Commons license and your intended use is not permitted by statutory regulation or exceeds the permitted use, you will need to obtain permission directly from the copyright holder.

CHAPTER 39

AI Applied to Intelligence and Security: Technologies and EC-Funded Initiatives

Domenico Frascà, Alessandro Zanasi, Maria Ustenko, Paola Fratantoni, and Giulia Venturi

INTRODUCTION

This chapter is focused on providing up-to-date information about the recent technological developments in artificial intelligence (AI), as well as insights on the relevance of these technologies for the intelligence communities (IC) and law enforcement agencies (LEAs). This work occurs in a delicate and crucial phase, such as the final approval and the entry into force of the European Union (EU) AI Act, the first EU regulation on AI and AI-based technologies. The global interest of scientists, professionals, organizations, and policymakers around AI technologies is rising continuously as AI adoption is getting ubiquitous, meaning it is being applied across a multitude of domains. AI is a generic term for multiple technologies and is an umbrella concept that embraces multiple technologies in the field of data analytics, while there is a subset of them, namely machine learning (ML) and deep learning (DL) techniques, which are intensively

D. Frascà • A. Zanasi • M. Ustenko (✉) • P. Fratantoni • G. Venturi
Zanasi & Partners, Modena, Italy
e-mail: maria.ustenko@zanasi-alessandro.eu

© The Author(s) 2025   485
I. Gkotsis et al. (eds.), *Paradigms on Technology Development for Security Practitioners*, Security Informatics and Law Enforcement, https://doi.org/10.1007/978-3-031-62083-6_39

gaining attention and being widely exploited. AI-based systems and applications used for improving and automation of crime investigations and solving may demand high computation and storage capabilities in order to ensure efficient and rapid management of the enormous amounts of data that intervene in the model training and testing activities. This is extremely relevant where data is entered in the AI system operation as a continuous flow of streaming data that needs to be analyzed. Therefore, big data and cloud technologies used to create information systems, databases, and diverse-associated services on top are in close relationships of AI adoption.

The focus of this chapter is not placed on the detailed technological explanation of each of the AI techniques, but rather in the description of the opportunities and usage scenarios of AI to facilitate and enhance security intelligence tasks. This work aims to be a source of inspiration and a reference for intelligence and security practitioners for further discussion on the potential opportunities and disadvantages of adopting AI in their daily activities.

The study descends from the research carried out in the first year of the NOTIONES project, which aims at providing intelligence and security practitioners with up-to-date information on technologies and research initiatives—specifically EC-funded projects—so that they can elaborate requirements and needs to be followed up by academic researchers and industrial technology providers. The technologies covered by the study include any kind of machine-based method, technique, mechanism, or system (equipment, platform, software, application) that constitutes, uses, or optimizes models to enable predictions, recommendations, classifications, and other tasks necessary in the security intelligence activities. Such kinds of AI technologies are key for the efficiency and accuracy of the intelligence tasks, and thus, the adoption of AI solutions is of paramount importance in the intelligence cycle [1].

The chapter is structured as follows. The first section presents the landscape of EU research projects on AI for civil security. The second section describes some relevant examples of AI-based solutions for intelligence and security practitioners as well as relative datasets. The third section presents the main challenges posed by the adoption of AI solutions. Finally, several recommendations on AI-based systems and applications used for security intelligence are provided.

EC-FUNDED RESEARCH PROJECTS ON AI FOR CIVIL SECURITY

In this section, the main initiatives of EU-funded research projects are reported, which support civil security and especially security and intelligence practitioners as well as LEAs. The list was obtained through desk research on the Community Research and Development Information Service (CORDIS). These initiatives include European research projects funded under the Horizon 2020 funding program, particularly under the secure societies framework [2]. Many current projects are dealing with explainability, trustworthiness, and ethical challenges of AI. Most of them have developed or are developing AI solutions oriented to law enforcement and/or involve direct collaboration of LEAs in these technologies. Below are some of the most relevant EU-funded research projects supporting LEA activities with regard to the use of AI technologies and its challenges for its adoption in the intelligence scenarios and use cases.

(a) *Artificial Intelligence Data Analysis (AIDA).* Closed in 2022, the AIDA project aimed to develop a framework for analyzing large volumes of data using AI technologies to improve the capabilities of LEAs to fight cybercrime and cyberterrorism [3]. Specifically, AIDA has developed a big data analysis and analytics framework equipped with a complete set of effective, efficient, and automated data mining and analytics solutions to deal with standardized investigative workflows, extensive content acquisition, information extraction and fusion, knowledge management, and enrichment through novel applications of big data processing, ML, AI, and predictive and visual analytics.

(b) *Deep AR Law Enforcement Ecosystem (DARLENE).* The DARLENE project aims to offer European LEAs a proactive security solution that will enable them to sort through massive volumes of data to predict, anticipate, and prevent criminal activities [4]. To achieve this, it aims to combine augmented reality (AR) and AI techniques in order to improve LEAs' decision-making and daily operations with regard to forensics and situational awareness.

(c) *Investigative, Immersive, and Interactive Collaboration Environment (INFINITY).* The primary goals of INFINITY are to revolutionize data-driven investigations through the use of AI, ML, and big data analytics to facilitate effectiveness of an investigation and utilize modern innovations in virtual reality (VR), AR, and visual analytics in order to facilitate a better intelligence cycle [5]. The INFINITY project will try to overcome the challenging task of dealing with enormous amounts of data in crime investigations of cybercrime, terrorism, and other hybrid threats. The project will build a collaborative platform between different LEAs relying on VR, AR, ML, and big data technologies. The INFINITY system for LEAs' operations addresses the whole intelligence cycle, including the generation of the required reporting and management of evidence admissible in court.

(d) *An Interoperable Multidomain CBRN System (NEST).* The NEST project will develop systems to provide threat indications and warnings, as well as guidance for facility security through appropriate information-sharing and analysis mechanisms [6]. Specifically, the aim of NEST is the creation of an Internet of Things (IoT) network of low-cost chemical, biological, radiological, and nuclear (CBRN) sensors in different physical infrastructures, and to leverage AI for the detection of CBRN threats and pandemic viruses. All threats and dangers will be displayed with the help of AR. The CBRN detectors would be low-cost sensors embedded in one unique detection equipment placed in the infrastructure or carried by security staff. The detectors will send data to the IoT platform that processes and merges data from internal and third-party services. Besides the use of AR to display hazards and CBRN threats, the NEST solution adopts AI for the generation of threat alerts and decision-making in security of facilities.

(e) *Artificial Intelligence Roadmap for Policing and Law Enforcement (ALIGNER).* The objective of the project is to bring together the main European actors in the field of AI applied to law enforcement and policing services [7]. The project will organize a series of workshops that gather different stakeholders with different points of view in order to focus, prioritize, and establish a roadmap of the most beneficial actions to cooperate on and research areas in the field of AI applied to law enforcement. Furthermore, the project will conduct a study on the policy and research needs as well as the AI capability needs of European LEAs. The project will also aid in the prevention of offensive AI by delivering a taxonomy of AI-powered crime. Finally, ALIGNER will assess and monitor AI technologies with potential for use by LEAs, together with their security, ethical, societal, and legal risks evaluation.

(f) *SusTainable Autonomy and Resilience for LEAs using AI against High-priority Threats (STARLIGHT)*. The STARLIGHT project aims to create a community that brings together LEAs, researchers, industry, and practitioners in the security ecosystem under a coordinated and strategic effort to bring AI into operational practices [8]. It is focused on improving the capacities and autonomy of LEAs in the use of AI tools. Two other objectives of the project are to protect own LEA's AI systems against attacks on them and improve LEAs' capacities to defend against attacks that use AI to be more effective; that is, against AI-powered crime and terrorist acts.

(g) *A European Positive Sum Approach Toward AI Tools in Support of Law Enforcement and Safeguarding Privacy and Fundamental Rights (popAI)*. popAI aims to address the concerns related to the use of AI-based technologies in the security domain [9]. To achieve this goal, popAI brings together security practitioners, AI scientists, ethics and privacy researchers, civil society organizations, as well as social sciences and humanities experts, aiming to boost trust in AI by increasing awareness and current social engagement, and delivering a unified European view and recommendations. The project also envisages the creation of an ecosystem and the structural basis for a sustainable and inclusive European AI hub for LEA.

It is worth noting that ALIGNER, STARLIGHT, and popAI are cluster projects that have been funded under the Horizon 2020 Artificial Intelligence calls (2020) and have approached the topic of AI-based technologies in the security domain by multiple perspectives (e.g., legal, ethical, and technological), facilitating knowledge sharing and nourishing joint activities.

AI TECHNOLOGIES AND TOOLS FOR INTELLIGENCE AND SECURITY

This section reports on the relevant examples of AI-based software tools and techniques that appear promising to support data processing and analysis activities performed by the intelligence and security practitioners. The technologies were selected based on a desk research on the open web and selected databanks such as *TheLens* [10], an integrated search engine on scholarly works and patents with an export functionality. The research was focused on technologies and tools that have been developed or are being developed to serve particular intelligence needs, including both commercial and open-source solutions. Furthermore, some relevant datasets were identified, which can be used for the training of the AI models. In this chapter, we report on the selected case study scenarios and their relative datasets, namely biometric recognition and video surveillance-based crime detection, violence detection, illegal trafficking detection, and crime prevention.

(a) *Biometric Recognition and Video Surveillance-Based Crime Detection.* Some AI-based commercial tools for face recognition, fingerprint identification, and surveillance are (1) "*Amazon Rekognition*," which can do facial analysis and facial search, and identify objects and scenes. It also offers face detection and analysis in videos, live, or stored [11]. (2) "*BioID*," which is a cloud-based online service providing biometric technology and includes an application programming interface (API) [12]. (3) "*Biometric Identification Services*" [13]. (4) "*Defendry*," which can autodetect hundreds of different kinds of guns and weapons, and monitor for intruders such as banned former employees, expelled students, and more, using the existing security cameras and/or Defendry EyesOn Cameras [14]. Some related AI-based datasets are "*FFHQ*," which consists of around 70,000 high-quality PNG images of human faces [15]; "*Labelled Faces*," which provides face photographs designed for studying the problem of unconstrained face recognition [16]; "*Google Facial*," which contains face image triplets along with human annotations that specify which two faces in each triplet form the most similar pair in terms of facial expression [17]; "*YouTube Faces DB*," which provides face videos designed for studying the problem of unconstrained face recognition in videos [18]; "*FVC2000_DB4_B*," with several hundred reference fingerprints of varying quality [19]; and "*SOCOFing*," which is a biometric fingerprint database designed for academic research purposes [20].

(b) *Violence Detection.* An interesting AI-based commercial tool for violence detection is "*Jarvis*," which is a customizable Video

Analytics Engine with state-of-the-art facial recognition technology and intelligent monitoring of objects, crowd (focused on violence detection), perimeters, and vehicles [21]. Examples of related datasets are *"RWF-2000,"* which is a large-scale video database for violence detection [22]; *"airtlab,"* which contains 350 video clips labeled "nonviolent" and "violent," to be used to train and test algorithms for violence detection in videos [23]; and *"XD-Violence,"* which contains more than 4700 untrimmed videos with audio signals and weak labels [24].

(c) *Illegal Trafficking Detection.* An AI-based tool funded by the European Union for firearms identification is *"iARMS"* (Illicit Arms Records and Tracing Management System). Police worldwide can record illicit firearms in the iARMS database and can search for seized persons to check if they have been reported as lost, stolen, trafficked, or smuggled [25]. Some of the relevant datasets for illegal trafficking detection are *"INTERPOL Open Databases,"* which is the only database at the international level with certified police information on stolen and missing objects of art [26]; and *"Global Human Trafficking,"* which contains information on almost 50,000 victims of human trafficking, including the reason, means of control, origin and destination, as well as other variables [27].

(d) *Crime Prevention.* Finally, an example of tool in crime prevention is *"PRECOBS"*: it generates forecasts using the most up-to-date crime data, which can be used by police authorities for operational and preventive purposes. Control centers and operational units receive temporal and spatial indications for situation-oriented operational planning [28]. Examples of datasets for crime prevention are *"FBI Crimes,"* which provides crime and policing analysis within the United States [29]; and *"London Crime,"* which covers the number of criminal reports that occurred in London by month, area, and major/minor category from 2008 to 2016 [30]. It is noteworthy that some of the tools presented, although categorized under one of the use cases, may serve more than one case. For example, tools used for face recognition may aid in different policy tasks such as identification of violent criminals, identity fraud crimes detection, or criminal pedophiles identification.

TRUSTWORTHINESS CHALLENGES

As in other fields of application, ensuring trustworthiness of the AI technologies used for law enforcement is one of the major challenges. The high sophistication of this type of technologies makes them prone to uncertainties of whether they may turn out against the purpose they were created for, that is, aid humans. Therefore, intense research is being carried out to ensure AI compliance with the proposed EU's AI Act [31], which calls for trustworthy AI solutions for Europe on the basis of the seven main requirements of trustworthy AI in the EU's HLEG on AI [32]. From a technical perspective, a trustworthy AI shall ensure three main characteristics. (a) Explainability of AI. Having means to guarantee the transparency and interpretability of the AI algorithm or model result is key to understand whether the system is getting against or twisted somehow from its originally designed objective. (b) Fairness of AI. Ensuring that the model or algorithm does not fall into bias or discrimination or stigmatization of humans. (c) Technical robustness of AI. The technical robustness includes reliability as well as security and safety of AI, that is, the fact that the system securely treats the data in storage, transit, and operation, and keeps the personal data private. The system shall also protect humans from any intentional harm. Although it is extremely challenging due to AI still being an emerging technology, growing faster in capabilities and implementation platforms, law enforcement is required to take steps to ensure that all these aspects are respected in the AI systems they adopt to facilitate their work.

CONCLUSION

In recent years, AI technologies have become extremely popular thanks to the horizontality of the applicability of AI in multiple domains. The wide variety of research projects that were funded in recent years by the European Commission about AI solutions for security, and the growing number of commercial solutions available to practitioners, demonstrate that the advances in ML, DL, and all sorts of algorithms for vast amounts of data processing are transforming intelligence, security, cybersecurity, law enforcement, and operations. Indeed, AI technologies may be adopted in almost all activities within the intelligence life cycle that require classification, identification, problem solving, decision-making, and prediction among others.

Despite all the benefits that AI can bring to intelligence and security practices, it is necessary to study and continuously monitor the potential downsides of the use of AI. The civil society needs to make sure that security practitioners do not trespass the limits of fundamental rights when using AI-based systems in intelligence gathering and processing, crime prevention, crime detection, case investigations, and criminal prosecution. In particular, it is necessary that practitioners closely follow trustworthy AI practices and recommendations, as well as ensure the acquisition and use of tested AI systems, trained with adequate unbiased data. As the AI standardization progresses and the research on more secure and resilient AI systems advances, their adoption with all guarantees will become easier.

Acknowledgments This work was supported by the NOTIONES (*iNteracting netwOrk of inTelligence and securIty practitiOners with iNdustry and acadEmia actorS*) project that has received funding from the Horizon 2020 Framework Programme of the European Union for Research and Innovation under grant agreement no. 101021853.

This work was also supported by the popAI (*A European Positive Sum Approach towards AI tools in support of Law Enforcement and safeguarding privacy and fundamental rights*) project that has received funding from the Horizon 2020 Framework Programme of the European Union for Research and Innovation under grant agreement no. 101022001.

REFERENCES

1. Ish, D., Ettinger, J., & Ferris, C. (2021). *Evaluating the effectiveness of artificial intelligence systems in intelligence analysis.* Retrieved July 18, 2023, from https://www.rand.org/pubs/research_reports/RRA464-1.html
2. European Commission. (2020, March 25). *Horizon 2020.* Retrieved July 24, 2023, from European Commission Decision C(2020)1862: https://ec.europa.eu/research/participants/data/ref/h2020/wp/2018-2020/main/h2020-wp1820-security_en.pdf
3. AIDA. (2018). *Artificial intelligence andd data analysis.* Retrieved July 24, 2023, from https://www.project-aida.eu/
4. DARLENE. (2023). *Deep AR law enforcement ecosystem.* Retrieved July 24, 2023, from https://www.darleneproject.eu/
5. INFINITY. (2023). *Investigative, immersive, and interactive collaboration environment.* Retrieved July 24, 2023, from https://h2020-infinity.eu/
6. NEST. (2023). *An interoperable multidomain CBRN system.* Retrieved July 24, 2023, from https://nest-h2020.eu/

7. ALIGNER. (2021). *Artificial intelligence roadmap for policing and law enforcement*. Retrieved July 24, 2023, from https://aligner-h2020.eu/
8. STARLIGHT. (2023). *SusTainable Autonomy and Resilience for LEAs using AI against High priority Threats*. Retrieved July 24, 2023, from https://www.starlight-h2020.eu/
9. PopAI. (2021). *A European Positive Sum Approach towards AI tools in support of Law Enforcement and safeguarding privacy and fundamental rights*. https://www.pop-ai.eu/
10. TheLens. (n.d.). Retrieved from https://www.lens.org/
11. Amazon. (2018). *Amazon Rekognition*. Retrieved July 18, 2023, from https://aws.amazon.com/rekognition/
12. BioID Web Service. (2020). Retrieved July 18, 2023, from https://www.bioid.com/bioid-web-service/
13. BIS. (2020). Retrieved July 18, 2023, from http://www.bioidentserv.com/web/
14. Defendry. (2022). Retrieved July 18, 2023, from https://defendry.com/
15. Github. (2023). *FFHQ*. Retrieved July 24, 2023, from https://github.com/NVlabs/ffhq-dataset
16. Kaggle. (2023). *Labelled Faces*. Retrieved July 24, 2023, from https://www.kaggle.com/datasets/ciplab/real-and-fake-face-detection
17. Google. (2023). *Google Facial*. Retrieved July 24, 2023, from https://research.google/resources/datasets/google-facial-expression/
18. YouTube. (2023). *YouTube Faces DB*. Retrieved July 24, 2023, from https://www.cs.tau.ac.il/~wolf/ytfaces/
19. Kaggle. (2023). *Fingerprint dataset*. Retrieved July 24, 2023, from https://www.kaggle.com/datasets/peace1019/fingerprint-dataset-for-fvc2000-db4-b
20. Kaggle. (2023). *Sokoto coventry fingerprint dataset*. Retrieved July 24, 2023, from https://www.kaggle.com/datasets/ruizgara/socofing
21. Jarvis. (2022). Retrieved July 18, 2023, from https://www.staqu.com/
22. Github. (2023). *RWF-2000*. Retrieved July 24, 2023, from https://github.com/mchengny/RWF2000-Video-Database-for-Violence-Detection
23. Github. (2023). *airtlab*. Retrieved July 24, 2023, from https://github.com/airtlab/A-Dataset-for-Automatic-Violence-Detection-in-Videos
24. Github. (2023). *XD-Violence*. Retrieved July 24, 2023, from https://roc-ng.github.io/XD-Violence/
25. INTERPOL. (2020). Retrieved July 18, 2023, from https://www.interpol.int/How-we-work/Databases/Illicit-Arms-Records-and-tracing-Management-System-iARMS
26. INTERPOL. (2023). *The stolen works of art database*. Retrieved July 24, 2023, from https://www.interpol.int/en/Crimes/Cultural-heritage-crime/Stolen-Works-of-Art-Database

27. Kaggle. (2023). *Global Human Trafficking.* Retrieved July 24, 2023, from https://www.kaggle.com/datasets/andrewmvd/global-human-trafficking
28. LogObject. (2022). Retrieved July 18, 2023, from https://logobject.com/en/solutions/precobs-predictive-policing/
29. Dolthub. (2023). *FBI Crimes.* Retrieved July 24, 2023, from https://www.dolthub.com/repositories/Liquidata/fbi-nibrs
30. Kaggle. (2016). *London Crime.* Retrieved July 24, 2023, from https://www.kaggle.com/datasets/jboysen/london-crime
31. European Commission. (2021, April 21). *Artificial Intelligence Act.* Retrieved July 18, 2023, from https://eur-lex.europa.eu/legal-content/EN/TXT/?uri=CELEX:52021PC0206
32. European Commission. (2019). *Ethics guidelines for trustworthy AI.* EC. Retrieved July 18, 2023, from https://ec.europa.eu/newsroom/dae/document.cfm?doc_id=60419

Open Access This chapter is licensed under the terms of the Creative Commons Attribution 4.0 International License (http://creativecommons.org/licenses/by/4.0/), which permits use, sharing, adaptation, distribution and reproduction in any medium or format, as long as you give appropriate credit to the original author(s) and the source, provide a link to the Creative Commons license and indicate if changes were made.

The images or other third party material in this chapter are included in the chapter's Creative Commons license, unless indicated otherwise in a credit line to the material. If material is not included in the chapter's Creative Commons license and your intended use is not permitted by statutory regulation or exceeds the permitted use, you will need to obtain permission directly from the copyright holder.

CHAPTER 40

Is Public Procurement a Hindering Factor in the Uptake of Innovation?

Jozef Kubinec

IPROCURENET AT THE HEART OF SECURITY INNOVATION

End users in the security sector often perceive public procurement as hindering innovation uptake. The iProcureNet project conducted an internal survey among the consortium on the main obstacles to innovation uptake. Several answers indicated that end users observe the procurement as an obstacle to obtaining innovative solutions for their needs.

To battle with the opinion that public procurement is a hindering factor of innovation uptake, the iProcureNet project envisaged creating an ecosystem of procurers, prescribers, legal advisors and other key stakeholders of security procurement to share and analyse procurement trends and needs, developing common and standardised practices from the technical, legal and financial perspectives, and open pathways for joint cross-border public procurement (JCBPP).

One of the activity lines is establishing pathways for JCBPP of innovative and new-to-market solutions, research services and commercial

J. Kubinec (✉)
Ministry of Interior of Slovak Republic, iProcureNet Project, Bratislava, Slovakia
e-mail: Jozef.Kubinec@minv.sk

© The Author(s) 2025
I. Gkotsis et al. (eds.), *Paradigms on Technology Development for Security Practitioners*, Security Informatics and Law Enforcement, https://doi.org/10.1007/978-3-031-62083-6_40

off-the-shelf (COTS) products in the field of internal security, that is, JCBPP of products or services acquired by the internal security sector such as border guard to perform their tasks.

Furthermore, iProcureNet dedicated its activities to promoting innovation in security procurement and supporting European procurers in JCBPP and innovation procurement. These actions should convince policymakers and end users that it is essential to turn the narrative into a more innovative procurement in the security sector, ultimately promoting the idea that procurement can catalyse innovation.

PROCUREMENT INSTRUMENTS TO PROMOTE INNOVATION

There are already well-established mechanisms to promote innovation in the case of solutions that are not yet on the market or not commercially available. These are pre-commercial procurement (PCP) and public procurement of innovative solutions (PPI). Except these, the chapter will identify other public procurement instruments that can promote innovation even in the case of procurement of COTS, such as preliminary market consultation, value engineering and functional specifications.

JOINT CROSS-BORDER PUBLIC PROCUREMENT

Joint cross-border public procurement (JCBPP) is regulated in art. 39 of Directive 2014/24/EU and art. 57 of the Directive 2014/25/EU. According to both directives, the JCBPP can take three different forms: (a) JCBPP by using a central purchasing body located in another member state; (b) JCBPP where several contracting authorities from different member states procure jointly the same subject of the tender based on the procurement done by lead procurer and (c) JCBPP where several contracting authorities from different member states procure through a joint entity which they set up for this purpose.

Joint cross-border public procurement (JCBPP) is an innovative way of procurement. A promising mechanism of efficient purchasing as well as a strategic tool for the positive use of purchasing power on the market, it enables the sharing of costs, securing economies of scale and developing innovation. [1]

Promoting competition, providing incentives for companies to enter new markets, as well as development, standardisation and sharing of good procurement practices are among other benefits of JCBPP [2].

On the other hand, JCBPP is a complex public procurement process with many issues that the contracting authorities should be aware of before deciding to participate. According to the European Commission team leader on innovation procurement, Ivo Locatelli, 'the challenges to be faced by contracting authorities can be legal, cultural, linked to the coordination effort required, using a foreign language in the procedure' [3].

Without a doubt, the benefits of JCBPP prevail against the perils that inevitably accompany it. Even though there are still few examples of JCBPP, it is 'gaining unexpected interest from a range of stakeholders: large cities, cross-border projects involving administrations, near borders, projects aiming at using public procurement to develop innovative products or services, inherently cross-border applications such as satellite services' [3].

The respondents to the online survey conducted by the iProcureNet project have chosen the following main benefits of JCBPP (Fig. 40.1).

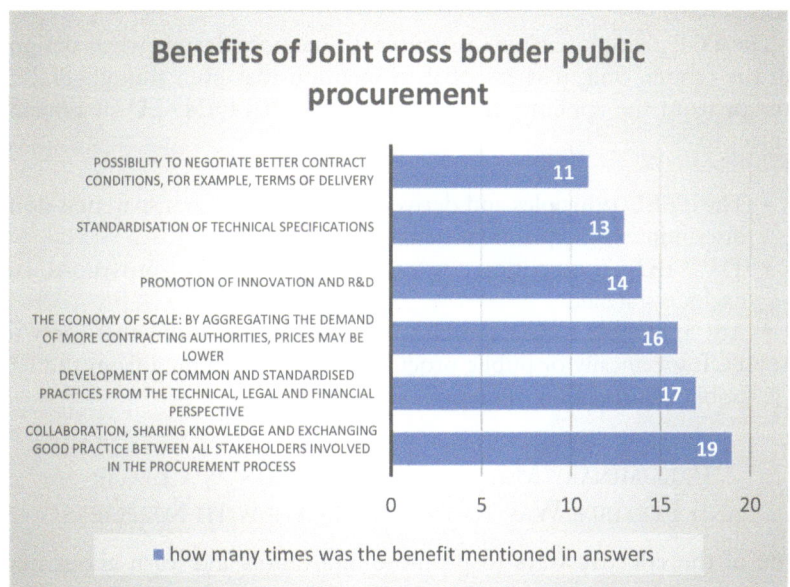

Fig. 40.1 Main benefits of JCBPP according to respondents to the iProcureNet survey

Pre-commercial Procurement and Public Procurement of Innovative Solutions

PCP can be described as a specific approach to procuring R&D services that involve competitive development in phases, risk–benefit sharing under market conditions and where there is a clear separation between the PCP and the deployment of commercial volumes of end products (potential follow-up PPI) [4].

On the other hand, public procurement of innovative solutions (PPIs) means procurement where contracting authorities act as a launch customer of goods or services that are not yet available on a large scale [4]. PPIs are part of public procurement tenders because they are done according to the rules of directives on public procurement.

The distinction is important because, in the case of PCP, more precisely, procurement of R&D, one is not obliged to follow the rules stated in Directive 2014/24/EU. However, in the case of PPI, the procurer has to follow the public procurement rules stated in Directive 2014/24/EU, specifically the rules of public procurement of the applicable legal regime (jurisdiction).

The PCP procedure allows procurers to be more flexible when designing the tender. Still, it is essential to mention that even though PCP is exempt from the application of the Directive 2014/24/EU, it remains subject to

- The TFEU principles and derived principles (e.g. transparency, non-discrimination, equal treatment)
- The EU competition rules include specific provisions to exclude state aid
- Any applicable national public procurement provisions that apply to PCP specifically or public procurements are exempted from the EU public procurement directives

Preliminary Market Consultation as a Legal and Flexible Way to Communicate with Suppliers

One of the effective ways to promote innovation and learn about new innovative solutions is to conduct open-market consultation or, as stated in article 40 of the Directive 2014/24/EU, preliminary market consultation (PMC). According to Voda and Jobse, specifically in innovation procurement, market consultation plays a crucial role because the innovation cycle is usually longer than the procurement cycle [5].

PMC can be described as a formalised dialogue between the contracting authority and other entities (economic operators, suppliers or independent experts), aiming to obtain answers to how the contracting authority's problems can be solved.

Another possible definition is that 'the concept of preliminary market consultation roughly encompasses a multi-faceted query whereby a contracting authority asks experts and market operators to offer their contributions to make up the object of the contract and to define the other feature of the procedure' [6].

You can conduct PMC anytime during the pre-tender phase of public procurement. In JCBPP, it is advised to do PMC after the collaboration agreement is drafted so that the collaborating parties know their roles in procurement. Also, conducting PMC before writing the final technical description is recommended as the conclusion from PMC will probably affect writing the technical description.

PMC is a cornerstone of innovation procurement, providing advance notice to the market of opportunities and the unmet needs of their customers, allowing both time and valuable insights to suppliers to direct their business plans. On the other side, buyers will understand the market's appetite, capacity and capability to meet their needs and the time frames involved. Experience from prior market engagement and consultations conducted is that it is welcomed by suppliers who find the process and access to customers beneficial.

The Reasons for Organising PMC

The reasons for organising PMC are numerous, and we are stating the ones we consider most crucial from the point of view of public buyers.

- The contracting authority does not know how it would be possible to ensure the realisation of its unmet need; for example, how to increase the capacities of the police force when analysing video streams from CCTV cameras
- Identification of the suitable suppliers on the market

- Obtaining information about existing technical solutions, their prices and conditions of performance
- Priming the market for new needs and demands of the buyer
- Promoting innovation in the market by giving innovative suppliers the possibility to present public buyers with their innovative solutions
- Verification of non-discriminatory qualification criteria and conditions for the performance of the contract
- Minimisation of imminent risks in the implementation and operation of the subject of tender
- Minimising the risk of review procedures within public procurement

It is essential to mention that suppliers can also benefit if they participate in the PMC organised by public buyers. To persuade the suppliers about the benefits of PMC must be of utmost importance to the public buyers to secure their presence on PMC; this is especially true due to the cost of suppliers they may have with participating in the PMC.

Suppliers can benefit from PMC in the following ways:

- A better understanding of the situation and problems of the public buyer and a better understanding of their need for innovation
- Suppliers have a legal way of 'influencing' the preparation of the tender
- Have an opportunity to present innovations, ideas and thoughts
- Can draw attention to discriminatory conditions without the need to use the review procedure
- Can decide faster whether the tender is interesting for the supplier and whether it is important to further deal with the prepared public tender

Within the iProcureNet project, we asked the respondents for their views on the benefits of conducting PMC. The answers are shown in Fig. 40.2.

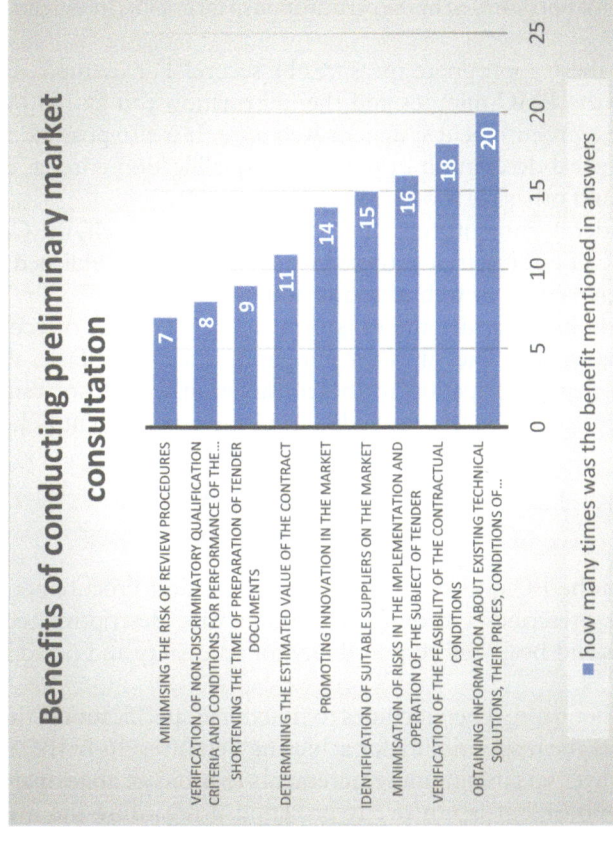

Fig. 40.2   Benefits of the PMC according to the respondents of the iProcureNet survey

Appropriate Measures in the Context of PMC

Article 41 of Directive 2014/24/EU states the rules concerning the prior involvement of candidates or tenderers specifying that

> the contracting authority shall take appropriate measures to ensure that competition is not distorted by the participation of that candidate or tenderer.

How can these appropriate measures be secure? For example, the CA publishes all the PMC minutes and the information you gather on your profile in the e-procurement system or web page. It is also possible to post the final technical description of the tender, qualification criteria, evaluation criteria and other parts of the tender document.

In the tender documents related to the public tender itself, the CA shall mention that all information gathered from the PMC is published in the e-procurement system or web page and add its link.

The CA should consider the risks of unequal treatment of economic operators and provide measures for mitigating such risks. First, the CA must ensure that participation in the consultation would not cause the economic operator to be later excluded from the procurement [7].

Functional Specification Versus Descriptive Requirements

According to the EC notice 'Guidance on Innovation Procurement', the public buyer prescribes the detailed solution with descriptive technical specifications and bears full responsibility for its quality and performance levels [8].

On the other hand, when it comes to functional specifications, the public buyer shifts the responsibility for achieving better results in the market. The public buyer sets minimum requirements to avoid an abnormally low-performing tender but is not overly prescriptive regarding the means of achieving a desired outcome. Economic operators enjoy openness and flexibility to reach the optimal performance [8].

Functional specifications should be preferred over technical specifications because they focus on long-term needs. It was mentioned in the online survey conducted by iProcureNet as a suggestion of good practice when referring specifically to PCP. Still, it can also be applied to public procurement tenders if the contracting authority wants to promote innovation. According to the EC Guidance Notice on Innovation Procurement, 'functional requirements are far more innovation-friendly'. This approach of using functional specifications was followed by 80% of the five EC-funded innovation procurement projects in the security sector mentioned in the assessment report on the performance of EC-funded innovation procurement projects in the security sector [8].

Does this mean that using descriptive requirements is unsuitable for attracting innovation? No. According to the mentioned assessment report, even descriptive requirements may attract innovative solutions, but 'this becomes easier to be achieved through the use of functional requirements that focus only on the description of the need, leaving the market free to come up, through competition, with solutions fit for the challenge in question' [8].

However, formulating sound functional requirements is not an easy task to do and presents a challenge. This challenge can be overcome by 'good knowledge of the market potential and most suitable technologies. Such knowledge is crucial for setting ambitious but realistic requirements and can be collected through a preliminary market consultation' [9].

The topic of the functional specification was also discussed on the second day of the 2021 Annual Conference of iProcureNet. Here are some of the most interesting conclusions:

- To be able to write functional specifications, you need special people who have strong technical knowledge and an excellent knowledge of the operational environment. Functional requirements are something in between. They explain what we need to buy without entering into technical details. They have to be understood by both end users and suppliers.
- The procuring organisation should establish a collaborative ecosystem to provide access to the right people knowledgeable about innovation and know what innovators are working on today. Short-term politics will not work. It must be a long-term collaboration, a trusted environment where you can establish dialogue and access information.
- The practice of using functional specifications varies from state to state. Using variant proposals in public tenders can be a perfect way to boost innovation. In the years before the 2004 EU Directive and before the 2008 crisis, variants were used a lot in Portugal. In Slovakia, we can see less use of functional specifications or variants because of the risks involved.

Promoting Innovation by Value Engineering

The iProcureNet project discussed on many occasions how we can promote innovation, even in the case of COTS products. One of the answers to this question is using value engineering. Value engineering consists of 'activities and actions that can be used during contract implementation to improve or preserve the functions of the innovative solution while reducing the costs'. Value engineering clauses typically incentivise vendors to continue improving their solution's quality/cost ratio by awarding part of the additional cost savings/quality improvements achieved after the contract signature to the vendor [10].

CAs should, when drafting the tender documents and the contractual clauses, consider using the value engineering clauses to promote innovation. In the procurement of COTS, the use of value engineering would be, for example, advisable in contracts where maintenance is a significant part of the value of the contract. For instance, in the case of tender for higher-value drones, the maintenance cost should be considered when evaluating supplier offers. Suppose the value engineering clauses are used in the tender for drones when the supplier presents how to make maintenance more practical and cheaper during the contract. In that case, the savings should be split between the contracting authority and the supplier.

The CA 'needs to announce the intention to use value engineering in the tender documents to ensure compliance with the principles of equal treatment, non-discrimination and transparency. Moreover, the procurement contract should clearly define the conditions for applying the value engineering approach to prevent unwarranted modifications to the procurement contract' [10].

For more information about value engineering, the CAs shall look into the Toolkit prepared by the EAFIP project, more specifically Module 2: an operational module addressed to public procurers aimed at clarifying the pre-requisites and key steps to design and implement an innovation procurement process (PCP and PPI).

Conclusions

In this chapter, we debated the role of public procurement in security research and presented several procurement instruments that can promote innovation uptake. The frustration of end users complaining about public procurement as an obstacle to innovation could be addressed by procurement departments by explaining these instruments. It is also vital to the success of innovation procurement that the end users are involved in these procurement instruments. For example, they should be part of the team responsible for conducting preliminary market consultations.

Acknowledgements This project received funding from the European Union's Horizon 2020 research and innovation programme under grant agreement no. 32875. Chapter article reflects only the authors' views, and the Research Executive Agency and the European Commission are not responsible for any use that may be made of the information it contains.

References

1. Heuninckx, B. Aggregated Procurement under Directive 2014/24/EU: Lessons from the Defence Sector, (2018) 27 P.P.L.R., 189; Racca, G. M., & Yukins, C. R. (2019). Introduction: The promise and perils of innovation. In G. M. Racca, & C. R. Yukins (Eds.), *Cross-border procurement, in Joint public procurement and innovation: Lessons across borders*. Bruylant. Available https://papers.ssrn.com/sol3/papers.cfm?abstract_id=3486897
2. Survey of JCBPP experiences conducted by Ministry of Interior of the Slovak Republic. (2019). Unpublished iProcureNet materials.
3. Locatelli, I. (2019). Process innovation under the new public procurement directives. In G. M. Racca & C. R. Yukins (Eds.), *Joint public procurement and innovation: Lessons across borders*. Bruylant. Available at https://www.research-gate.net/publication/347489725_Process_Innovation_Under_the_New_Public_Procurement_Directives_in_Joint_Public_Procurement_and_Innovation_Lessons_Across_Borders_GM_Racca_and_CR_Yukins_eds_Bruylant_2019/related
4. eafip Toolkit, Module 3, p. 7. Available at http://eafip.eu/toolkit/module-3-2/
5. Voda, O. P., & Jobse, C. (2016). Rules and boundaries surrounding market consultations in innovation procurement. *European Procurement & Public Private Partnership Law Review, 11*(3), 179–193.
6. Lopez, A. M. (2019). Preliminary market consultations in innovative procurement: A principled approach and incentives for anticompetetive behaviors. In G. M. Racca & C. R. Yukins (Eds.), *Joint public procurement and innovation: Lessons across borders* (p. 390). Bruylant.
7. see Art 24, Art 40 last sentence, Art 41, Art 57 (4) e & f of Directive 2014/24/EU.
8. EC notice – Guidance on Innovation procurement, European Commision, C (2021) 4320 final, 2021 available at https://ec.europa.eu/docsroom/documents/45975

9. Assessment report on the performance of the EC funded Innovation Procurement projects in the security field according to the EC Guidance Notice on Innovation Procurement. Available at https://ec.europa.eu/home-affairs/sites/homeaffairs/files/what-we-do/policies/industry-for-security/assessment_report_innovation_procurement_dg_home_final.pdf
10. eafip Toolkit, Module 2, p. 119. Available at https://eafip.eu/toolkit/

Open Access   This chapter is licensed under the terms of the Creative Commons Attribution 4.0 International License (http://creativecommons.org/licenses/by/4.0/), which permits use, sharing, adaptation, distribution and reproduction in any medium or format, as long as you give appropriate credit to the original author(s) and the source, provide a link to the Creative Commons license and indicate if changes were made.

The images or other third party material in this chapter are included in the chapter's Creative Commons license, unless indicated otherwise in a credit line to the material. If material is not included in the chapter's Creative Commons license and your intended use is not permitted by statutory regulation or exceeds the permitted use, you will need to obtain permission directly from the copyright holder.

CHAPTER 41

Policy Recommendations for Combating New Trends in Drug Trafficking

Freideriki Makri, Genny Dimitrakopoulou, Nikolaos Kapsalis, and George Kokkinis

INTRODUCTION

Organised criminal groups (OCGs) involved in drug trafficking are increasingly becoming poly-criminal in a range of criminal activities, while the profits from illegal drugs are used to fund other forms of criminal operations, and even, on some occasions, to finance terrorism. Perpetrators are constantly evolving their modus operandi to cope with law enforcement agencies' (LEAs) anti-crime operations. In the last years, OCGs have advanced their smuggling methods with the advantages that new information and communication technologies (ICTs) offer [1], leaving LEAs in a disadvantageous position.

Tackling drug trafficking is one of EU's priorities in the fight against serious and organised crime [2]. According to the latest European Drug Report, the availability of all types of drugs in Europe remains high while the diversity in the substances consumed today may pose a greater risk to

F. Makri (✉) • G. Dimitrakopoulou • N. Kapsalis • G. Kokkinis
Center for Security Studies (KEMEA), Athens, Greece
e-mail: f.makri@kemea-research.gr

© The Author(s) 2025
I. Gkotsis et al. (eds.), *Paradigms on Technology Development for Security Practitioners*, Security Informatics and Law Enforcement, https://doi.org/10.1007/978-3-031-62083-6_41

the health of consumers [3]. While cannabis remains the most widely used drug in Europe, opioids cause greater harm to societies since they may lead more easily to fatal overdoses (ibid.).

In this context, the Mediterranean and Black Sea network of practitioners (MEDEA) [4] analysed the relevant challenges and needs of practitioners and other stakeholders involved in combatting drug trafficking. MEDEA is an EU-funded Coordination and Support Action project, aiming to establish and further develop a regional network of practitioners and other security-related actors in the Mediterranean and Black Sea regions. Its practitioners are grouped into four thematic communities of practitioners (TCPs). The third TCP pertains to the 'fight against cross-border crime and terrorism' and focuses on the identification of new forms of terrorism—such as radicalisation, smuggling of illicit drugs, weapons and prohibited goods, as well as the exchange of security-related intelligence across EU Member States (MS), along with the aspiration to develop a roadmap for addressing operational capability gaps in data management. The following sections outline the methodology used in the context of TCP3 and offer the findings from the analysis as well as policy recommendations to overcome the challenges faced by LEAs in their anti-drug trafficking operations.

Methodology

MEDEA developed a realistic heroin trafficking scenario to study and analyse the challenges of practitioners. The scenario was focused on trafficking heroin (since it is Europe's most commonly used illicit opioid [3]) from Asia to the EU through the Balkan route, which is the main heroin trafficking corridor linking production countries to the European market. The Balkan route traverses the Islamic Republic of Iran (often via Pakistan), Turkey, Greece and Bulgaria across Southeast Europe to the Western European market, with an annual market value of $20 billion [5].

The scenario was jointly developed by first-line police officers involved in anti-drug smuggling operations. Other security stakeholders also contributed with their tacit knowledge. The scenario analysis first revealed several security gaps [6]. Next, these gaps were further analysed and processed to formulate the main challenges that the security practitioners are facing. Further interaction and processing of findings pave the way to recommendations that shall guide policy and decision-makers and benefit security practitioners.

CHALLENGES

Lack of Intelligence Exchange and Cooperation Between Security Stakeholders

One of the biggest challenges for practitioners is the lack of intelligence exchange, not only between EU MS, but also between the competent authorities at the national level, especially in countries where more than one LEA are involved. However, the cross-border nature of drug smuggling makes intelligence exchange from the early stages an imperative to fight it effectively. Complementary to the above, the cooperation of security stakeholders and implicated countries should be further enhanced to combat drug smuggling. This collaboration should concern all affected countries, be it EU MS, other European countries that act as transit and third countries (where drugs are produced or transited). Apart from a new framework that needs to be developed to frame and promote this cooperation, existing ones shall have to be amended so as to not impede cooperation with third countries. The EU Police Cooperation Code proposed by the Commission [7] is crucial in tackling current difficulties.

Some of the specific challenges of intelligence exchange are (i) differences in legal frameworks and standards: different countries may have distinct laws and regulations regarding the collection, processing and sharing of intelligence, especially when it involves personal data or sensitive information; (ii) lack of trust and reciprocity: intelligence cooperation requires a high level of trust between partners as they share sensitive information that may expose their sources and capabilities. However, trust may be undermined by divergent interests or priorities and (iii) technical and operational challenges: intelligence cooperation also involves technical and operational challenges, such as ensuring the compatibility, dependability and consistency of communication systems and platforms, and establishing common standards.

Lack of Advanced Technological Solutions in the Fields of Intelligence Analysis, Scanning of Freights and Lawful Interception (LI)

To set the basis, a few terms will be defined: lawful interception (LI) is the process of legally obtaining and monitoring communications of individuals or organisations, such as telephone calls, email and instant chat messages, or web browsing activities, under the authorisation of a court order. Intelligence analysis is the process of collecting, processing and analysing information gathered from several sources, such as signals intelligence (SIGINT), imagery intelligence (IMINT) or open-source intelligence (OSINT), to yield actionable intelligence for security practitioners. Scanning of freights is the process of inspecting and detecting the contents of cargo containers, or vehicles, by means of numerous technologies (scanners either X-ray or gamma-ray, radiation and chemical detectors, etc.).

Some needs and challenges are related to the existing technologies and current capabilities of respective LEAs. Investigations of organised crime require access to electronic communications data in order to detect existing links between OCG members. There is also a need for additional capabilities in SIGINT, IMINT and OSINT technologies, as well as in the surveillance tools deployed in land and sea routes used by criminals for drug smuggling. Advanced technologies and portable solutions are also needed to detect concealed drugs. More specifically, there is a need for an improved and coordinated monitoring and analysis of the threats posed by digitalisation, particularly the accessibility of illicit drugs via social media platforms, apps, Internet/darknet marketplaces, as well as the use of online payments (including cryptocurrencies) and encrypted digital communication together with the tools for facing those threats. ICT advancements, like 5G and end-to-end encrypted communication applications, provide criminals with a significant head start as they offer safer and difficult-to-decrypt communication channels, making existing tools and procedures followed for LI by LEAs increasingly outdated. Additionally, for easier detection of drug smuggling cases, increased monitoring-surveillance capabilities, detection devices and intelligence gathering-analytical tools by LEAs would be proven extremely beneficial.

Legal Obstacles

Several challenges have more legal grounds, instead of technological. The strict legal frameworks in place and the new EU regulations may impede LEAs from following the flow of money of drug smugglers that use cryptocurrencies [8]. The process of following the money flows of drug smuggling becomes more difficult due to the lack of commonly accepted terms at the international level, along with the absence of harmonisation of the legislative frameworks among states. Thus, following the money flows of drug smuggling may include differences in national laws and regulations regarding cryptocurrencies and money laundering; lack of international cooperation and coordination among LEAs and judicial authorities; challenges in tracing the transactions involved in the exchanges; difficulties in seizing and confiscating the proceeds of crime; and risks of violating data protection rights of legitimate users of cryptocurrencies.

POLICY RECOMMENDATIONS

Interconnect Existing EU Databases

The importance of information and intelligence exchange between MS with the aim of combatting serious crime is already known, and various steps have been taken to enhance the cooperation and information sharing [9]. Databases and frameworks to accompany their usage have been developed for enhancing data sharing, like the European Arrest Warrant, the European Criminal Records Information System and the Customs Advance Cargo Information System. The EU bodies should move towards the interconnection of the existing EU databases used by practitioners and the unification of their frameworks under one single framework that will constitute the basis of a more structured information sharing, taking into consideration the national legislations.

Expand the Mandate of Passenger Information Units (PIUs) to Other Modes of Transport

In 2016, the European Parliament adopted the directive on the collection, processing and transfer of Passenger Name Record (PNR) data of extra-EU flights that would be handled by the PIUs of each MS [10]. Since then, most of the MS have established their PIUs, and it has been recognised that they play an important role in the fight but also the prevention of serious crime including drug trafficking. One important aspect of the PIUs is that they can share intelligence with other PIUs but also with other national competent authorities, working as a link between the cooperation at national and EU levels. The expansion of the mandate of the PIUs to collect and handle data of other modes of transport, like railway and maritime data, seems very promising regarding the information exchange and crime prevention as they already count success stories with the data collected by air carriers. In fact, some MS have already expanded the obligations of the PIUs to other modes of transport even though this is not foreseen by the directive (ibid.). Research should be made to explore how this expansion is feasible from a legal, technological and organisational point of view.

Harmonise the Lawful Interception (LI) Process

At the EU level, the legal framework for LI is based on the non-binding Council Resolution of 1995 on the LI of telecommunications [11]. In 2006, the EU issued Directive 2006/24/EC [12] to define the time of data preservation and the type of information to be stored. However, no harmonised process among MS has been established for LI, and each communication service provider has to follow a different process each time an LEA initiates the process. A coordinated legislative initiative is required to harmonise the legislation between MS for requesting LI from telecom providers and software application companies, having in mind what is more convenient and less time-consuming for both LEAs and service providers.

Better Regulate the End-to-End Encryption (E2EE)

The rapid proliferation of 5G technology and E2EE poses significant challenges to LEAs. From a technical aspect, LI in 'traditional' voice communication is addressed by the 3GPP SA3-LI working group, which published a set of specifications (3GPP TS 33.127 version 16.6.0 Release 16) for the implementation of LI for 5G technology. E2EE, which is currently supported by many companies, poses a major challenge for LEAs since encrypted conversations cannot be decrypted. To address this issue, software application companies must support the option to disable E2EE after request by judicial authorities.

Develop a Framework for the Seizure of Cryptocurrencies

In January 2020, the 5th Anti-Money Laundering Directive (AMLD) [13] came into effect, bringing cryptocurrency exchanges under the scrutiny of EU legislation, on the grounds that they can be used for money laundering and terrorist financing. Since then, various proposals have been made to increase the transparency of the use of cryptocurrencies while protecting their users. This framework should include the steps to be followed for seizing, storing and handling seized crypto assets and should be implemented at EU level. It is recommended that the crypto assets be included in the proposal for a directive on asset recovery and confiscation, and further research to be done on the development and possible implications of an EU database for seized cryptocurrencies.

CONCLUSION

This chapter has discussed the main challenges and policy recommendations for combating drug trafficking in the EU based on the findings of the MEDEA project. The need for enhancing intelligence exchange and cooperation between security stakeholders at the national and EU levels, as well as for developing advanced technological solutions in the fields of intelligence analysis, scanning of freights and LI, was highlighted. Moreover, legal obstacles that hinder the effective fight against drug trafficking, such as the use of cryptocurrencies, end-to-end encryption and lack of harmonisation of the lawful interception process, were explained. A series of concrete measures to overcome these obstacles were proposed.

Acknowledgements The MEDEA project received funding from the European Union's Horizon 2020 research and innovation programme under grant agreement no. 787111. The support is gratefully acknowledged. The views expressed are purely those of the authors and may not in any circumstances be regarded as stating an official position of the European Commission.

REFERENCES

1. UNODC. (2020). *Drug trafficking – Module 3: Organized crime markets*. United Nations Office on Drugs and Crime. [Online]. Available: https://www.unodc.org/e4j/zh/organized-crime/module-3/key-issues/drug-trafficking.html. Accessed July 25, 2023.
2. EU Policy Cycle – EMPACT, 20 January 2022. [Online]. Available: https://www.europol.europa.eu/crime-areas-and-trends/eu-policy-cycle-empact
3. E. M. C. f. D. a. D. Addiction. (2023). European Drug Report 2023.
4. MEDEA Consortium. *Mediterranean practitioners' network capacity building for effective response to emerging security challenges*. [Online]. Available: https://cordis.europa.eu/project/id/787111. Accessed May 8, 2022.
5. UNODC. *Drug trafficking*. United Nations, [Online]. Available: https://www.unodc.org/unodc/en/drug-trafficking/index.html. Accessed July 25, 2023.

6. MEDEA Consortium. *Capability gaps.* [Online]. Available: https://www. medea-project.eu/capability-gaps/. Accessed June 12, 2023.
7. European Commission. (2021). *Proposal for a Council Recommendation on operational police cooperation, 2021/0415 (CNS).*
8. Europol. (2022). Cryptocurrencies key to tackiling organised crime – Europol and Basel Institute on Governance. In *6th Global conference on criminal finaces and cryptocurrencies,* Hague.
9. European Commission. *Migration and home affairs: Information exchange.* [Online].   Available: https://home-affairs.ec.europa.eu/information-exchange_en. Accessed July 25, 2023.
10. European Parliament and Council of the European Union. (2016). Directive (EU) 2016/681 of the European Parliament and of the Council of 27 April 2016 on the use of passenger name record (PNR) data for the prevention, detection, investigation and prosecution of terrorist offences and serious crime. *Official Journal of the European Union, 50,* 132–149.
11. Council of the European Union. *Council Resolution (C329/01) on the lawful interception of telecommunications.*
12. Parliament, EC. (2006). Directive 2006/24/EC of the European parliament and of the council on the retention of data generated or processed in connection with the provision of publicly available electronic communications services or of public communications. *Official Journal of the European Union, 105,* 54–63.
13. European Union. (2018). Directive (EU) 2018/843 of the European parliament and of the council of 30 May 2018 amending directive (EU) 2015/849 on the prevention of the use of the financial system for the purposes of money laundering or terrorist financing, and amending Directive. *Official Journal of the European Union, 156,* 1–32.

Open Access   This chapter is licensed under the terms of the Creative Commons Attribution 4.0 International License (http://creativecommons.org/licenses/by/4.0/), which permits use, sharing, adaptation, distribution and reproduction in any medium or format, as long as you give appropriate credit to the original author(s) and the source, provide a link to the Creative Commons license and indicate if changes were made.

The images or other third party material in this chapter are included in the chapter's Creative Commons license, unless indicated otherwise in a credit line to the material. If material is not included in the chapter's Creative Commons license and your intended use is not permitted by statutory regulation or exceeds the permitted use, you will need to obtain permission directly from the copyright holder.

CHAPTER 42

EU-CIP: European Knowledge Hub and Policy Testbed for Critical Infrastructure Protection

Emilia Gugliandolo, John Soldatos, and Habtamu Abie

THE CRITICAL INFRASTRUCTURE PROTECTION AND RESILIENCE ECOSYSTEM

In our contemporary and evermore interdependent and globalized world, critical infrastructure systems are the cornerstones of societies. They are complex, interrelated systems, networks, and services essential for everyday life, businesses and social activities, and underwrite the security of societies and communities [1].

Critical infrastructure in its essence means defining the infrastructure, processes, and systems that are crucial for the functioning of the wider

E. Gugliandolo (✉)
Engineering Ingegneria Informatica S.p.A, Rome, Italy
e-mail: Emilia.Gugliandolo@eng.it

J. Soldatos
INNOV-ACTS Limited, Nicosia, Cyprus

H. Abie
Norsk Regnesentral, Oslo, Norway

© The Author(s) 2025   517
I. Gkotsis et al. (eds.), *Paradigms on Technology Development for Security Practitioners*, Security Informatics and Law Enforcement, https://doi.org/10.1007/978-3-031-62083-6_42

social community. In the basic definition, the term "*Critical Infrastructure means an asset, system or part thereof located in Member States which is essential for the maintenance of vital societal functions, health, safety, security, economic or social well-being of people, and the disruption or destruction of which would have a significant impact in a Member State as a result of the failure to maintain those functions*" (EU Directive, 2008 [2]).

Critical infrastructures constitute the backbone of the functioning of our modern and interconnected societies. Major shock events of all types, from natural hazards to industrial accidents, terrorist or cyberattacks, have demonstrated the vulnerabilities of these critical systems. Their destruction, disruption, or interruption could lead to cascading effects across sectors and sometimes across national borders. Vulnerabilities of critical infrastructure to this range of hazards and threats call for increased attention to critical infrastructure security and resilience. Disaster risks, compounded by climate change, present a set of challenges for infrastructure resilience. In addition, the rise of hybrid threats and associated digital security risks calls for increased resilience of critical infrastructures to digital security incidents [3].

In the current landscape, critical infrastructure systems are becoming more complex and interconnected, making them more vulnerable to cascading failures and disruptions. This makes it even more critical for these systems to be resilient and able to adapt and recover from disruptions. Furthermore, disruptions to critical infrastructure systems are becoming more frequent and severe due to factors such as climate change, cyber threats, and pandemics. New security risks are emerging, and the number of cyberattacks against critical infrastructure assets is increasing.

The consequences to societies and communities are that existing vulnerabilities are greatly exacerbated, new ones are created, and social inequalities become more entrenched [4]. Critical infrastructure systems need to be resilient in the current landscape due to the growing complexity and interconnectivity of these systems, the increasing frequency and severity of disruptions, criticality of these systems, and high costs of disruptions. The resilience of critical infrastructure refers to its ability to withstand, adapt to, and recover from disruptions, stresses, and shocks while maintaining its essential functions and services. Resilience is a critical aspect of critical infrastructure, given the high likelihood of disruptions and the severe consequences of failures in these systems. Therefore, the importance of critical infrastructure resilience is high on the agenda of contemporary sustainable development as the "keystone" of nationally led processes of building resilience and actions. In that regard, a resilient society is based on multidimensional principles that contribute to sustainable and resilient development. Critical infrastructure resilience can be understood as the ability of these systems to anticipate, withstand, or absorb shocks and stresses, while adapting to new conditions that would result in a quick recovery and transformation as a way to better cope with chronic stresses and acute shocks in the future [4].

Critical Infrastructure Protection (CIP): A Complex Technological and Policy Landscape

To ensure the resilience and continuity of critical infrastructures, CI operators are deployed a host of different solutions, which range from security risk assessment and cybersecurity solutions to solutions that protect physical assets of critical infrastructures. The development of effective and innovative CIP solution is vital, given the growing sophistication of the CIs and the emerging challenges and threats that are recently faced by operators. For instance, in an increasingly digitally interconnected world, CI operators need to deal with much more sophisticated cybersecurity challenges than ever before. The latter include challenges (e.g., fake news and disinformation management) that were hardly considered in CIP solutions a few years ago. Most importantly, CI owners and operators are nowadays conducting business in a highly unstructured, volatile, and highly unpredictable environment, where asymmetric threats and other previously rare events are becoming the norm. This has been very evident during the last couple of years when several CI operators had to confront challenges like the COVID-19 pandemic outbreak, various large-scale supply chain disruptions following the pandemic, as well as the recent Ukrainian war.

In this landscape, CI operators are forced to design and implement novel solutions that are not just reactive in nature, but rather able to predict and anticipate potential disruptions and security threats. Recent technological advances yield the development of such solutions viable and more pragmatic than a few years ago. Nevertheless, the task of planning, designing, developing, deploying, and operating innovative solutions must take place in the scope of a complex landscape of

- *Multiple technological solutions,* such as security risk assessment, adaptive threat hunting, anomaly detection, biometric authentication, blockchain technology, cyber-physical threat intelligence, data loss prevention, digital twins, disinformation management, end-point protection, image analysis, impact assessment, machine learning, modeling of cascading effects, and pentesting solutions. There are also emerging technological solutions (e.g., generative AI [5]) that present both opportunities and threats to strengthening critical infrastructure protection.

- *A sea of different standards,* such as ISO/TC 292, ISO 27001 series, ISO 28000 series, and ISO 22000 facilities of standards, which cover many different aspects of CI protection aspects.
- *A considerable number of regulations and directives,* such as the NIS directive, GDPR regulation, AI Act of the Council of Europe, cybersecurity Act, and many more.
- *A multistakeholder environment,* including not only CI owners and CI operators, but also technology vendors, researchers, security experts, security integrators, regulatory experts, and policymakers, which view the development and deployment of innovative solutions from their own unique perspective.
- *A host of opportunities for combining* diverse technologies into more complex, integrated, and sophisticated solutions. For instance, these opportunities can be boosted by the amalgamation of advanced technologies such as big data, IoT, IIoT, artificial intelligence (AI), cloud computing, digital twins, edge computing, advanced analytics, robotics, cognitive computing, etc.

In this complex landscape, stakeholders need orientation regarding available solutions, gaps in current knowledge, limitations of the state of the art and the state of practice, as well as roadmaps for the development and deployment of innovative CIP solutions.

The EU-CIP Mission

EU-CIP is a 3-year Coordination and Support Action (CSA) that is funded by the European Commission. EU-CIP's vision is to establish a sustainable knowledge network of European CIP experts and stakeholders, which will provide knowledge, insights, foresights, and guidance regarding research and innovation opportunities in the CIP domain. Specifically, one of the main objectives of the project is to enhance Europe's analytical capability regarding research outcomes, technologies, and policies—foster data-driven evidence-based policy and innovation development. The activities that will lead to the accomplishment of this objective are conveniently called EU-CIP-ANALYSIS activities.

The EU-CIP-ANALYSIS activities will be implemented in a period of 3 years following the start of the project in October 2022. This chapter presents some of the findings of the EU-CIP-ANALYSIS, notably findings produced following state-of-the-art analysis and consultation with CIP stakeholders with the EU-CIP consortium and the European Cluster for Securing Critical Infrastructures (ECSCI) cluster, that is, a cluster comprising more than 32 of the most prominent European projects on security and critical infrastructure protection and protection.

Preliminary Findings on Capability Needs and Capability Gaps

The EU-CIP consortium members have identified the following CIP capabilities that are not adequately supported and covered by state-of-the-art solutions:

- *C1—Enhanced adaptability.* There is a need to enhance adaptability to new threats as the novel and sometimes asymmetric threats against CIs happen with higher frequency than ever before.
- *C2—Reduced response times.* EU-CIP experts identified a need for increasing the speed of response as a means of coping with highly volatile environments and minimizing the costs of potential damage.
- *C3—Increased transparency.* In an era where complex threats are handled by sophisticated technologies, there is a need for improving transparency of the solution for the stakeholders.
- *C4—Improved detection capabilities.* Improved detection over current state of the art based on solutions that (i) address novel threats like hybrid threats combining cybersecurity and physical security aspects; and (ii) provide improved analytics capabilities of detection and response tools that account for the rapid shifting in the IoT and 5G spaces, while offering improved informed decision-making.
- *C5—Improved risk and impact assessment capabilities* to address novel integrated, hybrid, and asymmetric threats against a broad range of cyber and physical assets.

- *C6—Better integration of Telco Security tools with information security management tools*: This integration should strive to avoid existing silos and fragmentation in security systems and capabilities.
- *C7—Solutions addressing cascading effects between different entities and states.* Such solutions must help prevent disruptions in supply chain services to ensure the continuity of business operations across different value chains.
- *C8—Transformation of proactive and adaptive protection tools and methods to incorporate real-time functionalities.* Such functionalities will boost the protection and resilience of CI through improved collection and analysis of real-time data in light of the ever-evolving and dynamically changing threat landscape.
- *C9—Better exploitation of information from critical sensors towards augmented situation awareness.* Input from critical sensors (e.g., cameras, human presence, luminosity, weather/environment parameters) must be better integrated and exploited within CIP systems toward improving situation awareness (e.g., about security personnel positions and CI assets state). The latter must be combined with proper revisions to security and emergency management processes.
- *C10—Risk prediction and anticipation,* leading to earlier detection of threats and subsequently enhancing resilience, monitoring, patrolling, decision support, and event management applications.
- *C11—Training, reskilling and upskilling.* There is a need for developing relevant skills and competencies in collaboration with the research and the academic community. Prominent examples of the required skills development include cybersecurity and cyber-resilience trainings.

Moreover, EU-CIP has also identified the following list of preliminary capability gaps (CG), which are partly linked to the above-listed capability needs:

- *CG1—Poor automation*: There is currently poor automation when it comes to achieving fast detection, protection, and recovery from cyberattacks. Specifically, there is a gap in closing the cybersecurity automation loop to automatically verify, diagnose, rectify, monitor, measure, and improve security controls. This gap is very evident when it comes to supporting heterogeneous technology and configurations in a systematic and justifiable manner. Fast-growing technology segments like artificial intelligence can greatly boost automation in processes like threat detection, attack anticipation, and fast enforcement of security policies, yet their use must adhere to the mandates and emerging regulations (e.g., the European AI Act).

- *CG2—Lack of proper control of interconnectedness*: CI operators are not currently offered strong control over interconnected assets and their dependencies. This makes it difficult to implement a holistic CIP approach that considers the dependencies of the various assets, as well as related cascading effects. To boost the interconnection of diverse assets and their security policies, there is a need for novel security knowledge modeling approaches, as well as for interconnection and interoperability across diverse security systems.
- *CG3—Poor alignment of resilience indicators*: Currently, CI operators deal with a host of resilience indicators, which are not aligned and, in several cases, diverse and noncompatible. This hinders the implementation of a structured CIR approach, while being a setback to interoperability across different interconnected critical infrastructures that support prominent value chains of our societies and economies.
- *CG4—Lack of agreed standards-based stress-testing procedures*: Along with poor alignment of indicators, there is also a lack of agreed stress-testing procedures that could foster the implementation of structured, standards-based CIR approaches.
- *CG5—Problems with the classification of IoT devices*: In recent years, there has been a proliferation of IoT devices in CIs. However, the identification and classification of IoT-based assets are lagging behind the output of the suppliers and do not scale to account for the increased diversity and complexity of the various classes of IoT devices. Currently, a "one-size-fits-all" approach to securing IoT devices is applied, which is inadequate to address the vulnerabilities and risks associated with the growing number of connected devices.
- *CG6—Scalability in the mitigation of distributed denial of service (DDoS) attacks*: DDoS attack mitigation is not scaling optimal given the increase in the bandwidth that is nowadays available to end users (e.g., multi-gigabit per second connectivity is now widely available to SMEs). Apart from bandwidth issues, there is also a need for mitigating measures for novel ways of performing DDoS attacks, such as ways based on IoT devices and AI bots.

- *CG7—Development and deployment of AI-based systems.* Despite recent advances in AI systems and technologies, the potential of AI in CIP/CIR solutions remains underexploited. Moreover, there is still poor awareness of AI capabilities among employees and CIP/CIR vendors. Likewise, existing solutions cannot deal effectively with AI-powered cyberattacks based on new adaptive AI technologies. As already outlined, generative AI systems (e.g., like ChatGPT) provide opportunities to increase protection (e.g., based on the intelligent identification of attack patterns), yet they also introduce new potential vulnerabilities and attacks (e.g., due to the generation of new attack patterns).

- *CG8—Lack of holistic security management systems.* There is a lack of holistic security management solutions, which are operationally applicable and do not need to be customized from scratch. Furthermore, there is a lack of solutions that are uniform and operations across a large number of different stakeholders.

- *CG9—Gaps in emergency management processes.* There are significant gaps in standardized processes in case of emergencies. For instance, there is a lack of universally agreed processes regarding the communication with external entities for emergency handling and with public authorities regarding security/safety incident-related communications.

- *CG10—Inability to cope with dynamically evolving threats.* There is a lack of dynamic and intelligent solutions that can deal with the ever-evolving and dynamically changing threats proactively and adaptively. Adaptive, AI-based solutions could provide the required intelligence, yet they are still not widely developed and used in the CIP/CIR domain. In the above direction, there is a need for evolving knowledge modeling and knowledge basis in directions that address the evolution of the threat landscape.

- *CG11—Poor awareness of modern CIP/CIR challenges.* There is generally a lack of awareness regarding contemporary CIP/CIR requirements, such as awareness of cascading events, preparedness against new natural and anthropic risks, and complex cyber resilience issues. Hence, there is a need for training security teams and developing new talent as part of a broader cultural shift that considers the latest CIP/CIR challenges such as evolving threats, novel forms of attacks and disruptions, and the need to consider interconnected infrastructures and cascading effects.

Beyond the above-listed general capabilities needs and gaps that are applicable to all CIP sectors, EU-CIP partners have expertise in different sectors and have identified capabilities needs and gaps linked to specific sectors.

PRELIMINARY INSIGHT INTO THE TRENDS AND STATE-OF-THE-ART TECHNOLOGIES THAT ADDRESS CAPABILITY GAPS

The technologies and tools that can help close the gaps in CIP/CIR and can help prevent, detect, and mitigate potential threats to CIs, such as cyberattacks, physical attacks, and natural disasters. The implementation of state-of-the-art technologies and tools can boost the resilience and continuity of critical infrastructure systems, protecting society from the potentially devastating consequences of their disruption or destruction.

The main findings (according to the priority) of the EU-CIP survey about the technologies that could help in alleviating existing capability gaps are:

- Cyber-physical threat intelligence
- Security risk assessment
- Impact assessment tools
- Digital twins
- Anomaly detection
- Modeling of cascading Effects
- Cybersecurity tools
- Malware detection
- Machine learning—pattern detection

In line with the identified gaps, technologies like cyber-physical threat intelligence, security risk assessment, and modeling of cascading effects were perceived as the most promising for mitigating the list of gaps. This is because the implementation of integrated security approaches and handling of cascading effects were perceived as some of the key capabilities that are currently missing in the CIP/CIR systems.

Some of the trending CIP/CIR technologies according to the internal survey of the project and the feedback of EU-CIP members were illustrated. The results of the trends survey are in line with the above-listed technologies. Specifically, cyber-physical threat intelligence represents one of the most important trends followed by collaborative threat intelligence and predictive security.

CONCLUSIONS

In an era characterized by unprecedented complexity and interconnectivity, safeguarding critical infrastructure systems is paramount to ensuring the continued functioning of modern societies and economies. The findings from the EU-CIP mission shed light on the pressing needs and gaps within the realm of critical infrastructure protection (CIP). These needs range from enhancing adaptability to improving response times, increasing transparency, and fortifying detection capabilities.

Addressing these needs is not only a technological imperative but also a multidimensional challenge that involves diverse stakeholders, standards, regulations, and policy considerations. As emerging threats and vulnerabilities reshape the CIP landscape, it is essential to foster a proactive approach that anticipates and mitigates potential disruptions. Leveraging state-of-the-art technologies, such as cyber-physical threat intelligence, security risk assessment, and machine learning, holds promise in closing existing capability gaps and enhancing resilience.

The preliminary results and the above considerations show that the project has already successfully addressed its ambitious goals. Further effort is needed and will be invested in order to improve the basis for (a) identifying gaps and priorities, (b) extracting knowledge, and (c) formulating recommendations, all of them needing to be data-driven, justified by the evidence gathered and transparent for the end users, as clearly confirmed in the interaction with the end users and resulting reports [6].

The EU-CIP mission serves as a pivotal initiative in advancing our understanding of CIP, providing a roadmap for innovation and policy development. The insights gleaned from this mission underscore the necessity for collaborative efforts among governments, industries, and academia to fortify critical infrastructure against evolving threats. By harnessing the power of innovation, aligning diverse perspectives, and fostering a culture of preparedness, we can bolster the security and resilience of critical infrastructure, ensuring its continued role as the foundation of our interconnected world.

Acknowledgments This project received funding from the European Union's Horizon Europe- research and innovation program under grant agreement no. 101073878. This chapter reflects only the authors', views and the Research Executive Agency and the European Commission are not responsible for any use that may be made of the information it contains.

References

1. Guidance notes on building Critical Infrastructure resilience in Europe and Central Asia, United Nations Development Programme. UNDP 2022.
2. European Council Directive 2008/114/EC on the identification and designation of European Critical Infrastructures and the assessment of the need to improve their protection. (2008). http://eurlex.europa.eu/LexUriServ/LexUriServ.do?uri=OJ:L:2008:345:0075:0082:EN:PDF
3. https://www.oecd-ilibrary.org/sites/76326acb-en/index.html?itemId=/content/component/76326acb-en
4. https://www.undp.org/albania/publications/guidance-notes-building-critical-infrastructure-resilience-europe-and-central-asia
5. Gupta, M., Akiri, C., Aryal, K., Parker, E., & Praharaj, L. (2023). From ChatGPT to ThreatGPT: Impact of generative AI in cybersecurity and privacy. *IEEE Access, 11,* 80218–80245. https://doi.org/10.1109/ACCESS.2023.3300381
6. Jovanović, A., Schernberg, H., & Sansavini, G. (Eds.). (2023). Disaster resilience and insurance, Springer Special Issue, https://link.springer.com/collections/jaagiajigj?utm_medium=referral&utm_source=rh_recs

Open Access   This chapter is licensed under the terms of the Creative Commons Attribution 4.0 International License (http://creativecommons.org/licenses/by/4.0/), which permits use, sharing, adaptation, distribution and reproduction in any medium or format, as long as you give appropriate credit to the original author(s) and the source, provide a link to the Creative Commons license and indicate if changes were made.

The images or other third party material in this chapter are included in the chapter's Creative Commons license, unless indicated otherwise in a credit line to the material. If material is not included in the chapter's Creative Commons license and your intended use is not permitted by statutory regulation or exceeds the permitted use, you will need to obtain permission directly from the copyright holder.

INDEX

© The Editor(s) (if applicable) and The Author(s) 2025
I. Gkotsis et al. (eds.), *Paradigms on Technology Development for Security Practitioners*, Security Informatics and Law Enforcement, https://doi.org/10.1007/978-3-031-62083-6